Mammals of Alabama

Philip Henry Gosse as a young man of twenty-nine, the year of his return to England from Alabama, painted by his brother, William Gosse. (1839, watercolor on ivory, courtesy of the National Portrait Gallery—London)

Philip Henry Gosse (1810–1888) was an English naturalist and illustrator who spent eight months of 1838 on the Alabama frontier, teaching planters' children in Dallas County and studying the native flora and fauna. Years after returning to England, he published the now-classic *Letters from Alabama: Chiefly Relating to Natural History*, with twenty-nine important black-and-white illustrations included. He also produced, during his Alabama sojourn, forty-nine remarkable watercolor plates of various plant and animal species, mainly insects, now available in *Philip Henry Gosse: Science and Art in "Letters from Alabama" and "Entomologia Alabamensis."*

The Gosse Nature Guides are a series of natural history guidebooks prepared by experts on the plants and animals of Alabama and designed for the outdoor enthusiast and ecology layman. Because Alabama is one of the nation's most biodiverse states, its residents and visitors require accurate, accessible field guides to interpret the wealth of life that thrives within the state's borders. The Gosse Nature Guides are named to honor Philip Henry Gosse's early appreciation of Alabama's natural wealth and to highlight the valuable legacy of his recorded observations. Look for other volumes in the Gosse Nature Guides series at http://uapress.ua.edu.

Mammals
OF ALABAMA

TROY L. BEST AND JULIAN L. DUSI

Published in Cooperation with the Alabama Wildlife Federation

The University of Alabama Press • Tuscaloosa

Typeface: Minion Pro
Cover photo: Raccoons are common in rural and urban areas
throughout Alabama. Photo courtesy of Joseph W. Hinton.
Design: Michele Myatt Quinn

Skull plates prepared and photographed by Scott A. Kincaid,
and maps prepared by Troy L. Best and Scott A. Kincaid.

∞

The paper on which this book is printed meets the minimum
requirements of American National Standard for Information
Sciences—Permanence of Paper for Printed Library Materials,
ANSI Z39.48-1984.

Library of Congress Cataloging-in-Publication Data

Best, Troy L.
  Mammals of Alabama / Troy L. Best and Julian L. Dusi.
    pages cm — (Gosse nature guides)
  "Published in cooperation with the Alabama Wildlife
Federation."
  Includes bibliographical references and index.
  ISBN 978-0-8173-5749-8 (quality paper : alk. paper) — ISBN
978-0-8173-8680-1 (e book) 1. Mammals—Alabama—
Identification. I. Dusi, Julian L. II. Title.
  QL719.A2B47 2014
  599.09761—dc23   2013018028

Published in cooperation with the Alabama Wildlife Federation

In the yard were some towering oaks, on which several Fox Squirrels (*Sciurus capistratus*) were frisking and leaping from bough to bough with great animation.

—Recorded by Philip Henry Gosse in *Letters from Alabama* on the first morning of his arrival in Dallas County

Contents

Rodents

Bats

Carnivores

Preface

This book is intended to provide an overview of the biology of wild mammals that have been documented to occur in Alabama, as well as extant species that previously inhabited the state, species that are of questionable or accidental occurrence in the state, and species that probably are present in the state. Because there are varying amounts of information about each species, our summaries are not uniform in length or detail. Each species account provides references to sources where we obtained information and where the reader can access additional facts about each species. For example, we often used information published in the journal *Mammalian Species*; accounts therein provide citations to the primary scientific literature. We have not included detailed taxonomic keys or technical descriptions, but readers should be able to identify species from their own experience, from images herein, and from descriptions in the accounts. Dental formulas and measurements of size and weight will aid in identification of some species. The dental formula describes the numbers of different kinds of teeth; e.g., i 3/3, c 1/1, p 4/4, m 3/3, total = 44. Letters designate incisors, canines, premolars, and molars, respectively. In shrews, "u" indicates unicuspid teeth. Numbers to the left of the diagonal line are the number of teeth of each kind on one side of the upper jaw; those to the right of the diagonal line indicate the number on one side of the lower jaw. To calculate the total number of teeth, all teeth in the dental formula are added together and then multiplied by two. In the size and weight sections of the species accounts, we have converted metric measurements in millimeters (mm), meters (m), grams (g), and kilograms (kg) to the nearest 0.1 inch, foot, ounce, or pound. In the text of the accounts, all measurements are metric. Generalized maps of the geographic range of each species in North America and Alabama are provided. In the sections on ecology, we include information on habitats, diet, predators, and other ecological data. Sections on behavior have information on daily and seasonal activity patterns, hibernation, migration, roosts or burrows, and other behaviors of interest. Life history includes information, when available, regarding breeding season, length of gestation, size of litter, number of litters, descriptions of newborns, age at sexual maturity, and estimated life span. For mammals, much mortality occurs early in life, so our estimates of average life spans actually apply to animals after they reach adulthood. Some listings of parasites, diseases, and disorders are lengthy, which usually indicates a species has been studied in more detail than others. Listings of parasites and diseases are intended to convey the breadth of interactions between mammals and disease-causing organisms, and to provide a starting point for readers who are interested in diseases and their transmission. Although it might appear so, we had no intention of providing a comprehensive list of parasites, diseases, or disorders for each species of mammal. We have listed current conservation status at state, federal, and international levels as warranted. In the comments sections, we present information that may be of special interest or that does not fit elsewhere. Derivation of scientific names provides an interesting glimpse into early perceptions of species, including names honoring early investigators of our mammalian fauna. References list sources for most of the information that appears in the accounts. However, we have added our own opinions, estimates, and original data in several accounts where information was not available in the scientific literature. We have enjoyed the decades we have spent studying the mammals of Alabama, and we hope you enjoy reading about Alabama's diverse mammalian fauna.

TROY L. BEST AND JULIAN L. DUSI

Acknowledgments

We are indebted to those who preceded us in studying the mammals of Alabama, especially A. H. Howell (1909, 1921) and D. C. Holliman (1963), for thorough reviews of what was known about our mammalian fauna. Also helpful were the annotated lists of mammals, species accounts, and discussions of conservation needs for wildlife in Alabama that were edited by R. H. Mount (1984, 1986), R. E. Mirarchi (2004), and R. E. Mirarchi et al. (2004a, 2004b). Several books about mammals provided data on life history, taxonomy, and useful summaries of the characteristics of orders and families (Anderson and Jones 1967; Hall 1981; Nowak 1999; Wilson and Reeder 2005; Feldhamer et al. 2007; Vaughan et al. 2011). In addition, numerous reviews have been published in *Mammalian Species*; we thank the authors of those accounts, which were instrumental in the preparation of this book. Numerous graduate and undergraduate students have contributed significantly to our understanding of the ecology and natural history of many species of mammals through their research. Analyses of DNA performed by M. C. Wooten verified the identification of the first specimen of the North American deermouse from Alabama (reported herein). J. D. Woodard obtained the first specimen of an American badger in Alabama (reported herein), which was deposited into the Auburn University Collection of Mammals by M. E. Sievering.

J. L. Hunt and L. A. McWilliams assisted in compiling derivations of scientific names, primarily from Jaeger (1955). S. C. Peurach and D. E. Wilson provided access to specimens in the United States National Museum of Natural History. S. A. Kincaid photographed and edited images of skulls and prepared images of distribution maps. L. A. Durden reviewed the information on ectoparasites. We thank J. A. Dellinger, G. A. Feldhamer, K. Geluso, M. J. Harvey, G. D. Schnell, and M. C. Wooten for assistance in acquiring images, and J. S. Altenbach, R. W. Barbour,

S. B. Castleberry, C. S. Clem, T. Dewey, C. M. Dunning, T. W. French, K. Geluso, L. M. Gilley, M. J. Harvey, J. W. Hinton, C. Jaworowski, M. L. Kennedy, D. B. Lesmeister, J. C. Medford, L. H. Moates, T. Murray, J. F. Parnell, J. S. Pippen, E. B. Pivorun, A. K. Poole, D. I. Stetson, J. Whitlock, C. Wemmer, and the United States Fish and Wildlife Service for allowing us to use their digital images. E. Motherwell and G. R. Mullen provided encouragement, many helpful suggestions, and extraordinary patience during the preparation of this book.

Alabama Counties

Mammals of Alabama

Introduction

Although naturalists visited Alabama beginning in the late 1700s, there are few scientific assessments of mammals for the state. Bartram (1791) traveled in Alabama and the Southeast during 1776 or 1777 and provided detailed descriptions of the flora and a few references to some of the mammals such as bats, moles, rabbits, squirrels, deer, bears, bobcats, and wolves. Reporting on his travels in Alabama during 1820, Hodgson (1824) mentioned deer, cougars, bears, and gray foxes. Gosse (1859) referred to several species in accounts of his visit to Alabama in 1838, many by recognizable common names, such as wild hogs, raccoons, opossums, bears, and wild cats (probably bobcats). Although most of the scientific names Gosse used are now outdated, many of the species he referred to are identifiable. For example, Gosse referred to fox squirrels as *Sciurus capistratus* (now *Sciurus niger*), rabbits as *Lepus americanus* (probably swamp rabbits, now *Sylvilagus aquaticus*), bats as *Lasiurus rufus* (now known as eastern red bats, *Lasiurus borealis*), yellow rats as *Mus floridanus* (probably eastern woodrats, *Neotoma floridana*), woodrats as *Arvicola hispidus* (probably hispid cotton rats, *Sigmodon hispidus*), and gray foxes as *Canis virginianus* (still known as gray foxes, but the scientific name is now *Urocyon cinereoargenteus*). The monumental works of Audubon and Bachman (1846) and Baird (1857) also had a few references to species of mammals occurring in Alabama.

Howell (1909, 1921) was the first to summarize what was known about our mammalian fauna; he presented a list of species, a historical review, and an annotated bibliography. Decades later, Holliman (1963) compiled records of mammals in Alabama in a doctoral dissertation. Lists and assessments of the status of species were presented by Mount (1984, 1986), and an updated list of species and conservation concerns were included in volumes edited by Mirarchi (2004) and Mirarchi et al. (2004*a*, 2004*b*).

None of these compilations provides a comprehensive listing of marine mammals that might occur along the Gulf Coast of Alabama. In Appendix 1, we present a list of cetaceans (whales and dolphins) that potentially occur in coastal Alabama. A proper assessment of the conservation status of most of these species in Alabama must await verification of their occurrence, data on their abundance, and additional information about their natural history.

Worldwide, there are 29 orders, 153 families, 1,229 genera, and 5,416 species of mammals (Wilson and Reeder 2005). Of these, there are 9 orders, 22 families, 51 genera, and 72 species that have occurred naturally in Alabama. Introduced species that have established wild breeding populations add another 3 families, 5 genera, and 6 species. The California sealion, which did not establish a breeding population, was introduced into Alabama by unknown means and represents an additional family, genus, and species. Of the 72 species of native mammals in Alabama, 4 became extinct in the state naturally (North American porcupine, jaguar, fisher, elk) and 3 were extirpated by humans (cougar, red wolf, bison). One species commonly assumed to occur in Alabama but for which there is no verified record is the eastern small-footed myotis. In this book, we present accounts for all 80 of these species of wild mammals.

Several species of mammals are of special conservation concern in Alabama and need additional study. The United States Fish and Wildlife Service lists 8 species as partially or totally endangered (West Indian manatee, oldfield deermouse, gray myotis, Indiana myotis, jaguar, cougar, jaguarundi, red wolf) and 1 as threatened (American black bear). An additional 2 are under review (Allegheny woodrat, Appalachian cottontail). The recent spread of white-nose syndrome (*Geomyces destructans*), a fungal infection of bats, has devastated populations of cave-dwelling bats in large areas of the central and northeastern United States. Because Alabama has large colonies of cave-dwelling bats, white-nose syndrome is likely to have a significant negative impact on endangered bats (Indiana myotis, gray myotis) as well as other species associated with caves (e.g., southeastern myotis, tri-colored bat).

Many of our mammals have benefited from game-management programs (e.g., our thriving population of white-tailed deer) and from a close association with humans (e.g., coyote, Virginia opossum, raccoon). Other species have become less common because of the modification and destruction of their habitat (e.g., Brazilian free-tailed bat, meadow jumping mouse), and some have been extirpated by overharvesting and predator-control efforts (cougar, red wolf, bison). While the impacts of wild mammals interacting with humans have varied from positive to negative, some new challenges face many species in our mammalian fauna. Destruction and fragmentation of natural habitats for roadways, homes, shopping centers, and other structures have increased the possibility for human-wildlife interactions. While many of these interactions are positive, others are negative or perceived to be negative. Negative perceptions, whether based in reality or not, can result in efforts to control or eliminate wildlife from some areas (e.g., nine-banded armadillos, bats, coyotes, raccoons, white-tailed deer).

The diverse mammalian fauna of Alabama varies in abundance from common to rare to extirpated to prehistoric to hypothetical. Continued research on this rich fauna is warranted for many reasons, including species recently reported for the first time in Alabama (American pygmy shrew, smoky shrew), the first records of 2 species reported herein (North American deermouse, American badger), ongoing searches for species that probably occur in the state (e.g., eastern small-footed myotis), and the seemingly incessant quest for information useful in developing management strategies. The goals of our book are to provide a brief historical overview, to present some of the methods used to study mammals, and to summarize information on each of the 80 species of mammals that occur, once occurred, or probably occur in Alabama. Perhaps our book will stimulate ongoing and new investigations into the biology of our varied and fascinating mammalian fauna.

Methods

Because of the nocturnal and secretive nature of most mammals, re-searching them is often difficult and usually requires dedicated, highly motivated, and well-trained individuals, including personnel of management and regulatory agencies, employees of public and private conservation organizations, and students and teachers at colleges and universities. In addition, members of the general public often become involved in research by providing information about species and by allowing access to their property. Information gained from research projects is used to make decisions regarding conservation and management of species, to train students to become teachers and researchers, and to inform the general public about mammals. Visits to classrooms, special programs

In the laboratory, students study variation in hair, bones, teeth, and skulls to learn about diversity and ecology of mammals.

In the field, students receive instructions on recording data in field notebooks and where and how to set traps to capture rodents and other mammals for study.

for the general public, presentation of information on the Internet, and preparation of educational materials such as posters, books, and popular articles are excellent ways to keep the public informed about the results of studies conducted on our mammalian fauna.

Research on most species of wild mammals in Alabama is regulated by laws and is conducted as authorized by appropriate permits issued by the United States Fish and Wildlife Service, the Alabama Department of Conservation and Natural Resources, and the research organization it-self (e.g., Institutional Animal Care and Use committees of Auburn University and The University of Alabama). Depending on the location of the research and the species studied, approval and permission may also be required from private landowners and from other federal and state agencies. Results may be presented in administrative reports submitted to funding and regulatory agencies, incorporated into theses and dis-sertations prepared as requirements for graduate degrees, published in peer-reviewed scientific journals, or some combination of these.

Because most species of mammals are not observed easily, a variety of methods are used to watch, capture, examine, and monitor them. We present only a few of the many methods because studies of mammals may include assessments of their ecology, behavior, parasites, populations, diet, interactions with other species, growth and development, anatomy,

Traps are placed at sites where small mammals might encounter them. Once opened and baited with rolled oats, traps are checked a few hours later and captured animals are removed for study.

physiology, diseases, biochemical composition, energetic requirements, reproduction, genetics, and myriad other aspects of their biology.

Rodents and other small mammals can be captured for study in small live traps baited with rolled oats or oatmeal. Larger species (e.g., eastern fox squirrel, woodchuck, swamp rabbit) and medium-sized species (e.g., raccoon, Virginia opossum) can be captured in larger live traps baited with sardines, apples, whole pecans, or other foods. Even larger live traps baited with scent lures, sardines, dog and cat food, and other food and nonfood items are used to capture larger species (e.g., gray fox, red fox, bobcat). Once captured, animals may be held in hand, placed into a suitable container for examination, or immobilized with drugs until species, sex, and reproductive condition are determined, measurements of size and weight are made, samples of tissues, blood, and parasites are obtained, ear tags or radiotransmitters are attached, or a combination of these. Under some circumstances, shooting, steel leghold traps, snap-back mouse and rat kill traps, and other kinds of kill traps may be used to secure specimens for research. Roadkill is also an important source of information about the distribution, abundance, and life history of mammals.

Mammals to be retained as voucher specimens are prepared as standard museum study skins, skulls, skeletons, or a combination of these.

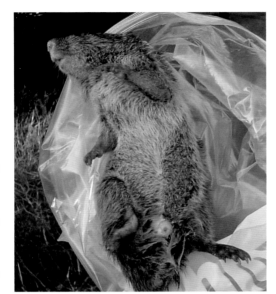

Upon capture, small mammals, such as this hispid cotton rat, are removed from traps and examined to determine species, sex, age, reproductive condition, and other information as needed.

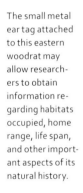

The small metal ear tag attached to this eastern woodrat may allow researchers to obtain information regarding habitats occupied, home range, life span, and other important aspects of its natural history.

Samples of tissues, ectoparasites, internal parasites, and feces may also be saved. These materials are deposited into appropriate research collections, where they are available for study by researchers from throughout the world. Voucher specimens serve as a permanent and verifiable record of the occurrence of a species at a particular place and time. Taxonomic identifications, refinements of distribution maps, estimates of abundance, information on diet, health, and parasitic fauna, and assessments of sexual, age, geographic, and genetic variation provide a basis for developing sensible management plans and reasonable conservation strategies. In addition to being used for research, these voucher specimens are used in a variety of educational activities, including oral presentations to students

Researchers unfurl a 20 x 2-meter mist net over a pond in northern Alabama. Most species of bats are attracted to sites with open water and abundant insects.

of all ages and the production of field guides, popular articles, television programs, and other resources that inform the general public about our diverse fauna.

Bats represent about 20% of our mammalian fauna. These animals are unique among mammals in their ability to fly, which provides a challenge in capturing them for study. Mist nets, which are fine-meshed nylon nets similar to gigantic hair nets, can be placed across a stream or pond, across a passageway in the woods, under a tree in a pasture, in front of a cave, or elsewhere and are effective in capturing bats unharmed. Although bats can detect the fine mesh of the mist net, naive bats often attempt to fly through the net and become entangled. Once a bat has been captured in

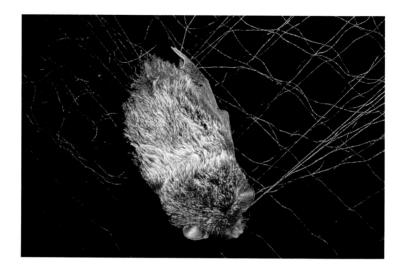

A silver-haired bat captured unharmed in a mist net.

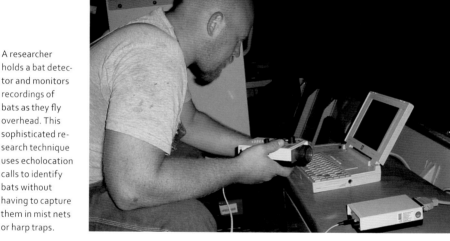

A researcher holds a bat detector and monitors recordings of bats as they fly overhead. This sophisticated research technique uses echolocation calls to identify bats without having to capture them in mist nets or harp traps.

a mist net, examined, and released, a researcher is unlikely to capture the same bat again because bats quickly learn to avoid the nets.

When researchers need to determine species and abundances of bats in an area without actually capturing them, one method is to use a bat detector to record echolocation calls of passing bats. Sophisticated computer programs then analyze these recordings to determine the number and species of bats present. This technique is especially useful in conducting surveys to determine the presence of endangered species in specific areas because it does not disturb routine activity patterns.

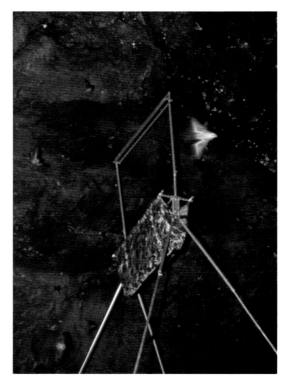

Harp traps are especially successful at capturing bats at openings of caves and mines. Captured bats fall gently into a collecting bag beneath the frame of the harp trap.

As with some other mammals that have predictable behaviors (e.g., elk, white-tailed deer), bats have certain behaviors that can be used to capture them for study. Because all bats that occur in the New World have a well-developed ability to echolocate, they are able to avoid flying into objects and can detect, pursue, and capture prey rapidly and with great precision. If a thin monofilament fishing line or wire is placed into its environment, the bat will usually detect and avoid it. If numerous lines or wires are placed vertically in a row at 2.5-cm intervals, the bat will detect the wires, turn onto its side, fly through the space between the wires, and immediately return to horizontal flight. A harp trap, which gets its name from the vertical, harp-like arrangement of its wires, takes advantage of the echolocation and agility of bats. If a second set of wires 2.5 cm apart is placed about 10 cm behind the first set of wires, the bat will detect the first set of wires, turn, fly through, return to a horizontal position, and strike the second set of wires. At this point, the bat gently falls unharmed into a bag attached to the bottom of the harp trap, where researchers can examine it.

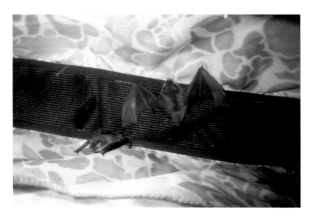

Gray myotis cannot fly out of the collecting bag beneath a harp trap. Species, age, sex, reproductive condition, and other information can be recorded and bats can be released unharmed within a few minutes of capture.

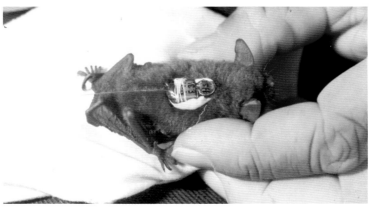

This gray myotis has a radiotransmitter attached to its back with nontoxic surgical glue. Three of these radiotransmitters weigh less than a penny. After about 2-3 weeks, the radiotransmitter and glue fall off of the bat.

Radiotransmitters are especially useful in monitoring the timing and duration of activity, assessing movement patterns, determining habitats occupied, and estimating the size and variation in home ranges of mammals. Some radiotransmitters weigh less than a dime, transmit signals 2–3 kilometers, and operate for about 3 weeks; others are much larger, transmit signals to orbiting satellites, and can function for 2 or more years. In addition to providing information on movement patterns, some radiotransmitters are designed to send a signal when a female is giving birth, to relay information about body temperature, and to notify researchers if the animal dies.

Digital game cameras are popular among sportsmen, the general public, and researchers. These cameras are effective for remotely monitoring wildlife, whether scouting for a trophy game animal, gathering information on the diversity and abundance of mammals and other wildlife, or just having fun capturing images of critters in the backyard or woods.

Researchers can use radiotelemetry to monitor direction and strength of radiotransmissions with a radio-receiver and antenna. Data from multiple locations or receivers can be used to determine foraging habitat, time of activity, and size of home range.

A female elk with a radiotransmitter that is capable of sending signals to a satellite. Researchers are able to monitor her movements from a remote location.

Various scented lures and food items can be used to attract species that might go undetected otherwise. Many people are surprised at the diversity of wildlife captured on digital cameras in Alabama; images often include Virginia opossums, eastern gray squirrels, raccoons, gray foxes, coyotes, and white-tailed deer.

Whether by casual observation of mammals as we travel in cities or

This researcher is checking a digital game camera at a monitoring site baited with corn, sardines, and dog food.

Raccoons are frequently attracted to cameras baited with any type of food. Specific kinds of scent lures, certain foods, or aluminum pie pans can be effective baits, depending on the species of interest.

in the countryside, by use of sophisticated equipment, or by trapping, it is amazing how many kinds of mammals we can encounter in Alabama. An early morning drive in any part of the state is likely to provide a good look at white-tailed deer and other species of wildlife. This abundance and diversity is part of our natural heritage, and we have a responsibility to conserve it for future generations. In the accounts that follow are summaries of what is known about the diverse mammalian fauna of Alabama. For convenience, we also include a table for conversions of metric measurements (Appendix 2) and a glossary of some common terms used in mammalogy.

Examination of the prepared skin and skull of this voucher specimen verifies that it is a meadow jumping mouse; that it was collected 18 October 2010 at Skyline Wildlife Management Area, Walls of Jericho, 34°59.896′N, 86°05.622′W, Jackson County, Alabama; that it was an adult male with testes 4 mm long by 2 mm wide; that standard measurements were taken; and that this was the first time this species was recorded from northeastern Alabama.

SPECIES ACCOUNTS
Class Mammalia

When we list characters that differentiate mammals from other organ-isms, we usually begin the list with the presence of hair and mammary glands. These characters are certainly enough to distinguish mammals from all other organisms, but numerous other traits are unique to this group and have allowed mammals to become successful inhabitants of the world for about 220,000,000 years. Having red blood cells that lack a nucleus and having a muscular diaphragm between the thoracic and abdominal body cavities have allowed for a more efficient exchange of respiratory gases (oxygen and carbon dioxide), which supports the higher metabolism and constant body temperature of most mammals. Well-developed facial muscles provide communication within and be-tween species; sweat glands provide evaporative cooling, removal of waste materials, and olfactory information; sebaceous glands lubricate and pro-tect hairs; and each hair has an erector muscle (arrector pili) that allows it to be raised for thermoregulation or communication. Even the 3 bones in the middle ear (stapes, incus, and malleus), the epiglottis, and the way the jaw articulates with the skull are unique characters of mammals. Ex-pressions of these traits vary greatly among mammals and, coupled with the myriad attributes shared with other organisms, make mammals es-pecially diverse in form and function, in the habitats they occupy, and in their behavior.

American Opossums

Order Didelphimorphia

This order ranges from southeastern Canada to southern Argentina and contains 1 family, 17 genera, and 87 species. Usually, the braincase is small, the rostrum is long, and there is a prominent sagittal crest. Members of Didelphimorphia occupy a wide range of habitats, but they inhabit primarily tropical or subtropical areas, where they may be locally abundant. The presence of 10 upper and 8 lower incisors distinguishes this order and family from all other mammals that are native to Alabama.

Family Didelphidae

Of the 17 genera and 87 species in this family, only 1 species occurs in the United States. All species have an opposable and clawless hallux (big toe). A marsupium is present in some species; in others, there is a fold of skin that protects the mammae, and in other species, there is no marsupium. The tail is usually long and prehensile. The family occurs only in North and South America. One species (Virginia opossum) is present in Alabama.

Virginia Opossum
Didelphis virginiana

IDENTIFICATION This is a medium-sized, relatively robust species. Hair may be pale or dark, with long, dark or pale guard hairs. The rostrum is narrow. Most of the tail is sparsely haired and darker on the proximal one-third than toward the end. The skull has a long rostrum, a small braincase, and an unusually large sagittal crest.

DENTAL FORMULA i 5/4, c 1/1, p 3/3, m 4/4, total = 50. No other species of mammal in Alabama has more than 44 teeth.

SIZE AND WEIGHT Average and range in size of 15 specimens from Alabama:

> total length, 607 (350–810) mm / 24.3 (14.0–32.4) inches
> tail length, 247 (142–312) mm / 9.9 (5.7–12.5) inches
> hind foot length, 55 (33–60) mm / 2.2 (1.3–2.4) inches
> weight, 989 (454–1,547) g / 34.6 (15.9–54.1) ounces

Average weight elsewhere in the geographic range is 2,800 g (98.0 ounces) for males and 1,900 g (66.5 ounces) for females. Males are larger than females.

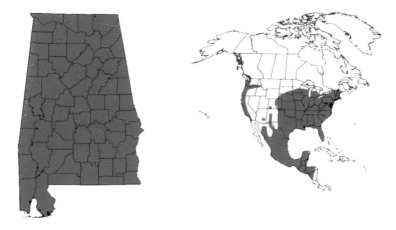

Distribution of the Virginia opossum in Alabama and North America.

DISTRIBUTION Statewide in Alabama. Geographic range of the Virginia opossum includes southeastern Canada, the eastern United States, the Pacific coastal region, a few scattered locations in the southwestern United States, and most of Mexico into Central America. The subspecies *D. v. virginiana* occupies most of eastern North America, including most of northern Alabama, and *D. v. pigra* occurs in southern Alabama.

ECOLOGY River bottoms and swamps are favored habitats, but Virginia opossums occur in all terrestrial habitats in the state. In suburban areas, hedges, piles of wood, fencerows, stream banks, and other habitats provide cover. Virginia opossums often occupy burrows excavated by other species. They are opportunistic omnivores and consume a wide range of foods including insects, carrion, fruits, garden vegetables, foods discarded by humans, and food that is left for pets. Predators include owls, coyotes, and domestic dogs. Once of moderate economic importance, the Virginia opossum was hunted for its fur and as food. Unfortunately, Virginia opossums are a common roadkill on our streets and highways.

BEHAVIOR Activity is usually nocturnal and begins at dusk and continues throughout the night. Death feigning, opening the mouth, baring the teeth, hissing, growling, and screeching are common defensive responses. During heat stress, Virginia opossums spread saliva on their body for evaporative cooling, and their sparsely haired tail also aids in cooling. When the animals are cold, shivering generates heat, and vasoconstriction of blood vessels, piloerection of hairs, and avoiding exposure to low temperatures conserve heat. The Virginia opossum is a strong and

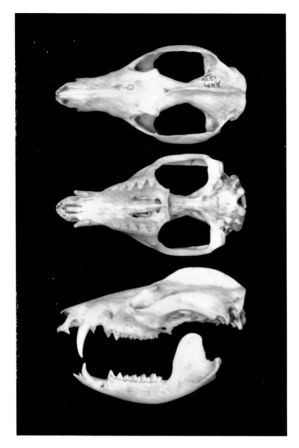

Dorsal, ventral, and lateral views of the cranium, and lateral view of the mandible of a male Virginia opossum. Greatest length of cranium is 111.1 mm.

slow swimmer on the surface of the water and can easily swim 100 m, but it can also swim underwater as an escape behavior. Virginia opossums wash their faces with their forefeet, licking them in a catlike fashion. When young are in their pouch, females lick their marsupium extensively. Except during mating, adults avoid other adults or are aggressive toward each other.

LIFE HISTORY Peak periods of reproduction are late January through late March and mid-May to early July. The scrotum is pendent and anterior to the hemipenis. Within the marsupium, there are usually 13 nipples (range is 9–17). There is a branched uterus with 2 lateral vaginae that receive sperm and a median vagina that serves as the birth canal. Birth of 18–21 altricial young occurs about 13 days after copulation. Because there are only about 13 teats and all of them are rarely used, average size of litter is 7–9. Two litters are usually produced each year. At birth, young weigh about 0.16 g and are 14 mm long; they quickly attach to a nipple. As they grow, the nipple enlarges and elongates in the mouth and detachment is unlikely. Young remain attached to the nipple for 50–65 days and after release continue to nurse until weaned, about 30–40 days later. Populations are mostly young animals, but there is considerable mortality in the first year of life, and turnover in populations is high. Average life span is probably 2–5 years, but some may live more than 7 years.

PARASITES AND DISEASES Ectoparasites include mites (*Androlaelaps, Archemyobia, Didelphilichus, Eulaelaps, Eutrombicula, Glycyphagus, Haemogamasus, Laelaps, Listrophorus, Macrocheles, Marsupialichus, Ornithonyssus, Pygmephorus, Walchia, Zibethacarus*), ticks (*Amblyomma, Dermacentor, Ixodes*), fleas (*Ctenocephalides, Echidnophaga, Orchopeas, Polygenis, Pulex, Xenopsylla*), and flies (*Lucilia*). Endoparasites include acanthocephalans (*Hamanniella*), cestodes (*Mesocestoides*), nematodes (*Anatrichosoma, Capillaria, Cruzia, Dipetalonema, Gnathostoma, Longistriata, Physaloptera, Trichinella, Viannaia*), and trematodes (*Brachylaeme, Brachylaima, Echinostoma, Fibricola, Mesostephanus, Phagicola, Rhopalias, Stictodora*). Diseases include canine distemper (*Morbillivirus*), canine hepatitis (*Mastadenovirus*), Chagas disease (*Trypanosoma*), eastern equine encephalomyelitis (*Alphavirus*), encephalomyocarditis (*Cardiovirus*), histoplasmosis (*Histoplasma*), leptospirosis (*Leptospira*), Lyme disease (*Borrelia*), murine typhus (*Rickettsia*), rabies (*Lyssavirus*), ringworm (fungal dermatophytes), Saint Louis encephalitis (*Flavivirus*), salmonella (*Salmonella*), toxoplasmosis (*Toxoplasma*), and tularemia (*Francisella*).

CONSERVATION STATUS Lowest conservation concern in Alabama.

COMMENTS *Didelphis* is from the Greek *di*, meaning "two," and *delph*, meaning "womb," referring to the paired uteri; *virginiana* refers to Virginia, origin of the type specimen. The subspecific name *pigra* is from the Latin *pigrus*, meaning "lazy."

REFERENCES Rausch and Tiner (1949), Hamilton (1958), Lumsden and Zischke (1962), Barr (1963), McManus (1974), Forrester (1992), Oliver et al. (1999), Best (2004*a*), Whitaker et al. (2007).

Dugongs, Manatees, and Sea Cows

Order Sirenia

Sirenia is derived from the word *siren*, the sea nymph of mythology, and dugongs and manatees may have been the basis of myths regarding mermaids. Geographic range of the order includes coastal regions of the western Pacific and Indian oceans from eastern Africa to southern Japan and Australia. Now extinct, Steller's sea cows (*Hydrodamalis gigas*) once occurred in the Bering Sea. Sirenians also occur along coastal regions of the western and eastern Atlantic Ocean from about 30°N to about 20°S latitude. The order contains 2 families, 3 genera, and 5 species. These are the only completely aquatic mammals that are herbivores. Forelimbs are paddle-like flippers, there are no hind limbs, and the large, horizontally flattened tail forms a broad paddle. Sirenians are closely related to elephants. One family (Trichechidae) occurs in Alabama.

Manatees

Family Trichechidae

Trichechidae contains 1 genus and 3 species. In addition to our West Indian manatee, the West African manatee (*Trichechus senegalensis*) also occupies freshwater and saltwater habitats (coastal ocean, bays, estuaries, and other shallow-water habitats) along the western coast of Africa, and the Amazonian manatee (*T. inunguis*) occupies only freshwater habitats throughout the Amazon and Orinoco drainages of South America. The distal end of the tail of manatees is rounded (spatulate) compared to the concave tail of dugongs. One species (West Indian manatee) occurs in Alabama.

West Indian Manatee
Trichechus manatus

IDENTIFICATION A large, streamlined aquatic mammal with flippers instead of front legs (nails are present), no hind legs, and a flattened, spatulate tail. Color is grayish, there may be dark or pinkish splotches dorsally and ventrally, and algae, barnacles, and other incrustations may be on the skin. Appears hairless, but there are scattered 30–45-mm hairs on the body and stiff, thick bristles on the upper lip.

DENTAL FORMULA In adults, no incisors or canines are present. At any time, manatees have 5–7 teeth on each side of the upper and lower jaws. These cheek teeth are replaced consecutively from the rear of the jaw as anterior teeth are worn down and lost. Up to 30 teeth may move through each quadrant of the jaw during the lifetime of a West Indian manatee. This adaptation is in response to the constant abrasion and wear of teeth caused by the coarse aquatic vegetation that is on sandy and muddy substrates.

SIZE AND WEIGHT Range in body length is 2.5 m (8.3 feet) to more than 4.5 m (14.9 feet), with corresponding weights of 200 kg (440 pounds) to

more than 1,600 kg (3,520 pounds). Average length is 3–4 m (9.9–13.2 feet) and average weight is less than 500 kg (1,100 pounds).

DISTRIBUTION Mobile Bay and adjacent waterways in southern Alabama northward to Claiborne Lock and Dam on the Alabama River. The West Indian manatee occurs southward from Florida and the Gulf of Mexico through the Caribbean Sea to northeastern Brazil, but it has been reported as far north as New Jersey and as far inland along the Mississippi River as Memphis, Tennessee.

ECOLOGY Present from May to November in Mobile Bay, the Mobile-Tensaw Delta, and adjacent waterways of Alabama. West Indian manatees have no known predators, but sharks could potentially prey on them. Scars from boat propellers and keels are evidence that humans may have a significant impact on populations. Although they have been reported to consume fish, West Indian manatees are nearly completely herbivorous. They usually forage on vegetation in shallow, coastal freshwater and saltwater habitats, but they may move inland up to 800 km along river chan-

Submerged with nostrils closed near a warm-water spring, this captive West Indian manatee is consuming lettuce provided by her caretakers.

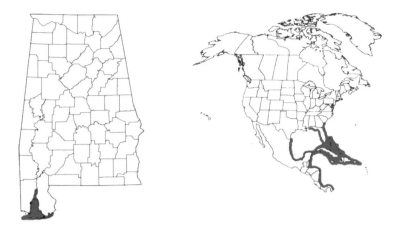

Distribution of the West Indian manatee in Alabama and North America.

nels. About 6–8 hours/day are spent foraging at depths of 1–3 m. Foraging is divided into 1–2-hour bouts followed by 2–4 hours of resting. In captivity, West Indian manatees consume 30–50 kg of food/day. They are well adapted to diving and may dive as deep as 10 m. Heart rate gradually decreases from 50–60 beats/minute to about 30 beats/minute during an 8-minute dive. Causes of death associated with humans include collisions with boats; drowning in fishing nets; ingestion of fishing hooks, fishing lines, and wire; and entrapment in locks, dams, and large pipes. West Indian manatees have been hunted for their skins, meat, oil, and bones, which were used as ivory. Although they have never been hunted commercially, killing by fishermen and boaters has led to small populations of this species.

BEHAVIOR Usually, West Indian manatees are solitary, paired (often a female and her young), or in small groups; larger groups may form near warm springs in inland waterways during winter. Acoustical abilities are excellent. Visually, they appear to be far sighted. The extreme density of their bones, their sluggish behavior, and their extremely low rate of oxygen consumption may be related to the unusual structure of their thyroid gland or to hypothyroidism. West Indian manatees rest 2–12 hours/day and even longer on cold days. Dorsoventral undulations of the tail move manatees forward; they also use their tail as a rudder for steering, banking, and rolling. They hold their flippers close to their sides.

LIFE HISTORY There are 2 mammae; 1 on the body beneath each flipper. The placenta is similar to that of elephants (Proboscidea) and hyraxes

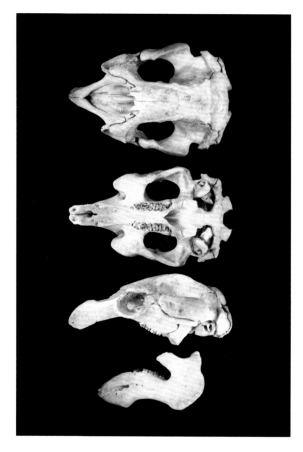

Dorsal, ventral, and lateral views of the cranium, and lateral view of the mandible of a West Indian manatee of unknown sex. Greatest length of cranium is 370.2 mm.

(Hyracoidea). Testes are abdominal and seminal vesicles are large. West Indian manatees become sexually mature at 8–10 years old. Breeding is year-round. When receptive, the female is promiscuous. Usually, 1 young is born following a gestation of about 13 months; sometimes there are twins. At birth, young weigh 11–27 kg and are about 1 m long. Young may nurse for 1–2 years. The reproductive potential is low; thus, losses from disease, boats, or other causes are difficult for a population to overcome. Longevity in the wild is unknown, but 1 captive animal lived more than 28 years. West Indian manatees do not breed in captivity.

PARASITES AND DISEASES Algae (*Compsopogen, Navicula, Zygnema*), cyanobacteria (*Lyngbya*), suckerfish (*Echineis*), and a variety of invertebrates may be on the skin, including the copepod *Harpacticus* and the turtle barnacle *Chelonibia*. Endoparasites include fungi (*Blastomyces,*

Fusarium), protozoans (*Eimeria*, *Toxoplasma*), cestodes (*Anoplocephala*), nematodes (*Heterocheilus*, *Plicalolabia*), and trematodes (*Chiorchis*, *Cochleotrema*, *Moniligerum*, *Nudacotyle*, *Opisthotrema*). West Indian manatees seem to be highly susceptible to pneumonia, other bronchial disorders, and a variety of microbial infections (*Aeromonas*, *Bacillus*, *Clostridium*, *Enterobacter*, *Klebsiella*, *Pasteurella*, *Proteus*, *Pseudomonas*, *Salmonella*, *Streptococcus*, *Vibrio*).

CONSERVATION STATUS Because of their habit of floating or swimming near the surface, West Indian manatees are especially vulnerable to boat propellers, vandalism, and poachers. Although West Indian manatees are relatively fast swimmers over short distances, boaters must take care to avoid collisions in areas where they occur. Highest conservation concern in Alabama, listed as endangered by the United States Fish and Wildlife Service, and listed on the International Union for Conservation of Nature and Natural Resources Red List of Threatened Species as vulnerable.

COMMENTS *Trichechus* is from the Greek *trichos*, meaning "hairs," and *manatus* is from the Haitian *manati*, meaning "big beaver." When sailors first saw manatees, some believed they were seeing mermaids.

REFERENCES Husar (1978), D. S. Hartman (1979), Forrester (1992), Odell (2003), Best (2004*a*), Lewis (2004), Pabody et al. (2009), R. H. Carmichael et al. (in litt.).

Armadillos

Order Cingulata

Present only in North and South America, this order contains 1 family, 9 genera, and 21 species. Because members have low basal metabolic rates and poor thermoregulatory abilities, they generally occur in warm regions. They do not enter torpor and they burrow extensively. Fossil species of this order include giant armadillos (family Glyptodontidae), some of which were 3 m long and weighed more than 1,100 kg.

Family Dasypodidae

The 9 genera and 21 species have a hard, armor-like carapace made of ossified dermal scutes that cover the head, body, and tail to varying degrees. Dorsally, these scutes are arranged as shields and bands separated by flexible skin with hair. Ventrally, there are scutes on the legs, but the soft skin on the venter is covered with hair. Species vary in size from about 120 g (pink fairy armadillo, *Chlamyphorus truncatus*) to 60 kg (giant armadillo, *Priodontes maximus*). One species (nine-banded armadillo) occurs in Alabama.

Nine-banded Armadillo
Dasypus novemcinctus

IDENTIFICATION A medium-sized mammal with 4 toes on the forefeet, a tapered rostrum, and a carapace with 8–11 (usually 9) bands of scutes. The carapace is divided into a scapular shield across the shoulders, a pelvic shield across the hips, and usually 9 bands that are bordered by flexible skin between the 2 shields and at the juncture of the head and tail. The tail is covered by 12–15 rings of scutes that decrease in size to a terminal end covered by irregular scutes.

DENTAL FORMULA No incisors or canines are present, but there are 8 small, peg-like teeth on each side of the upper and lower jaws.

SIZE AND WEIGHT Average and range in size of 3 adults from Alabama:
    total length, 730 (720–740) mm / 29.2 (28.8–29.6) inches
    tail length, 335 (330–340) mm / 13.4 (13.2–13.6) inches
    hind foot length, 72 (70–75) mm / 2.9 (2.8–3.0) inches
    weight, 4.6 (4.2–5.0) kg / 10.1 (9.2–11.0) pounds

Distribution of the nine-banded armadillo in Alabama and North America.

The ear is leathery and about 40 mm (1.8 inches) long. Males average larger (5.5–7.7 kg / 12.1–16.9 pounds) than females (3.6–6.0 kg / 7.9–13.2 pounds).

DISTRIBUTION Statewide in Alabama. Until the 1980s, nine-banded armadillos were known only from southern Alabama, but by the mid-1990s, they had extended their range across the state and into central Tennessee. The nine-banded armadillo occurs from the central United States to some Caribbean islands, through Central America, and into Argentina and Uruguay.

ECOLOGY Insects are the most common items in the diet, but nine-banded armadillos will also eat millipedes, centipedes, earthworms, snails, eggs, amphibians, reptiles, birds, baby mammals, carrion, fruits, and berries. A unique salivary bladder, surrounded by skeletal muscle, stores large amounts of highly viscous saliva. This bladder opens into the oral cavity, and the thick saliva coats the tongue and aids in feeding. The nine-banded armadillo usually digs its own burrows, which may be shared with cottontails, hispid cotton rats, striped skunks, and Virginia opossums, as well as a variety of invertebrates, such as spiders, mites, and camel crickets (*Ceuthophilus*). Cougars may eat nine-banded armadillos, but their only significant predators are humans. Humans have used nine-banded armadillos as food and have made their skins into boots, bags, and other items. Although nine-banded armadillos consume many

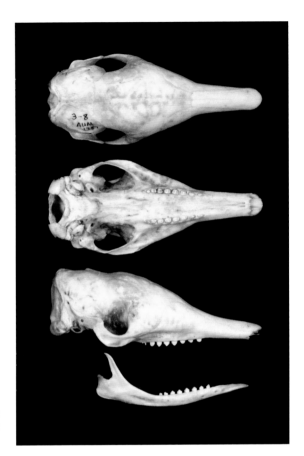

Dorsal, ventral, and lateral views of the cranium, and lateral view of the mandible of a male nine-banded armadillo. Greatest length of cranium is 96.2 mm.

harmful insects, they are often not welcome because of the damage they can cause to vegetable gardens and flower beds.

BEHAVIOR There is no apparent territoriality or aggressive behavior toward other nine-banded armadillos, but sick or injured individuals may be mutilated or cannibalized. They spend up to 17 hours/day sleeping in a burrow; at dusk, they exit and immediately begin to forage. As the night progresses, nine-banded armadillos move into more open habitats. On cool, cloudy, or rainy days, they may begin activity earlier in the day. When they forage, they use their keen sense of smell to search for food. Nine-banded armadillos move slowly along with their nose to the ground, stopping to root with their nose and dig small holes in the ground or leaf litter when they detect food. They may make a low, grunting sound while rooting and digging. Nine-banded armadillos can swim for considerable distances with a dog-paddling style, and they can also cross narrow bodies of shallow water by walking along the bottom. When alarmed, nine-banded armadillos arch their backs and jump straight up. They will also arch their backs when chased into a burrow, making it nearly impossible to pull them out.

LIFE HISTORY Nine-banded armadillos are believed to pair for the breeding season and the male and female will share the same burrow. Females have 4 mammae. Instead of a true vagina, they have a urogenital sinus that serves as a vagina and urethra. Testes descend no farther than the pelvis and there is no scrotum. Because of the carapace and ventral position of the genitalia, copulation occurs with the female lying on her back. First breeding is at about 1 year of age, with ovulation occurring from June to August. Following fertilization, the developing blastocyst moves to the uterus in 5–7 days, but it is about 4 months before implantation occurs. In March or April, usually identical quadruplets are born, but the number of newborns may be 1–7. At birth, young are fully formed, with eyes open. Within a few hours, they are walking, and in a few weeks, they accompany their mother on foraging expeditions. Even after weaning, the young may stay with their mother for several months.

PARASITES AND DISEASES Ectoparasites include mites (*Echimyopus, Marsupialichus, Ornithonyssus*), ticks (*Amblyomma*), and fleas (*Echidnophaga, Polygenis*). Endoparasites include protozoans (*Toxoplasma*), acanthocephalans (*Hamanniella, Oncicola, Travassosia*), cestodes (*Oochoristica*), nematodes (*Ascarops, Aspidodera, Delicata, Macielia, Mazzia, Moennigia, Physocephalus, Pintonema*), and trematodes (*Brachylaemus*). Diseases and disorders include Chagas disease (*Trypanosoma*), a leprosy-like disease (*Mycobacterium*), leptospirosis (*Leptospira*), nocardiosis (*Nocardia*), Saint Louis encephalitis (*Flavivirus*), salmonellosis (*Salmonella*), schistosomiasis (*Schistosoma*), and tularemia (*Francisella*). Renal disease is common. Because they are the only mammals other than humans that suffer from lepromatid leprosy, nine-banded armadillos have played a significant role in medical research concerning this disease.

CONSERVATION STATUS Lowest conservation concern in Alabama.

COMMENTS *Dasypus* is from the Greek *dasys* and *podos*, meaning "hairy foot." Armadillos do not have hairy feet; when Linnaeus named the animal he may have meant "rough-footed"; *novemcinctus* is from the Latin *novem* and *cinct*, meaning "nine-banded."

REFERENCES Taber (1939), Fitch et al. (1952), Talmage and Buchanan (1954), Wolfe (1968), McBee and Baker (1982), Forrester (1992), Best (2004*a*), Whitaker et al. (2007), Frey and Stuart (2009).

Rodents

Order Rodentia

This is the largest order of mammals and includes 33 families, 481 genera, and 2,277 species. Rodents occupy a wide range of habitats and occur naturally worldwide, except in New Zealand, Antarctica, and some oceanic islands. Most rodents are small (20–100 g); the largest is the South American capybara (*Hydrochoerus hydrochaeris*), which weighs up to 50 kg. Rodents are distinguished from other mammals by a single pair of upper and lower incisors that are ever growing and used for gnawing. They are important components of the diet of many predators, they serve as hosts for vectors and as reservoirs of disease, and they may be serious agricultural pests throughout the world. Eight families of rodents have been reported to occur in Alabama.

Squirrels

Family Sciuridae

This large and diverse family includes 51 genera and 278 species. Squirrels occur worldwide, except for some deserts and polar regions, southern South America, Australia, and Madagascar. Sciurids have well-developed and sharply pointed postorbital processes and are divided into 2 groups; the tree and ground squirrels, and the flying squirrels. Tree squirrels are generally diurnal and arboreal, and they live in cavities or build nests among branches; ground squirrels and marmots are terrestrial and burrow. Flying squirrels are distinguished by furred webbing (patagia) between their front and hind limbs that facilitates gliding. There are 5 species of squirrels in Alabama (eastern gray squirrel, eastern fox squirrel, southern flying squirrel, woodchuck, and eastern chipmunk).

Eastern Gray Squirrel
Sciurus carolinensis

IDENTIFICATION A gray, medium-sized tree squirrel that has a long bushy tail with white-tipped guard hairs, a whitish venter, and peg-like premolars in the upper jaw.

DENTAL FORMULA i 1/1, c 0/0, p 2/1, m 3/3, total = 22.

SIZE AND WEIGHT Average and range in size of 25 males and 25 females of *S. c. carolinensis*, respectively, from counties outside of Mobile County, Alabama:

> total length, 445 (413–476), 450 (401–474) mm / 17.8 (16.5–19.0), 18.0 (16.0–19.0) inches
>
> tail length, 206 (186–225), 209 (188–230) mm / 8.2 (7.4–9.0), 8.4 (7.5–9.2) inches
>
> hind foot length, 62 (55–69), 62 (57–66) mm / 2.5 (2.2–2.8), 2.5 (2.3–2.6) inches
>
> ear length, 30 (23–34), 31 (25–34) mm / 1.2 (0.9–1.4), 1.2 (1.0–1.4) inches

Nest of leaves (a drey) constructed by an eastern gray squirrel.

Average and range in size of 32 specimens of *S. c. fuliginosis* from Mobile County, Alabama:

  total length, 430 (380–477) mm / 17.2 (15.2–19.1) inches
  tail length, 195 (140–223) mm / 7.8 (5.6–8.9) inches
  hind foot length, 62 (57–66) mm / 2.5 (2.3–2.6) inches
  ear length, 31 (27–33) mm / 1.2 (1.1–1.3) inches
  weight of adults, 300–710 g / 10.5–24.9 ounces

There is no sexual dimorphism in size or color.

DISTRIBUTION Statewide, with the subspecies *S. c. carolinensis* present in all but extreme southwestern Alabama, where *S. c. fuliginosis* occurs. The species ranges throughout most of eastern North America from central Saskatchewan, Canada, southward to the central Gulf Coast of Texas.

ECOLOGY Eastern gray squirrels are most common in forests of oaks (*Quercus*), hickories (*Carya*), and walnuts (*Juglans*) where there is a diverse woody understory. When they climb, their hind feet can be reversed 180° from forward to permit a headfirst descent. Diet includes more than 100 kinds of plants and considerable animal matter. Common foods include nuts, buds, fruits, and flowers of trees such as oaks and hickories.

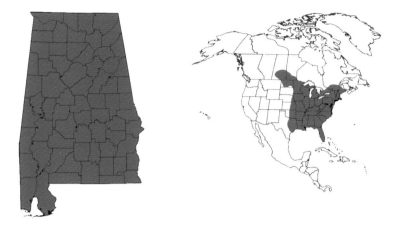

Distribution of the eastern gray squirrel in Alabama and North America.

Other items consumed are insects, bones, bird eggs, nestlings, frogs, and carrion. Predators are numerous and include snakes, hawks, owls, bobcats, long-tailed weasels, gray and red foxes, coyotes, and domestic dogs and cats. Eastern gray squirrels are an important game species and thousands are harvested each year for food. Although often considered pests at bird feeders, they are highly ranked by nature watchers.

BEHAVIOR Eastern gray squirrels are active throughout the day, but peaks in activity occur in early morning and late afternoon. These squirrels are scatterhoarders; they carry nuts in their mouth and bury them in shallow holes. Males dominate females and adults dominate juveniles. Residents are especially aggressive toward immigrants; combat is rare but may result in torn ears, broken tails, and other wounds. They use cavities most often in winter and leaf nests in summer. Leaf nests (or dreys) often consist of a platform of twigs on a tree limb, a compacted base of decaying matter, an outer shell of twigs and leaves, and usually a lining of shredded material. These nests may average 47 liters in size. Males may disperse long distances; there is a record of a male moving 100 km.

LIFE HISTORY Reproductive longevity is usually more than 10 years. The 8 mammae become black at the first pregnancy. Males begin following females 5 days before estrus, and as many as 34 males have been observed following a female at this time. When receptive, females will copulate with multiple males. Copulation lasts less than 30 seconds. Following a gestation of 44 days, 2–3 young are born (range is 1–8). In years

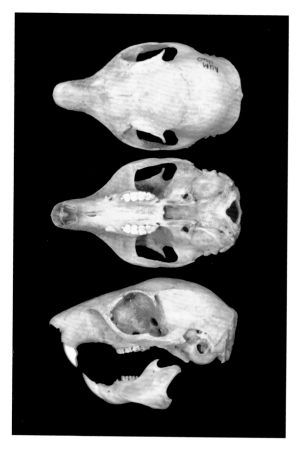

Dorsal, ventral, and lateral views of the cranium, and lateral view of the mandible of an eastern gray squirrel of unknown sex. Greatest length of cranium is 59.7 mm.

with a good mast crop, 2 litters may be produced. Neonates weigh 13–18 g, they are naked except for vibrissae on the rostrum, and they make lip-smacking noises and squeaking calls. At about 21 days, hair begins to grow on top of the tail, and teeth begin to erupt; at 21–28 days, ears open; and at 24–42 days, eyes open. By 4 weeks old, youngsters make a variety of calls. At 7–10 weeks old, young are weaned. By 8–9 months, they attain adult weight. Females may give birth at 5.5 months of age, but most do not reproduce until 15 months old. Males reach sexual maturity at 10–11 months. Average annual mortality of adults is 42–57%. Maximum longevity in the wild is 9 years for males and 12.5 years for females. In captivity, a female lived more than 20 years.

PARASITES AND DISEASES Ectoparasites include mites (*Androlaelaps, Cheyletus, Echimyopus, Eulaelaps, Eutrombicula, Haemogamasus, Laelaps,*

Leptotrombidium, Macrocheles, Ornithonyssus, Microtrombicula, Neo-trombicula, Parasecia, Walchia), ticks (*Amblyomma, Dermacentor, Ixodes*), lice (*Enderleinellus, Hoplopleura, Neohaematopinus*), and fleas (*Echidnophaga, Hoplopsyllus, Orchopeas, Pulex*). Eastern gray squirrels may also be infested with larval flies (*Cuterebra, Phaenicia*). Endoparasites include protozoans (*Eimeria, Hepatozoon, Trypanosoma*), acanthocephalans (*Moniliformis*), cestodes (*Catenotaenia, Hymenolepis, Raillietina, Taenia*), and nematodes (*Ascaris, Boehmiella, Capillaria, Citellinema, Dipetalonema, Heligmodendrium, Physaloptera, Rictularia, Strongyloides, Syphacia, Trichostrongylus*). Diseases and disorders include acute fatal toxoplasmosis (*Toxoplasma*), California encephalitis (*Orthobunyavirus*), carcinosarcoma mammary tumors, human ECHO virus (*Enterovirus*), leptospirosis (*Leptospira*), mange, pox virus–induced fibromas of the skin, plague (*Yersinia*), Q fever (*Coxiella*), ringworm (fungal dermatophytes), Saint Louis encephalitis (*Flavivirus*), tetanus (*Clostridium*), rabies (*Lyssavirus*), and tularemia (*Francisella*).

CONSERVATION STATUS Lowest conservation concern in Alabama.

COMMENTS *Sciurus* is from the Latin *sciurus*, meaning "a squirrel," and *carolinensis* is Latin for "of Carolina." The subspecific name *fuliginosis* is from the Latin *fuliginis*, meaning "sooty."

REFERENCES Rausch and Tiner (1948), Sharp (1959), Forrester (1992), Koprowski (1994*b*), Best (2004*a*), Whitaker et al. (2007), Whitaker and Mumford (2009).

Eastern Fox Squirrel
Sciurus niger

IDENTIFICATION This is the larger of the 2 species of tree squirrels in Alabama. Compared to the eastern gray squirrel, the eastern fox squirrel is more than 20% larger, is more orange-tawny, usually has black pelage on the head and face, and does not have peg-like premolars in the upper jaw.

DENTAL FORMULA i 1/1, c 0/0, p 1/1, m 3/3, total = 20.

SIZE AND WEIGHT Average and range in size of 14 specimens from Alabama:

> total length, 593 (526–716) mm / 23.7 (21.0–28.6) inches
> tail length, 303 (262–380) mm / 12.1 (10.5–15.2) inches

hind foot length, 73 (57–78) mm / 2.9 (2.3–3.1) inches

ear length of 7 specimens, 30 (27–33) mm / 1.2 (1.1–1.3) inches

weight of 5 specimens, 839 (584–1,047) g / 29.4 (20.4–36.7) ounces

There is no sexual dimorphism in color or size.

DISTRIBUTION Statewide in Alabama; the subspecies *S. n. niger* occurs in southeastern counties and *S. n. bachmani* is present throughout the remainder of the state. The eastern fox squirrel ranges from southern Saskatchewan and Manitoba, Canada, throughout the eastern United States (except New England) to the Rio Grande Valley of Texas, and into northeastern Mexico. Eastern fox squirrels have been widely introduced into the western United States.

ECOLOGY A variety of deciduous and mixed forests are occupied, but eastern fox squirrels are most common in small forests with an open understory. Densities are greatest in forests with trees that produce foods that can be stored in winter, such as oaks (*Quercus*), hickories (*Carya*), walnuts (*Juglans*), and pines (*Pinus*). Seeds of trees are consumed throughout the year, but buds, flowers, and fruits are also consumed when available. More than 100 kinds of plants serve as food, including oaks, hickories, walnuts, pines, sweet gums (*Liquidambar*), dogwoods (*Cornus*), mulberries (*Morus*), hawthorns (*Crataegus*), and maples (*Acer*). Eastern fox squirrels also eat insects, fungi, birds, eggs, and carrion. They may consume agricultural crops but rarely cause a significant reduction in the harvest. Predators are numerous and include snakes, hawks, owls, Virginia opossums, long-tailed weasels, gray and red foxes, coyotes, bobcats, and domestic dogs and cats.

BEHAVIOR Up to 9 nests may be used during a year. Leaf nests are used most often in warmer months and cavities in trees and burrows of other animals are used most often in winter. Leaf nests (or dreys) average 34 cm in diameter and are made of a platform of twigs placed on limbs, a shell of leaves and twigs, and often a lining of shredded or woven materials. Intraspecific threats are characterized by an upright posture with the tail over the back, accompanied by a tail flick and rapid approach by the aggressor. Submissive animals usually retreat. The most common vocalization is a series of barks. Chatter barks are given when startled, teeth

Distribution of the eastern fox squirrel in Alabama and North America.

chatters indicate restlessness or mild stress, a two-part scream is emitted by animals in distress, and a high-pitched whine is given during mating chases. Swimming is accomplished with a dog-paddling motion, and eastern fox squirrels are good sprinters on the ground.

LIFE HISTORY Females are capable of producing litters for more than 12 years. At the onset of her 1-day estrus, males begin following the female. When receptive, the female mates with multiple males. Copulation lasts less than 30 seconds. The 8 mammae turn black during the first pregnancy and remain that color. Gestation is 44–45 days. Usually, 2–3 young are born and only 1 litter is produced each year. Neonates are naked except for vibrissae on the rostrum, they weigh 13–18 g, and they are 50–60 mm long. By 3 weeks, downy hair grows on the dorsum of the tail, ears open, and lower incisors erupt. In week 5, hair grows on the ventrum of the tail, eyes open, and upper incisors erupt. Weaning occurs at 8–12 weeks. Molt to adult pelage occurs at 75–90 days of age. Males reach sexual maturity at 10–11 months; functional testes are pendent in the scrotum. Females may first bear young at 8 months, but most do not reproduce until more than 15 months old. Maximum longevity is 8.3 years for males and 12.6 years for females; a captive female lived for 13 years.

PARASITES AND DISEASES Ectoparasites include mites (*Androlaelaps, Atricholaelaps, Echimyopus, Eulaelaps, Eutrombicula, Haemogamasus, Laelaps, Listrophorus, Trombicula*), ticks (*Amblyomma, Dermacentor,*

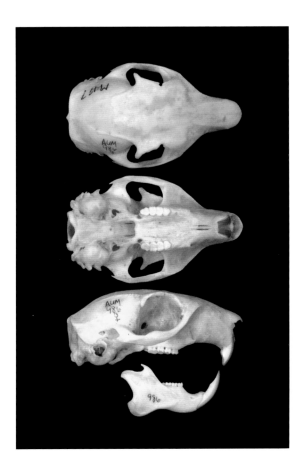

Dorsal, ventral, and lateral views of the cranium, and lateral view of the mandible of a female eastern fox squirrel. Greatest length of cranium is 67.0 mm.

Haemaphysalis, Ixodes), lice (*Enderleinellus, Hoplopleura, Neohaematopinus*), and fleas (*Ceratophyllus, Ctenocephalides, Echidnophaga, Hoplopsyllus, Leptosylla, Orchopeas*). Eastern fox squirrels may be parasitized by larval botflies (*Cuterebra*). Endoparasites include protozoans (*Eimeria, Trypanosoma*), acanthocephalans (*Macracanthorhynchus, Moniliformis*), cestodes (*Bothriocephalus, Catenotaenia, Choanotaenia, Cisticircus, Hymenolepis, Mesocestoides, Raillietina, Taenia*), and nematodes (*Ascaris, Boehmiella, Capillaria, Citellinema, Enterobius, Heligmodendrium, Physaloptera, Rictularia, Strongyloides, Trichinella, Trichostrongylus*). Diseases include California encephalitis (*Orthobunyavirus*), coccidiosis, leptospirosis (*Leptospira*), mange (*Cnemidoptes, Notoedres, Sarcoptes*), plague (*Yersinia*), rabies (*Lyssavirus*), tularemia (*Francisella*), and western equine encephalitis (*Alphavirus*).

CONSERVATION STATUS Several subspecies appear threatened by loss of suitable habitat. Low conservation concern in Alabama.

COMMENTS *Sciurus* is from the Latin *sciurus*, meaning "a squirrel," and *niger*, meaning "dark" or "black." The subspecific name *bachmani* honors John Bachman (1790–1874), an American naturalist.

REFERENCES Raush and Tiner (1948), Koprowski (1994*a*), Forrester (1992), Best (2004*a*), Whitaker et al. (2007).

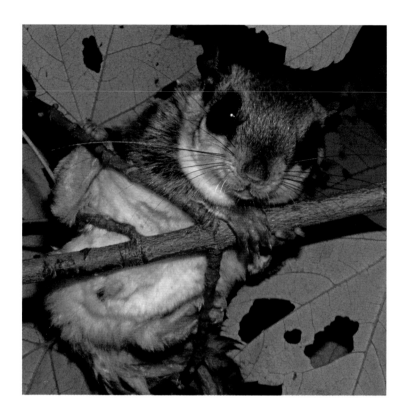

Southern Flying Squirrel
Glaucomys volans

IDENTIFICATION Front and hind legs are connected from wrist to ankle by a loose fold of furred skin (patagium). The densely haired tail is dorso-ventrally flattened, its sides are nearly parallel, and it is rounded at the tip. Pelage is dense, soft, and silky. Color is gray-brown on the back and head, and creamy white on the underside.

DENTAL FORMULA i 1/1, c 0/0, p 2/1, m 3/3, total = 22.

SIZE AND WEIGHT Average and range in size of 19 specimens from Alabama:

　　total length, 223 (199–248) mm / 8.9 (8.0–9.9) inches
　　tail length, 99 (82–115) mm / 4.0 (3.3–4.6) inches
　　hind foot length, 30 (27–35) mm / 1.2 (1.2–1.4) inches

An orphan southern flying squirrel drinking milk from a medicine dropper.

ear length of 7 individuals, 18 (16–20) mm / 0.7 (0.6–0.8) inch

weight of 5 specimens, 60.5 (35.1–75.7) g / 2.1 (1.2–2.7) ounces

There is no sexual dimorphism.

DISTRIBUTION Statewide in Alabama. Southern flying squirrels occur from northern Minnesota and southeastern Canada southward across the eastern United States. There are also isolated populations from western Mexico southward into Central America.

ECOLOGY Habitats are temperate and subtemperate pine-hardwood forests. Nests may be constructed on branches with sticks and leaves and lined with finely shredded bark, but they are often in cavities 4.5–6 m above the ground (range is 1.2–12 m). Openings into the nesting cavity are usually 40–50 mm in diameter, which is large enough to allow southern flying squirrels to enter and small enough to exclude tree squirrels. Interiors of nests are usually lined with finely divided inner bark, but moss, lichens, and feathers are used occasionally. In addition to cavities and woodpecker holes that are used in forested habitats, attics, barns, and birdhouses are common nesting sites in Alabama. Several secondary nests may be constructed near the primary nest and used for refuge, feeding, and defecation. Foods include a variety of plant matter, such as nuts, seeds, fruits, buds, blossoms, fungi, and bark. Southern flying squirrels may be the most carnivorous squirrels in North America; they consume insects and other invertebrates, eggs, nestlings (birds and mice), and

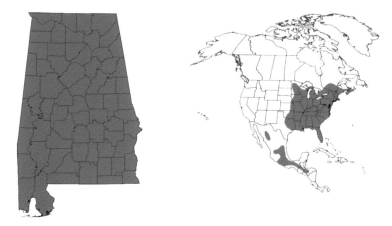

Distribution of the southern flying squirrel in Alabama and North America.

carrion when available. In autumn, oak and hickory trees are important sources for acorns and nuts, which are stored in the nest, in cracks and crevices, in woodpecker holes, and even in the ground. Predators include owls, hawks, snakes, domestic cats, bobcats, long-tailed weasels, and raccoons.

BEHAVIOR This is the only squirrel in Alabama that is active primarily at night. Although southern flying squirrels may run up trees, along branches, or across roofs, gliding in a descending curve from one tree to another is the primary means of moving longer distances. Glides are usually 6–9 m, but downslope glides of up to 90 m have been documented. During a glide, these squirrels are capable of making 90–180° turns to avoid obstacles. As they approach a tree at the end of a glide, they lower their tail, raise their head and shoulders, and slacken their gliding skin, which acts like an air brake. They then land gently on all four feet and run rapidly to the other side of the tree, possibly to avoid detection by predators. Although not considered to be true hibernators, southern flying squirrels may become torpid during cold weather. When disturbed during torpor, these squirrels may take up to 40 minutes to awake. In winter, 2 or more may huddle together in cavities or abandoned birdhouses. Adults make several distinct birdlike calls, including a chuck-chuck note, a high-pitched "tseet," and a soft sneeze-like call. Young produce high-pitched squeaks, some of which are ultrasonic.

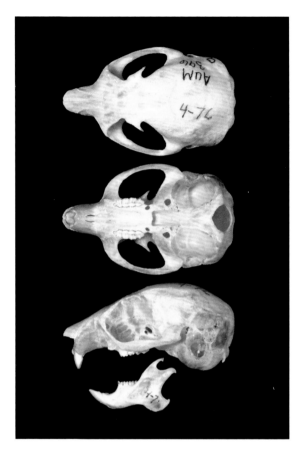

Dorsal, ventral, and lateral views of the cranium, and lateral view of the mandible of a female southern flying squirrel. Greatest length of cranium is 35.0 mm.

LIFE HISTORY Reproductive peaks for southern flying squirrels occur from April to May and from August to September, but it is not known if females produce more than 1 litter each year. Sexual maturity is reached 8–12 months after birth. Gestation is 40 days. In spring or summer, 3–4 young (range is 2–7) are born that weigh 3–5 g. They are pink and nearly hairless, and their eyes and ears are closed. The patagium is clearly visible at the time of birth. Ears open, hair begins to appear, and eyes open within 1–3 weeks. Weaning occurs 6–8 weeks after birth. Young may remain with their mother for weeks after weaning. Average life span is probably about 5 years, but some may live 10 years in the wild.

PARASITES AND DISEASES Ectoparasites include mites (*Androlaelaps, Euhaemogamasus, Haemogamasus, Psorergates, Trombicula*), lice (*Ender-*

leinellus, Hoplopleura, Neohaematopinus), and fleas (*Conorhinopsylla, Epitedia, Leptopsylla, Orchopeas, Peromyscopsylla, Polygenis*). Southern flying squirrels may be parasitized by larval botflies (*Cuterebra*). Endoparasites include protozoans (*Eimeria, Trypanosoma*), acanthocephalans (*Moniliformis*), cestodes (*Raillietina*), and nematodes (*Capillaria, Citellinema, Enterobius, Syphacia, Strongyloides*). The first record of rabies (*Lyssavirus*) in rodents was in a southern flying squirrel. Lice or fleas on southern flying squirrels may harbor the bacterium responsible for epidemic or louse-borne typhus (*Rickettsia*). Nutritional deficiencies have led to osteomalacia and alopecia (loss of hair) in animals kept in captivity.

CONSERVATION STATUS Lowest conservation concern in Alabama.

COMMENTS *Glaucomys* is from the Greek *glaukos*, meaning "gray" or "silvery," and *mys*, meaning "mouse"; *volans* is Latin for "to fly."

REFERENCES Dolan and Carter (1977), Forrester (1992), Best (2004*a*), Whitaker et al. (2007).

Woodchuck
Marmota monax

IDENTIFICATION This is the largest member of the family Sciuridae in Alabama. It has a relatively large body and short legs that are well adapted for digging. Overall color is dark brown interspersed with yellowish, white, black, and brown hairs, giving the pelage a frosted appearance. The tail is short and dark.

DENTAL FORMULA i 1/1, c 0/0, p 2/1, m 3/3, total = 22.

SIZE AND WEIGHT Average and range in size of 4 specimens from Alabama:

     total length, 607 (560–636) mm / 24.3 (22.4–25.4) inches
     tail length, 144 (142–155) mm / 5.8 (5.7–6.2) inches
     hind foot length, 88 (82–94) mm / 3.5 (3.3–3.8) inches
     weight of 2 specimens, 2.9 (2.7–3.0) kg / 6.4 (5.9–6.6) pounds

DISTRIBUTION Northern Alabama. Woodchucks occur from central Alaska southward to northern Idaho, across most of southern Canada to the Atlantic Ocean, and southward from Manitoba and North Dakota into Louisiana, Mississippi, Alabama, and Georgia.

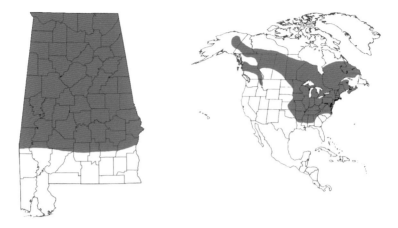

Distribution of the woodchuck in Alabama and North America.

ECOLOGY Generally, woodchucks occupy lowland habitats between woodlands, where they hibernate, and fields, where they breed and forage. They also live near fencerows, vegetated gullies, banks of streams or lakes, near human-made structures, and other sites suitable for constructing burrows. During wildfires, most woodchucks survive by staying inside their burrows, where they avoid exposure to flames and extreme heat. Woodchucks are true hibernators and rely solely on body fat to get them through winter. There is an annual cycle of weight gain and loss. Depending on age and sex, individuals may gain 12–20 g/day in summer and lose 2–9 g/day during hibernation and arousal in spring. Diet consists of a variety of vegetation and invertebrates, including beetles, snails, grasshoppers, and leaves of hackberry (*Celtis*), maple (*Acer*), peach (*Prunus*), and mulberry (*Morus*) trees. Predators include humans, domestic dogs, foxes, coyotes, long-tailed weasels, American minks, American black bears, bobcats, owls, hawks, and rattlesnakes.

BEHAVIOR Although active primarily during the day, woodchucks may also be active at night, especially when food is in short supply. In northern Alabama, woodchucks are often seen foraging, basking, scratching, or grooming along railroad embankments or along highway rights-of-way. Interactions between males and females end after copulation and females and young do not associate after weaning and dispersal. However, interactions between individuals may include greeting nuzzles, visual threats (arched back, erect tail that may be flipped up and down, open mouth exposing incisors), teeth chattering, physical attacks and fighting,

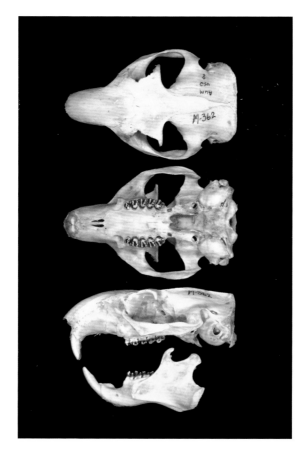

Dorsal, ventral, and lateral views of the cranium, and lateral view of the mandible of a female woodchuck. Greatest length of cranium is 97.8 mm.

or subtle movements of the head and body. During the breeding season in early spring, males fight with other males to establish dominance hierarchies and to defend hibernacula of females. Woodchucks are vocal animals and may squeal, chatter, bark, or give a loud, shrill whistle. Burrows often have up to 11 entrances, with an apron of excavated soil at the openings. In addition, plunge holes are dug up to the surface from inside the burrow; thus, an apron of soil is not present at these inconspicuous openings. These plunge holes can be used to escape into or out of the burrow. Burrows and trails between burrows are scent-marked, indicating territoriality. Woodchucks are good swimmers and may climb trees to forage or escape from predators. Sites used for hibernation include hedgerows, steep inclines on rocky ground, woods, haystacks, and sites with adequate drainage and southern exposure. During hibernation, dens may be shared with eastern cottontails, Virginia opossums, raccoons, or striped skunks. Typically, adult males emerge from hibernation before females and subadults.

LIFE HISTORY Mating occurs soon after emergence from hibernation. Gestation is 31–32 days and a litter of 1–9 (average is 4) is born in spring. At birth, average weight is 27 g, ears are closed, pinnae of ears are erect, eyes are closed and appear as darkened areas on the face, and teeth have not erupted. Newborns are nearly hairless, wrinkled, and pinkish. Within the first 3 weeks after birth, the skin has become pigmented, short black hair has appeared, and the young begin to crawl. At about 4 weeks, eyes open and feeding on solid foods begins. By 5–6 weeks, weaning, whistling, and teeth chattering occur, and young begin to emerge from the burrow. Young are fully grown in 2–3 years. Males become sexually mature at about 2 years of age, and most females have their first litter when 2 years old. Life span is 4–6 years, but some have lived nearly 10 years in captivity.

PARASITES AND DISEASES Ectoparasites include mites (*Androlaelaps, Eulaelaps, Macrocheles*), ticks (*Ixodes*), lice (*Enderleinellus*), and fleas (*Ceratophyllus*). Woodchucks may be parasitized by larval botflies (*Cuterebra*). Endoparasites include protozoans, nematodes (*Ascaris, Citellina, Citellinema, Obeliscoides*), and trematodes (*Quinqueserialis*). Diseases and disorders include arteriosclerosis, fibroma, hepatitis, hepatoma, rabies (*Lyssavirus*), Rocky Mountain spotted fever (*Rickettsia*), and tularemia (*Francisella*). Woodchucks have been used as laboratory animals in medical studies.

CONSERVATION STATUS Lowest conservation concern in Alabama.

COMMENTS *Marmota* is from the French *marmotte*, meaning "mountain mouse"; *monax* is from the Greek *monachos*, meaning "solitary" or "a monk."

REFERENCES Hamilton (1934), Raush and Tiner (1948), Chandler and Melvin (1951), Grizzell (1955), Young and Sims (1979), Kwiecinski (1998), Best (2004*a*), Whitaker et al. (2007).

Eastern Chipmunk
Tamias striatus

IDENTIFICATION A small squirrel with a dark brown stripe along the middle of the back from head to rump and 3 prominent stripes along each side of the back. Color varies from the dark brown and white stripes on the back to mixtures of gray, brown, and reddish orange on the head, sides, and rump. Venter is creamy white. There are pale stripes above and below the eyes.

DENTAL FORMULA i 1/1, c 0/0, p 1/1, m 3/3, total = 20.

SIZE AND WEIGHT Average and range in size of specimens from Alabama:
 total length of 21 specimens, 230 (175–311) mm / 9.2 (7.0–12.4)
  inches
 tail length of 18 specimens, 89 (67–124) mm / 3.6 (2.7–5.0)
  inches
 hind foot length of 21 specimens, 35 (27–38) mm / 1.4 (1.1–1.5)
  inches
 weight of 6 specimens, 104.2 (65.8–125.7) g / 3.7 (2.3–4.4)
  ounces

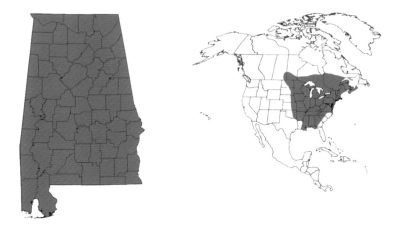

Distribution of the eastern chipmunk in Alabama and North America.

DISTRIBUTION Statewide in Alabama. Geographic range of the species is from southern Saskatchewan eastward across southeastern Canada and southward into Louisiana, Mississippi, Alabama, and northern Florida. Eastern chipmunks are absent from much of the coastal plain in the southeastern United States.

ECOLOGY Eastern chipmunks inhabit primarily open brushy habitats in mature deciduous forests, but they are common in urban areas that provide adequate cover and resources. Diet includes a variety of fungi, plants, invertebrates (especially insects), and vertebrates (frogs, snakes, birds, mice). The internal cheek pouches are useful in collecting and transporting food items, which may include seeds, nuts, and acorns that are stored in burrows for use in winter or even during the following spring and summer. When full, cheek pouches are about the size of the head. Predators include domestic cats, bobcats, coyotes, gray and red foxes, long-tailed weasels, snakes, and hawks. Although usually harmless, eastern chipmunks may damage lawns, flower beds, and retention walls by burrowing; patios, insulation, and electrical fixtures by gnawing; and gardens, seedlings, and flower bulbs by harvesting for food.

BEHAVIOR Eastern chipmunks are active only during the day. They defend the area around their burrow with threats, chases, and fights. Both sexes vocalize. Most vocalizations are believed to be alarm calls, but they may announce territories or the presence of predators. The most prominent call is a high-pitched chipping sound that may be repeated 130 times/minute for more than 10 minutes. Sometimes several chipmunks

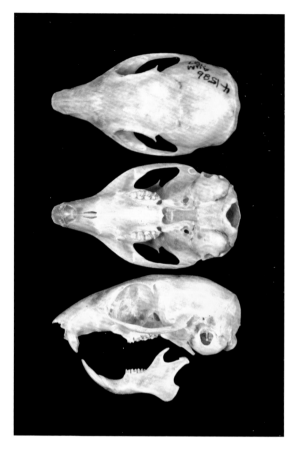

Dorsal, ventral, and lateral views of the cranium, and lateral view of the mandible of a female eastern chipmunk. Greatest length of cranium is 41.4 mm.

form a chorus of these chipping calls. During the breeding season, males actively inspect burrow openings and core areas of other chipmunks, and they congregate in the home range of females in estrus. When 2–3 males compete for a female, the dominant male mates with her, but when 7–8 males are competing, the dominant male becomes so involved in chasing his rivals that the female usually mates with 1 or more of the subdominant males. Burrows do not have an apron of dirt at the openings, so they are usually inconspicuous; during excavation, dirt is carried away. Burrows range in complexity from a single tunnel to tunnels 10 m long that have 1 or more large storage galleries, many smaller galleries, and short side pockets where food or debris is stored. Burrows are 40–60 mm in diameter; usually, only 1 is actively used and there are plugged openings that frequently reach the surface. Nests are made of chewed or crushed leaves. Eastern chipmunks do not hibernate. From late autumn to early spring, they stay underground in various degrees of torpor and consume stored food, but some venture aboveground in favorable weather.

LIFE HISTORY Two litters may be born each year; most mating occurs from February to April and from June to July. Estrus may last only about 6.5 hours, although females may appear to be receptive to males for 3–10 days. Gestation is 31–32 days and litters average 4–5 young (range is 1–9). Newborns are pink and nearly hairless, eyes and ears are closed, and they weigh 2.5–5 g. By 1 month old, young weigh about 30 g, their eyes and ears are open, and they are well furred. Young first emerge from the burrow 5–7 weeks after they are born. Within 2 weeks following emergence, young establish their own home range. By about 12 weeks, they are fully grown and have fully formed dentition. By 1 year, dentition is well worn. Sexual maturity is reached in spring or summer following their first winter. Average life span is probably 2–3 years, but some have lived more than 8 years in the wild.

PARASITES AND DISEASES Ectoparasites include mites (*Androlaelaps, Aplodontopus, Dermacarus, Echinonyssus, Eucheyletia, Euschoengastia, Eulaelaps, Haemogamasus, Laelaps, Lepidoglyphus, Lynxacarus, Macrocheles, Pygmephorus, Sciurocoptes, Trombicula, Xenoryctes*), ticks (*Dermacentor, Haemaphysalis, Ixodes*), lice (*Hoplopleura*), and fleas (*Ceratophyllus, Ctenophthalmus, Epitedia, Megabothris, Monopsyllus, Nosopsyllus, Orchopeas, Peromyscopsylla, Stenoponia, Tamiophila*). Larval botflies (*Cuterebra*) are common in warmer months and they can cause anemia, splenomegaly (enlargement of the spleen), and elevated counts of white blood cells, but usually not death. Endoparasites include protozoans (*Eimeria, Trypanosoma*), acanthocephalans (*Macracanthorhynchus, Moniliformis*), cestodes (*Cladotaenia, Hymenolepis, Paruterina, Taenia*), nematodes (*Capillaria, Rictularia, Spirura, Trichostrongylus, Trichuris*), and trematodes (*Postharmostomum, Scaphiostomum*). Diseases include California encephalitis (*Orthobunyavirus*).

CONSERVATION STATUS Lowest conservation concern in Alabama.

COMMENTS *Tamias* is Greek for "a storer," referring to the food-storing habits of chipmunks; *striatus* is Latin for "striped."

REFERENCES Snyder (1982), Best (2004*a*), Jaffe et al. (2005), Whitaker et al. (2007).

Beavers

Family Castoridae

This family contains 1 genus and 2 species. Well adapted to a semiaquatic life, beavers are about 1 m long and may weigh more than 30 kg. Hind feet are large and digits are webbed. The tail is broad, flat, scaly, and mostly hairless. The family occurs from northern Alaska and Canada to northern Mexico in North America and across most of northern Europe and Asia in the Old World. One species (American beaver) occurs in Alabama.

American Beaver
Castor canadensis

IDENTIFICATION This is the largest rodent in North America. American beavers are tan to dark reddish brown and they have prominent, ever-growing incisors and webbed toes on the hind feet. The tail is dorsoventrally flattened, scaled, and relatively hairless.

DENTAL FORMULA i 1/1, c 0/0, p 1/1, m 3/3, total = 20.

SIZE AND WEIGHT Range in size:
    total length, 1,000–1,200 mm / 40–48 inches
    tail length, 258–365 mm / 10.3–14.6 inches
    tail width, 90–200 mm / 3.6–8.0 inches
    hind foot length, 156–205 mm / 6.2–8.2 inches
    ear length, 23–29 mm / 0.9–1.2 inches
    weight, 11–26 kg / 24.2–57.2 pounds
There is no sexual dimorphism.

DISTRIBUTION Statewide in Alabama. Geographic range of the species is from northwestern Alaska eastward and southward across most of Canada and southward across the United States into northern Mexico.

ECOLOGY American beavers occupy forested areas near significant bodies of water. This species can close its mouth behind its incisors, allowing it to transport logs and branches without taking in water. Using their incisors, American beavers cut down trees and use them as construction materials for dens and dams. They construct watertight dams from mud and wood, sometimes across fast-flowing streams. Dams may be long and cause large areas to be flooded. Dens (lodges) also are made of wood and mud and are usually surrounded by open water, but they may join dry land when the water level is low. Dens have 1 or 2 openings from under the water into a chamber a few centimeters above water level. Chambers are usually about 2–3 m wide and up to 1 m tall. American beavers also construct canals that are about the same width and depth as their body; these canals are especially useful for transporting parts of trees across areas with shallow water. Diet includes leaves, twigs, and bark of most species of woody plants that occur near water, as well as cattails (*Typha*), water lilies (*Lotus*), and many other kinds of aquatic vegetation. Caches of logs and branches may be stored underwater for use as food in winter. Modifications of the environment by American beavers have a large effect on which other species can live in that ecosystem; these species include a

Lodges of American beavers are used for refuge and as a place to rear young.

Dams constructed by American beavers provide access to food, afford protection from predators, and create a critical habitat for many other species.

wide variety of microbes, plants, invertebrates, and vertebrates. In Alabama, the only significant predator of American beavers is humans, but domestic dogs, coyotes, American minks, North American river otters, and long-tailed weasels may kill young American beavers.

BEHAVIOR American beavers live in colonies of up to about 8 related individuals and are active primarily from late afternoon to early morning. They are excellent swimmers. In addition to other adaptations for swimming (such as webbed hind feet and flattened tail), their nostrils and ears close, and a membrane protects their eyes and allows them to see as they swim underwater The sound of running water stimulates them to build or repair dams. During the day, all members of the colony rest, either together in a single den, in subgroups in the den, or in a burrow dug into the bank. Interactions between American beavers may include mutual grooming, nose touching, and play-like wrestling. They communicate by posturing, vocalizing, tail slapping, and scent-marking piles of mud in their home range. They scent-mark using castoreum, a yellowish secretion of the large castor sac, in combination with urine. Humans use castoreum as a lure for trapping animals, in some perfumes, as a food ad-

Distribution of the American beaver in Alabama and North America.

ditive, and as a supposed aphrodisiac. Tail slapping is probably a danger signal to other American beavers, or it may frighten predators, or both. Aggression is characterized by hissing, grunting, and teeth grinding.

LIFE HISTORY American beavers mate for life, but if 1 of the pair dies, the survivor will find a new mate. Breeding usually occurs from January to February and gestation is about 107 days. In May or June, 3–4 young are born, but births may occur during any month. At birth, the body is fully furred, eyes are at least partially open, incisors have erupted, and weight is usually 340–630 g. Weaning occurs when young are 6–8 weeks old. Dispersal of up to 16 km occurs at about 2 years of age, but some have dispersed up to 110 km from where they were born. Both sexes may attain sexual maturity at about 1.5 years old, but most American beavers breed first at about 2–3 years old. Average life span is probably 5–7 years, but a life span of more than 20 years has been documented.

PARASITES AND DISEASES Ectoparasites include mites (*Androlaelaps, Ornithonyssus, Prolabidocarpus, Schizocarpus, Zibethacarus*), ticks (*Dermacentor, Ixodes*), larval flies (*Cochliomyia*), and beetles (*Platypsyllus, Leptinillus*). Endoparasites include protozoans (*Eimeria*), nematodes (*Capillaria, Castorstrongylus, Travassosius*), and trematodes (*Renifer, Stephanoproraoides, Stichorchis*). Diseases include tularemia (*Francisella*), which can be spread in water or by ticks.

CONSERVATION STATUS Historically, American beavers were a primary stimulus for the exploration of North America. By the late 1800s, popu-

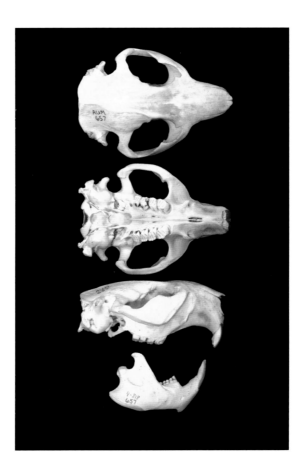

Dorsal, ventral, and lateral views of the cranium, and lateral view of the mandible of a male American beaver. Greatest length of cranium is 131.4 mm.

lations had been greatly reduced due to hunting and trapping. By the late 1900s, the species had been widely reintroduced, some populations had recovered, and some populations had grown to the point that they had to be controlled because they were considered to be pests. Lowest conservation concern in Alabama.

COMMENTS *Castor* is from the Greek *kastōr*, meaning "beaver"; *canadensis* is Latin for "of Canada."

REFERENCES Erickson (1944*b*), Jenkins and Busher (1979), Best (2004*a*), Whitaker et al. (2007).

Pocket Gophers

Family Geomyidae

Represented by 6 genera and 40 species, this is 1 of the 2 families of mammals with external, fur-lined cheek pouches; the other family is Heteromyidae, which contains kangaroo rats and pocket mice. Pocket gophers are 13–40 cm long and highly specialized for subterranean life; the body is stocky, the neck is short, the eyes and ears are small, the forefeet have long and powerful claws, and the sparsely haired tail is short and serves as a tactile organ. Pocket gophers occur from central British Columbia and Saskatchewan, Canada, across North America southward to northern Colombia, except north of Georgia and east of the Mississippi Valley. One species of pocket gopher (southeastern pocket gopher) is present in Alabama.

Southeastern Pocket Gopher
Geomys pinetis

IDENTIFICATION This medium-sized rodent has external fur-lined cheek pouches that open on either side of the mouth, small eyes, reduced pinnae, front claws that are elongated for digging, a nearly hairless tail, and a stout, muscular-appearing body. The front surface of each large upper incisor has 1 wide and 1 narrow groove. Color is brown on the back, paler brown on the shoulders, sides, and flanks, and grayish on the underside.

DENTAL FORMULA i 1/1, c 0/0, p 1/1, m 3/3, total = 20.

SIZE AND WEIGHT Average and range in size of 6 males from Alabama:
total length, 220 (182–261) mm / 8.8 (7.3–10.4) inches
tail length, 72 (54–86) mm / 2.9 (2.2–3.4) inches
hind foot length, 28 (19–31) mm / 1.1 (0.8–1.2) inches
ear length of 4 specimens, 4 (3–6) mm / 0.2 (0.1–0.2) inch
weight of 4 specimens, 159.4 (104.8–233.4) g / 5.6 (3.7–8.2) ounces
Males average larger than females.

DISTRIBUTION Most of southern and eastern Alabama. Geographic range of the species includes southern Alabama and Georgia and extends southward across most of northern Florida.

ECOLOGY Southeastern pocket gophers are associated with rolling topography, sandy soils, longleaf pines (*Pinus palustris*), and turkey oaks (*Quercus laevis*). They also inhabit highway rights-of-way, parks, lawns, golf courses, orchards, cemeteries, and pasturelands. Burrowing serves to mix and aerate the soil. Burrows provide shelter for a variety of organisms including insects, toads, frogs, salamanders, snakes, and small mammals. Diet is primarily roots, tubers, bulbs, and stems of plants, but southeastern pocket gophers may forage on the surface for grasses, forbs, and sedges, which they can collect and transport back to the burrow in their external cheek pouches. Potential predators of southeastern pocket gophers are snakes, owls, coyotes, long-tailed weasels, American minks, and eastern spotted skunks.

BEHAVIOR Southeastern pocket gophers are solitary, territorial, and aggressive. Except for mating and when young are present, only 1 individ-

Mounds of soil pushed to the surface by southeastern pocket gophers during excavation of underground passageways are evidence of their presence in an area.

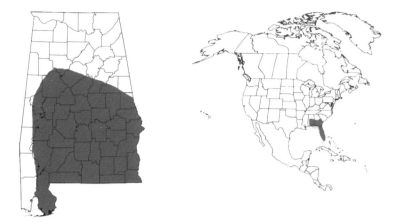

Distribution of the southeastern pocket gopher in Alabama and North America.

ual inhabits a burrow. Southeastern pocket gophers are active all year, at all times of the day, and they do not hibernate. Long hairs on the snout, scattered hairs on the tail, and a well-developed sense of smell provide sensory information in their dark, subterranean environment. Although southeastern pocket gophers may venture onto the surface in search of food or to disperse, they spend almost all of their life underground. Southeastern pocket gophers are excellent swimmers, and air trapped in their fur allows them to float, potentially enabling them to disperse across open water and to survive floods. During construction of their subterranean burrows, they push mounds of soil to the surface from side tunnels, which are then plugged with soil. On the surface, these mounds of soil are usually about 30 cm wide, 15 cm tall, and 2–4 m apart, and they form a crooked line of 8–10 mounds across the ground. Main burrows are usually 30–60 cm below the surface and about 6–8 cm in diameter, and they may extend more than 100 m with many branches and side tunnels.

LIFE HISTORY Breeding occurs throughout the year. Each year, up to 2 litters of 1–3 young are born after a gestation of about 20–30 days. Young are born tail first with eyes, ears, and cheek pouches closed. Newborns are nearly hairless and their teeth are not erupted; they weigh about 6 g and are about 50 mm long. Young are weaned and disperse within 1–2 months of birth. Females reach sexual maturity at 4–6 months of age. Life span is unknown but is probably 3–4 years.

PARASITES AND DISEASES Ectoparasites include mites (*Androlaelaps,*

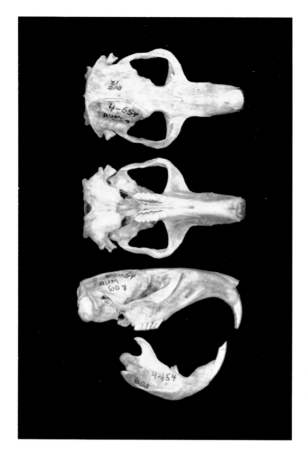

Dorsal, ventral, and lateral views of the cranium, and lateral view of the mandible of a male southeastern pocket gopher. Greatest length of cranium is 53.1 mm.

Geomylichus), ticks (*Amblyomma*), lice (*Geomydoecus*), and fleas (*Echidnophaga*). Endoparasites include nematodes (*Capillaria, Mastophorus*).

CONSERVATION STATUS The southeastern pocket gopher is a species of special concern in Alabama. One subspecies in the Southeast is now extinct and other populations of this species may be threatened. High conservation concern in Alabama.

COMMENTS *Geomys* is from the Greek *ge*, meaning "earth," and *mys*, meaning "mouse"; *pinetis* is Latin for "of the pines."

REFERENCES Best and Hart (1976), Pembleton and Williams (1978), Forrester (1992), Best (2004*a*), Whitaker and Hamilton (1998), Whitaker et al. (2007).

Jumping Mice and Jerboas
Family Dipodidae

The 16 genera and 51 species in this family occur in North America, Europe, Asia, and across much of Africa. Many species have an elongated tail and hind limbs adapted for saltatorial (hopping or jumping) locomotion, but some have a shorter tail and less elongated hind feet. Habitats include deserts, semiarid plains, grasslands, marshes, and coniferous and deciduous forests. One species (meadow jumping mouse) occurs in Alabama.

Meadow Jumping Mouse
Zapus hudsonius

IDENTIFICATION A yellowish mouse with coarse hair, a dark band of hair along the back from head to rump, paler sides, a creamy white to yellowish underside, long hind feet, and a tail that is sharply bicolored (brownish above and yellowish below) and longer than the body. Each upper incisor is strongly grooved.

DENTAL FORMULA i 1/1, c 0/0, p 1/0, m 3/3, total = 18.

SIZE AND WEIGHT Average and range in size of 18 specimens from Alabama:

    total length, 192 (175–225) mm / 7.7 (7.0–9.0) inches
    tail length, 114 (97–127) mm / 4.6 (3.9–5.1) inches
    hind foot length, 27 (24–29) mm / 1.1 (1.0–1.2) inches
    ear length of 13 specimens, 11 (8–13) mm / 0.4 (0.3–0.5) inch
    weight of 12 specimens, 16.3 (10.5–23.0) g / 0.6 (0.4–0.8) inch
There is no sexual dimorphism.

DISTRIBUTION Northeastern Alabama. Meadow jumping mice occur from western Alaska across most of southern Canada and the northern

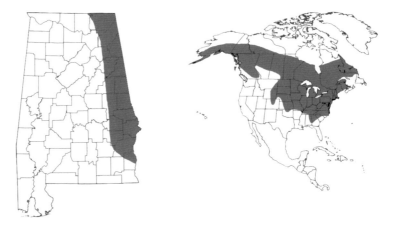

Distribution of the meadow jumping mouse in Alabama and North America.

Great Plains, throughout most of the eastern United States, and as far south as east-central Alabama.

ECOLOGY Meadow jumping mice occur primarily in fallow, moist, and woody agricultural fields, but also in thick vegetation along ponds, streams, and marshes, and in wooded areas when herbaceous growth is adequate. Diet includes seeds (especially of grasses), insects (mostly beetles and larval moths), fungi (*Endogone*), nuts, fruits, and berries. Predators probably include coyotes, gray and red foxes, long-tailed weasels, owls, and red-tailed hawks (*Buteo jamaicensis*).

BEHAVIOR Meadow jumping mice are generally solitary, but they are not aggressive toward other individuals. They easily climb grasses and shrubs, and they are excellent swimmers on the surface and under water. Burrows are not used extensively, but they are used for nests and hibernation. Although active mostly at night, some meadow jumping mice may be active during daylight hours. They do not store food. When feeding on grasses, meadow jumping mice sit on their haunches and use their forefeet to manipulate the food. They cut off the fruiting head of a grass, strip all the parts beginning at one end, and eat only the seeds. Seeds that fall to the ground are not retrieved. Meadow jumping mice increase their weight significantly before entering hibernation, which usually takes place in an underground burrow. They hibernate as long as, or longer than, most other hibernating mammals (early autumn to early spring). Vocalizations

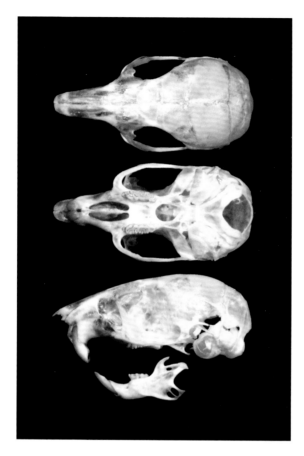

Dorsal, ventral, and lateral views of the cranium, and lateral view of the mandible of a male meadow jumping mouse. Greatest length of cranium is 21.1 mm.

are few but may include chirps and clucking noises. Meadow jumping mice can also produce a noise by vibrating their tail rapidly against a surface. When attempting to avoid a threat, they may make several leaps, stop abruptly, and then remain motionless. Meadow jumping mice often clean their face, feet, and tail. They grasp their tail in their forepaws and pass it completely through their mouth, whereas they clean their faces and feet using their forepaws. Nests are usually made of grasses or leaves and located in protected sites such as clumps of grass, hollow logs, under objects, or underground.

LIFE HISTORY Breeding begins shortly after hibernation and continues through the warmer months. Gestation is 17–21 days. Two or three litters of about 4–6 young may be born each year. At birth, neonates have short hairs (vibrissae) on the rostrum; otherwise, they are hairless and pink, but they can emit high-pitched squeaking sounds. Ears are closed. Eyes are closed and appear as dark spots. Average and range of measurements (in mm) of 19 newborns were: total length, 34 (30–39); tail length, 9 (7–11); hind foot length, 5 (3–6). Average and range of weight of 14 newborns from 3 litters was 0.8 g (0.7–1.0 g). By 1 week, vibrissae have grown noticeably, the tail becomes bicolored, pinnae unfold and become tipped with black, and claws appear. During days 9–13, vibrissae grow to about 8 mm long, teeth erupt, and yellowish hairs appear on the back and sides. By 3 weeks, hair covers the body and ears begin to open. At about 4 weeks, the eyes open. Young weigh 2–4 g by day 10, 4–8 g by day 20, 8–11 g by day 30, and reach 14–15 g by about day 60. By 90 days, weight is about 20 g. Sexual maturity may be reached 2–3 months following birth. Life span is probably about 2–3 years, but meadow jumping mice have lived 5 years in captivity.

PARASITES AND DISEASES Ectoparasites include mites (*Androlaelaps, Dermacarus, Echinonyssus, Eulaelaps, Euschoengastia, Eutrombicula, Glycyphagus, Haemogamasus, Laelaps, Listrophorus, Macrocheles, Neoschongastia, Neotrombicula, Ornithonyssus, Orycteroxenus, Pygmephorus, Radfordia, Trombicula, Xenoryctes*), ticks (*Dermacentor, Ixodes*), and fleas (*Corrodopsylla, Ctenophthalmus, Megabothris, Orchopeas, Stenoponia*). Meadow jumping mice may be parasitized by larval botflies (*Cuterebra*). Several kinds of bacteria (*Bacillus, Bacteroides, Escherichia, Klebsiella*), parasitic diplomonads (*Hexamita*), and protozoans (*Eimeria*) occur in the intestine. Endoparasites include cestodes (*Choanotaenia, Hymenolepis, Mesocestoides, Taenia*), nematodes (*Citellinoides, Longistriata, Mastophorus, Rictularia, Spirocerca, Subulura*), and trematodes (*Echinostoma, Notocotylus, Plagiorchis, Quinqueserialis, Schistosomatium*).

CONSERVATION STATUS High conservation concern in Alabama.

COMMENTS *Zapus* is from the Greek *za* and *podos*, probably referring to the large hind feet; *hudsonius* refers to Hudson Bay, Canada, where the type specimen was obtained.

REFERENCES Whitaker (1972), Scott and French (1974), Best (2004*a*), Whitaker et al. (2007).

New World Rats and Mice, Voles, Hamsters, and Relatives

Family Cricetidae

Cricetidae, with 130 genera and 681 species, is the second most diverse family of mammals after Muridae. The family is distributed throughout North and South America, most of Eurasia, and parts of eastern Africa. Members inhabit virtually every terrestrial environment, including polar, temperate, desert, and tropical regions of the world. Many species are mouselike and ratlike, their bodies may be rounded or slender, and they may weigh from about 8 g (pygmy mouse, *Baiomys*) to 2 kg (common muskrat, *Ondatra*). In some species, there is no sexual dimorphism, but in others, males are larger than females, and in yet others, females are larger than males. Usually, there are 4 well-developed incisors, 2 upper and 2 lower, and a total of 12 molariform teeth. In Alabama, there are 13 species of cricetid rodents.

Prairie Vole
Microtus ochrogaster

IDENTIFICATION Tail is short; fur is long, coarse, and grayish brown; sides and underside are paler than the back; and ears are mostly hidden by hair. The only other mouse in Alabama with a tail that is shorter than the body is the woodland vole, which is smaller and often has a brownish or reddish back.

DENTAL FORMULA i 1/1, c 0/0, p 0/0, m 3/3, total = 16.

SIZE AND WEIGHT Average and range in size of 15 specimens from Alabama:

     total length, 142 (124–157) mm / 5.7 (5.0–6.3) inches
     tail length, 31 (26–37) mm / 1.2 (1.0–1.5) inches
     hind foot length, 19 (16–26) mm / 0.8 (0.6–1.0) inch
     ear length of 10 specimens, 9 (6–14) mm / 0.4 (0.2–0.6) inch
     weight of 15 specimens, 41.7 (26.7–59.0) g / 1.5 (0.9–2.1) ounces

There is no sexual dimorphism.

DISTRIBUTION Northern Alabama. The species ranges from central Alberta, Canada, across the Great Plains and Midwest to central Oklahoma, Arkansas, and northern Alabama. A disjunct, and now extinct, population once occupied southeastern Texas and southwestern Louisiana.

Distribution of the prairie vole in Alabama and North America.

ECOLOGY A wide variety of prairie habitats are occupied including grasslands, riparian corridors, and agricultural areas; all of these habitats allow the development of well-used runways through the dense vegetation. Prairie voles construct burrows and store seeds and plant material in underground caches. Their home range is usually 0.11–0.22 hectares. Abundance varies widely in response to available moisture and vegetation. Diet consists of a wide variety of stems and leaves and insects, and these voles have been documented preying on tri-colored bats in a cave. Prairie voles may become pests, especially in orchards and gardens. They are a significant problem where people plant trees; injury to stems has been reported in 14 species of hardwoods and 16 species of conifers, with pines being the most adversely affected. Nearly every predator that occupies the same range as the prairie vole eats this species, including snakes, hawks, owls, foxes, bobcats, and coyotes. Hispid cotton rats occupy the same types of habitats as prairie voles, and they compete for space rather than for food. Hispid cotton rats may exclude prairie voles from some sites and may even kill and eat them. When hispid cotton rats are present, prairie voles may be less abundant and have altered sex ratios, decreased survival, and changes in spatial distribution and movements.

BEHAVIOR Daytime activity decreases in summer when temperatures are high. In winter, when temperatures are low at night, nocturnal activity decreases. Young produce ultrasonic sounds that enable parents to find them; compared to young, adults produce few ultrasonic sounds. However, ultrasonic sounds produced by males may communicate gender

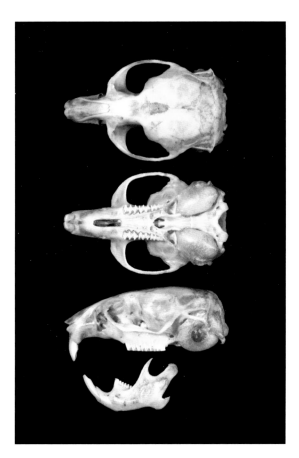

Dorsal, ventral, and lateral views of the cranium, and lateral view of the mandible of a male prairie vole. Greatest length of cranium is 24.9 mm.

and announce availability for reproductive activity. Nonsocial behaviors include locomotion, maintenance, and comfort movements, as well as digging, swimming, constructing runways, and building nests. Runways are usually bare and depressed into the surface due to repeated use. Each runway consists of a long, crooked central path with several branches to the side. Food and nesting chambers are up to 200 mm in diameter, with tunnels varying from very short to several meters. Typically, nests consist of coarse dried grass on the outside, with finely shredded grass in the center. Often, nests are constructed under boards or logs.

LIFE HISTORY Prairie voles seem to be monogamous, unlike most other mammalian species. Three pairs of mammae are present. Reproduction occurs throughout the year, with peaks from May to October. Ovulation occurs about 10.5 hours after copulation. Gestation is 20–23 days and results in an average of 4 young/litter. At birth, young have a crown-to-rump length of 30–35 mm and a weight of 3.5 g; they are hairless, their eyes and ears are closed, and their digits are 75% fused. At 1–2 days, incisors emerge; at 2–3 days, fur appears and pinnae unfold; at 4–5 days, crawling begins; at 5–10 days, eyes open; and at 10–14 days, solid food is consumed. Young are weaned at 2–3 weeks old and are fully grown by 2 months of age. Females may become pregnant by 6 weeks of age. Males are capable of reproduction within 40–60 days after birth. Average life span in the wild is about 12 months, but 2 individuals lived in captivity for 27 and 35 months.

PARASITES AND DISEASES Ectoparasites include mites (*Androlaelaps, Echinonyssus, Eulaelaps, Euschoengastia, Glycyphagus, Haemogamasus, Laelaps, Listrophorus, Myobia, Myocoptes, Neotrombicula, Ornithonyssus, Orycteroxenus, Pygmephorus, Radfordia, Trichoecius*), ticks (*Dermacentor, Ixodes*), lice (*Hoplopleura*), and fleas (*Ctenophthalmus, Epitedia, Hystrichopsylla, Megabothris, Monopsyllus, Nearctopsylla, Nosopsyllus, Orchopeas, Peromyscopsylla, Rhadinopsylla, Stenoponia*). Endoparasites include protozoans (*Eimeria, Trypanosoma*), acanthocephalans (*Moniliformis*), cestodes (*Andrya, Choanotaenia, Cladotaenia, Hymenolepis, Paranoplocephala, Taenia*), nematodes (*Capillaria, Syphacia, Trichuris*), and trematodes (*Quinqueserialis*).

CONSERVATION STATUS Moderate conservation concern in Alabama.

COMMENTS *Microtus* is from the Greek *mikros*, meaning "small," and *ot*, meaning "ear"; *ochrogaster* is from the Greek *ōchra*, meaning "yellow-ochre," and *gastēr*, meaning "belly."

REFERENCES Rausch and Tiner (1949), Timm (1985), Stalling (1990), Best (2004a), Whitaker et al. (2007).

Woodland Vole
Microtus pinetorum

IDENTIFICATION A small rodent with a short tail. Eyes and ears are small. Fur is soft, dense, and variable in color, from reddish brown to dark brown on the back and from whitish to gray on the underside. The only other mouse in Alabama with a tail that is shorter than the body is the prairie vole, which is larger and usually has a grayish-brown back.

DENTAL FORMULA i 1/1, c 0/0, p 0/0, m 3/3, total = 16.

SIZE AND WEIGHT Average and range in size of 25 specimens from Alabama:

> total length, 120 (105–142) mm / 4.8 (4.2–5.7) inches
> tail length, 20 (16–26) mm / 0.8 (0.6–1.0) inch
> hind foot length, 15 (13–18) mm / 0.6 (0.5–0.7) inch
> ear length, 8 (5–10) mm / 0.3 (0.2–0.4) inch
> weight of 22 specimens, 26.6 (17.6–38.9) g / 0.9 (0.6–1.4) ounces

There is no sexual dimorphism.

DISTRIBUTION Except for southern areas, this species is present nearly statewide in Alabama. Geographic range of the woodland vole is from

Distribution of the woodland vole in Alabama and North America.

southeastern Canada across the eastern United States and southward into northern Florida. There is an isolated population in south-central Texas.

ECOLOGY A wide range of habitats are occupied throughout the geographic range from beech (*Fagus*), maple (*Acer*), and oak (*Quercus*) forests with varying depths of leaf litter to grassy fields with many shrubs, patches of brambles (*Rubus*) and briers (*Smilax*), and mats of honeysuckles (*Lonicera*). Diet is roots and stems of grasses in summer, fruits and seeds in autumn, and roots and bark in winter. Common foods are fungi (*Endogone*), sprouts of clover (*Trifolium*), roots of morning glories (*Convolvulus*) and docks (*Rubus*), acorns (*Quercus*), and stems of grasses (*Echinochloa*, *Panicum*). Woodland voles can cause significant damage to crops of potatoes, peanuts, nursery stock, planted seeds, and especially to trees in orchards, where they may strip roots or girdle trunks as they remove bark for consumption. Predators include owls, hawks, and snakes as well as Virginia opossums, gray and red foxes, coyotes, bobcats, and long-tailed weasels.

BEHAVIOR Woodland voles are active throughout the day and night. Vocalizations include chatterings that seem to be associated with alarm and fighting. There is no apparent territoriality; burrows and nests may be occupied communally by animals of multiple ages and sexes. Woodland voles are semifossorial and spend most of their time in burrows and runways they construct on the surface. Digging involves a shoveling motion with the head and neck to loosen soil, which is scratched out with the forefeet, pushed backward using a sideward and backward motion of the

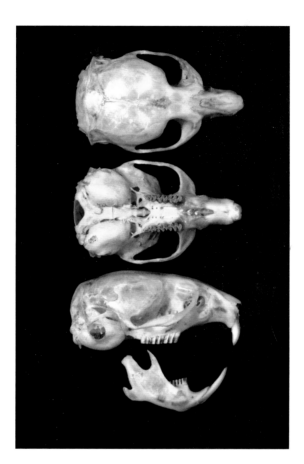

Dorsal, ventral, and lateral views of the cranium, and lateral view of the mandible of a male woodland vole. Greatest length of cranium is 22.7 mm.

forefeet, and swept farther back with the hind feet. After excavating about 30 cm of the burrow, the vole turns around and pushes the accumulated soil and debris to the surface with its head. This soil is deposited beneath the leaf litter into piles about 10 cm wide and 5–8 cm tall. Most burrowing is 5–10 cm below the surface, but some burrows may be 30 cm deep. Diameter of burrows is about 3.0–3.5 cm and they may incorporate old runways of eastern moles. In areas with a thick layer of leaves or humus, surface runways may be constructed to connect burrows. Nests made of leaves and grasses are constructed in burrows, where they are usually globular, or under fallen logs, where they may be saucer shaped.

LIFE HISTORY Breeding may occur at any time during the year. Each year, 1–4 litters are born, with an average of about 2 litters/year. Most young are born from March to July. Following a gestation of 20–24 days, 1–6 young are born (average is 2–3 young/litter). Newborn woodland voles have a sparse coat of fuzzy hairs on the back and well-developed vibrissae on the rostrum, and they can pull themselves about with short blunt claws on the forelimbs. At birth, average weight is 1.9–3.2 g, total length is 39–48 mm, eyes and ears are closed, and they emit low squeaks and suckling noises when disturbed. Newborns quickly attach to a nipple and are difficult to dislodge. By day 3, the back is grayish, underparts are pink, total length is 43–54 mm, and average weight is 3.7 g. By day 5, the body is well covered by fine hairs and only the limbs and genital regions are pinkish. By days 7–9, weight is 5.0–6.4 g, smooth brown hair appears on the back, the underparts are covered with grayish hairs, the eyes open, and the young crawl vigorously. By 11–12 days, ears open, young are very active, and weight is 7.0–8.0 g. Young are weaned 17–21 days following their birth. By 3 weeks, juvenile pelage is replaced by adult pelage, total length is 69–98 mm, and weight is 7.5–9.6 g. Although they are still growing, they attain sexual maturity at 6–8 weeks of age. Life span is probably about 1 year.

PARASITES AND DISEASES Ectoparasites include mites (*Androlaelaps, Atricholaelaps, Bakerdania, Dermacarus, Eucheyletia, Eulaelaps, Euschoengastia, Glycyphagus, Haemogamasus, Laelaps, Lepidoglyphus, Leptotrombidium, Listrophorus, Myocoptes, Neotrombicula, Ornithonyssus, Orycteroxenus, Pseudopygmephorus, Psorergates, Pygmephorus, Radfordia, Trichoecius*), ticks (*Amblyomma, Dermacentor, Ixodes*), lice (*Hoplopleura, Polyplax*), fleas (*Atyphloceras, Ceratophyllus, Ctenophthalmus, Doratopsylla, Epitedia, Hystrichopsylla, Megabothris, Orchopeas, Peromyscopsylla, Rhadinopsylla, Stenoponia*), and beetles (*Leptinus*). Woodland voles may be parasitized by larval botflies (*Cuterebra*). Endoparasites include protozoans (*Sarcocystis*), acanthocephalans (*Moniliformis*), cestodes (*Catenotaenia, Cladotaenia, Hymenolepis, Taenia*), and nematodes (*Capillaria, Oxyuris, Trichinella, Trichuris*). Diseases and disorders include lesions in the stomach, gastritis, and skin disorders.

CONSERVATION STATUS Low conservation concern in Alabama.

COMMENTS *Microtus* is from the Greek *mikros*, meaning "small," and *ōtos*, meaning "ear"; *pinetorum* is Latin for "of pine-woods."

REFERENCES Smolen (1981), Timm (1985), Best (2004*a*), Whitaker et al. (2007).

Common Muskrat
Ondatra zibethicus

IDENTIFICATION This is a brownish, medium-sized rodent with a laterally compressed, nearly hairless, scaly tail. The underside is pale brown, hind feet are partially webbed, ears are small and rounded, and eyes are relatively small.

DENTAL FORMULA i 1/1, c 0/0, p 0/0, m 3/3, total = 16.

SIZE AND WEIGHT Average and range in size of 13 specimens from Alabama:

total length, 543 (425–594) mm / 21.7 (17.0–23.8) inches
tail length, 243 (198–255) mm / 9.7 (7.9–10.2) inches
hind foot length, 86 (69–87) mm / 3.4 (2.8–3.5) inches
ear length of 8 specimens, 21 (15–26) mm / 0.8 (0.6–1.0) inch
weight of 3 specimens, 1,084 (425–1,556) g / 2.4 (0.9–3.4) pounds

There is no sexual dimorphism.

DISTRIBUTION Nearly statewide in Alabama, except for the southeastern region. Geographic range of the species includes nearly all of Alaska, Canada, the contiguous United States, and northernmost Mexico.

Distribution of the common muskrat in Alabama and North America.

ECOLOGY Common muskrats can inhabit most aquatic habitats within their geographic range, including ponds, streams, lakes, and marshes. They may also inhabit human-made ponds at strip mines, power plants, and fish-rearing facilities, as well as irrigation ponds and canals. The underfur is waterproof, and air trapped there gives common muskrats buoyancy and provides insulation. The lips close behind the incisors, allowing common muskrats to gnaw while under water. The small forefeet are used for digging and manipulating food and construction materials. In addition to serving as a rudder during swimming, the laterally flattened tail may also act to remove heat from the body during exercise and in hot weather. Populations generally follow a 5–10-year cycle. Common muskrats are primarily herbivores and eat a variety of plants, but they may also consume animal matter such as crayfish, mollusks, or turtles. Burrowing activities of common muskrats may cause extensive damage to riverbanks, canals, and agricultural areas, which may lead to attempts to control or eradicate local populations. Predators include humans, eagles, hawks, owls, raccoons, and American minks.

BEHAVIOR Although primarily nocturnal, common muskrats are also active during daylight hours. Sounds made include squeaks, high-pitched whines, and teeth chattering. Both males and females actively defend territories. Females are more aggressive than males in defending their territories and often kill intruders. Common muskrats are most aggressive before and during the breeding season. Dominance hierarchies are determined by fighting. Optimum sites for burrows are occupied by older or larger individuals in the population. Common muskrats may construct

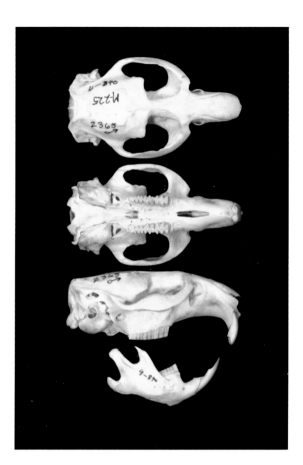

Dorsal, ventral, and lateral views of the cranium, and lateral view of the mandible of a male common muskrat. Greatest length of cranium is 67.3 mm.

conical houses or dig burrows into banks. They usually begin constructing their main houses on a firm substrate and use dominant emergent vegetation in the area. They build houses above water level with several underwater tunnels as the only entrances. One or more nest chambers lined with fresh plant material are usually near the center of the house. Feeding houses are usually smaller than main houses. When several common muskrats huddle in a house, temperatures rise, increasing survival during winter. During the breeding season, common muskrats deposit scent around dens, houses, defecation sites, and along trails.

LIFE HISTORY Breeding occurs throughout the year, with a peak in winter. Common muskrats mate while partially submerged. Each year, 2–6 litters are born in nest chambers inside their houses. Potentially, subsequent litters can be born at monthly intervals due to an immediate postpartum estrus. Following a gestation of 20–25 days, 4–8 young are born. Newborns are pink and almost hairless, weigh about 21 g, and have closed eyes and a rounded tail. Females care for the young until weaning, but a male will care for them if the female dies. By 6–7 days, incisors erupt. By 14–16 days old, young are covered with soft fur, their eyes are open, they are active, and they can swim. Young are weaned about 4 weeks after birth. The tail becomes laterally compressed during the second month. Most common muskrats become sexually active the first spring following their birth. Life span is 3–4 years.

PARASITES AND DISEASES Ectoparasites include mites (*Androlaelaps, Dermacarus, Eutrombicula, Hirstionyssus, Ichoronyssus, Laelaps, Listrophorus, Marsupialichus, Myocoptes, Radfordia, Schizocarpus, Zibethacarus*), ticks (*Ixodes*), and fleas (*Ceratophyllus*). Endoparasites include protozoans, acanthocephalans (*Polymorphus*), cestodes (*Andrya, Anomotaenia, Cladotaenia, Cysticercus, Hymenolepis, Monoecocestus, Taenia*), nematodes (*Ascaris, Capillaria, Dirofilaria, Dracunculus, Heligmosomum, Hepaticola, Litomosoides, Nematospiroides, Rictularia, Strongyloides, Trichostrongylus, Trichinella, Trichuris*), and trematodes (*Alaria, Allassogonoporus, Amphimerus, Echinochasmus, Echinoparyphium, Echinostoma, Fibricola, Levinseniella, Mediogonimus, Metorchis, Notocotylus, Nudacotyle, Opisthorchis, Paragonimus, Paramonostomum, Plagiorchis, Pseudodiscus, Psilostomum, Ptyalincola, Quinqueserialis, Schistosomatium, Urotrema, Wardius*). Diseases include adiaspiromycosis (*Chrysosporium*), epizootic chlamydiosis (*Chlamydophila*), hemorrhagic disease, leptospirosis (*Leptospira*), pseudotuberculosis (*Yersinia*), ringworm (fungal dermatophytes), salmonellosis (*Salmonella*), tularemia (*Francisella*), Tyzzer's disease (*Clostridium*), and yellow-fat disease.

CONSERVATION STATUS Lowest conservation concern in Alabama.

COMMENTS The common name is derived from the conspicuous odor produced by secretions from scent glands near the lower base of the tail in both sexes. *Ondatra* is a Native American name for the muskrat; *zibethicus* is from the Greek *zibeth*, meaning "musty-odored."

REFERENCES Macy (1933), Meyer and Reilly (1950), Chandler and Melvin (1951), Beckett and Gallicchio (1967), Gash and Hanna (1973), McKenzie and Welch (1979), Willner et al. (1980), Best (2004a), Whitaker et al. (2007).

Eastern Woodrat
Neotoma floridana

IDENTIFICATION A medium-sized, nocturnal rat with soft, brownish-gray pelage on the back; white or creamy white fur on the underside of the body; large, somewhat bulging eyes; relatively long ears; and a long tail covered with short dark hairs on top and short whitish hairs beneath. The tail is sharply bicolored. Externally, this species appears to be identical to the Allegheny woodrat. However, the eastern woodrat can be distinguished from the Allegheny woodrat by shape of the palate (Allegheny woodrats have a maxillovomerine notch on the posterior palate, which is rarely present in eastern woodrats) and by genetic evaluation. Present knowledge of the distribution of these species in Alabama indicates that except for 3 specimens of Allegheny woodrats from south of the Tennessee River near Muscle Shoals in Colbert County, the eastern woodrat is the only species of woodrat that occurs south of the Tennessee River.

DENTAL FORMULA i 1/1, c 0/0, p 0/0, m 3/3, total = 16.

SIZE AND WEIGHT Average and range in size of 12 specimens from Alabama:

> total length, 331 (230–392) mm / 13.2 (9.2–15.7) inches
> tail length, 158 (108–184) mm / 6.3 (4.3–7.4) inches
> hind foot length, 36 (33–39) mm / 1.4 (1.3–1.6) inches
> ear length of 9 specimens, 27 (24–29) mm / 1.1 (1.0–1.2) inches
> weight of 5 specimens, 208.8 (149.3–251.2) g / 7.3 (5.2–8.8)
>> ounces

Males average larger than females.

A midden constructed by the eastern woodrat. This pile of branches and other woody materials provides shelter and protection from predators.

DISTRIBUTION South of the Tennessee River in Alabama. Geographic range of the species is from eastern Colorado to the Atlantic Ocean and southward through most of Florida.

ECOLOGY Although the geographic range is primarily in eastern deciduous forest, western populations occur in grasslands of the Great Plains. Habitats occupied include lowland hardwood forests, marshes, mountainous areas, coastal plains, swamps, and grasslands. Woodrat houses

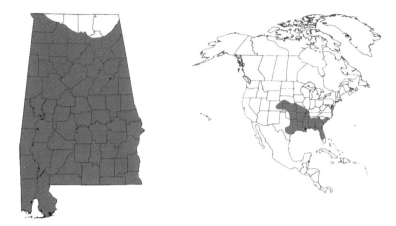

Distribution of the eastern woodrat in Alabama and North America.

(middens) are used as shelter by many kinds of animals including toads, snakes, lizards, shrews, white-footed deermice, eastern cottontails, and Virginia opossums. Although diet is quite variable among locations and seasons, eastern woodrats consume mushrooms, acorns (*Quercus*), pecans (*Carya illinoiensis*), seeds of honey locusts (*Gleditsia triacanthos*), beechnuts (*Fagus grandifolia*), and fruits and bark of sumacs (*Rhus*), redbuds (*Cercis canadensis*), and dogwoods (*Cornus asperifolia*). They collect and store acorns and other foods in their houses beginning in autumn. Common predators include snakes, great horned owls (*Bubo virginianus*), eastern spotted skunks, long-tailed weasels, coyotes, and bobcats.

BEHAVIOR Eastern woodrats are usually solitary except during the breeding season and when young are present. They construct large houses that provide escape from predators, protection from environmental extremes, and a place for 2 or more nests. Nests are usually about 200 mm in diameter and made of dried grasses, shredded bark, and sometimes feathers. Primary construction materials for houses are branches, twigs, and leaves, but bones, cattle dung, aluminum cans, and shotgun shells also are collected and added to the houses if they are available. These houses may be 2–4 m across, 1 m tall, and built under rocky outcrops, in brush piles, at the base of large trees, in or near caves, in trees, or in abandoned buildings and motor vehicles. Houses may be maintained, modified, and occupied by generations of woodrats. Several trails lead to and from houses but are usually less than 10 m long, indicating that eastern woodrats may forage relatively close to their houses. The senses of smell and hearing are keen, and the vibrissae provide sensory information using touch. Both

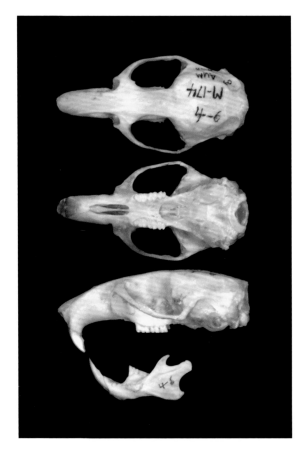

Dorsal, ventral, and lateral views of the cranium, and lateral view of the mandible of a male eastern woodrat. Greatest length of cranium is 52.5 mm.

sexes have a ventral abdominal gland, but in males it becomes well developed during the breeding season and produces an oily secretion that stains the pelage a dark brown. Males may mark their houses with the secretion, which may help females to find mates.

LIFE HISTORY Breeding may occur at any time of the year in Alabama but is most probable from late winter to early spring. During the mating season, a male approaches the female, attempts to mate, drums his hind feet, and sniffs her perineal region. If the female is receptive to these behaviors, mating occurs. Each year, 1–2 litters are born. Following a gestation of 33–38 days, a litter of 1–7 young (average is 3) is born. Newborns are nearly naked, with sparse fur and 5-mm vibrissae, but the back and top of the tail are darker than the rest of the body. At birth, the young have claws, their eyes and ears are closed, their incisors are erupted, they are 87–96 mm long, and they weigh 12–14 g. They occasionally make suckling and squeaking noises. By 9 days, the ears open, and by 15–21 days, the eyes open. Young are weaned at about 4 weeks old. Females reach sexual maturity in 5–6 months, and some have a litter during the year of their birth. Males are sexually mature by February–March of the year following their birth. Life span is probably at least 2 years in the wild.

PARASITES AND DISEASES Ectoparasites include mites (*Androlaelaps, Cheyletus, Echinonyssus, Eulaelaps, Eutrombicula, Glycyphagus, Hypoaspis, Listrophorus, Myocoptes, Ornithonyssus, Pseudoschoengastia, Pygmephorus, Trombicula*), ticks (*Dermacentor, Haemaphysalis, Ixodes*), and fleas (*Conorhinopsylla, Epitedia, Orchopeas, Polygenis, Xenopsylla*). Larval botflies (*Cuterebra*) up to about 15 mm in diameter and 25–30 mm long are conspicuous parasites and may be embedded in the skin of the throat, chest, or abdomen. Endoparasites include protozoans (*Eimeria, Isospora*), cestodes (*Andrya, Taenia*), nematodes (*Boehmiella, Capillaria, Longistriata*), and trematodes. Diseases include plague (*Yersinia*) and toxoplasmosis (*Toxoplasma*).

CONSERVATION STATUS Moderate conservation concern in Alabama.

COMMENTS *Neotoma* is from the Greek *neos*, meaning "new," and *tomos*, meaning "a cut" or "slice." The specific name *floridana* is Latin for "of Florida," where the type specimen was obtained.

REFERENCES Wiley (1980), Forrester (1992), Hayes and Richmond (1993), Best (2004*a*), Whitaker et al. (2007).

Allegheny Woodrat
Neotoma magister

IDENTIFICATION A medium-sized rat with soft, gray to brownish-gray pelage on the back; white or creamy white fur on the underside of the body; large, somewhat bulging eyes; relatively long ears; and a long, sharply bicolored tail covered with short dark hairs on top and short whitish hairs beneath. Externally, this species appears to be identical to the eastern woodrat. However, the Allegheny woodrat can be distinguished from the eastern woodrat by the shape of the palate (Allegheny woodrats have a maxillovomerine notch on the posterior palate, which is rarely present in eastern woodrats) and by genetic evaluation.

DENTAL FORMULA i 1/1, c 0/0, p 0/0, m 3/3, total = 16.

SIZE AND WEIGHT Average and range in size of adult Allegheny woodrats from across the geographic range:

> total length of 91 specimens, 397 (311–451) mm / 15.9 (12.4–18.0) inches
>
> tail length of 91 specimens, 180 (147–210) mm / 7.2 (5.9–8.4) inches
>
> hind foot length of 92 specimens, 42 (35–46) mm / 1.7 (1.4–1.8) inches
>
> ear length of 42 specimens, 28 (23–34) mm / 1.1 (0.9–1.4) inches

A midden constructed by the Allegheny woodrat at the base of a cliff. Middens are constructed primarily of leaves, twigs, and branches, but almost any other object may be gathered and deposited on them (e.g., bottle caps, shotgun shells, aluminum cans).

Average weight of 48 adult males and 36 adult females was 357 and 337 g (12.5 and 11.8 ounces), respectively. Males are larger than females.

DISTRIBUTION Present knowledge of the distribution of this species in Alabama indicates that the Allegheny woodrat is the only species of woodrat that occurs north of the Tennessee River. The only occurrence of Allegheny woodrats south of the Tennessee River in Alabama is from 16 km southeast of Muscle Shoals in Colbert County. Geographic range of the species is from northern Alabama into northern Pennsylvania and New Jersey. Historically, the species occurred as far northeastward as New York and Connecticut.

ECOLOGY Regardless of geographic area or type of vegetation, habitats occupied include rocky areas such as boulder fields, caves, or cliff faces. Types of forests occupied include northern hardwood, mixed mesophytic, and mixed oak-pine (*Quercus-Pinus*). In West Virginia, Allegheny woodrats occur on rocky sandstone outcrops with steeper and wider slopes, less accumulation of leaf litter, and fewer understory trees than are present on unoccupied rocky outcrops. In Kentucky, Allegheny woodrats occupy rocky outcrops on steep slopes with high densities of trees. Throughout their range, they seem to prefer an understory with diverse vegetation. Habitats occupied usually provide year-round access to water in lakes or streams, and seepage water in caves and rocky outcrops. Diet includes a variety of vegetative and fruiting parts of woody plants including black birch (*Betula lenta*), American chestnut (*Castanea dentata*), dogwood (*Cornus florida*), apple (*Malus pumila*), bear oak (*Quercus ilicifolia*), black cherry (*Prunus serotina*), rhododendron (*Rhododendron*

maximum), and mountain ash (*Sorbus americana*). Also important in the diet are fronds of ferns (*Polypodium*), fungi, leaves of greenbriers (*Smilax*), acorns (*Quercus*), hollies (*Ilex*), and lichens. Predators include great horned owls (*Bubo virginianus*), other owls, timber rattlesnakes (*Crotalus adamanteus*), other snakes, bobcats, gray foxes, striped skunks, eastern spotted skunks, and long-tailed weasels.

BEHAVIOR Allegheny woodrats are nocturnal. Although usually not vocal, young may emit squeaks and adults may vocalize during fights or when males follow females before mating. Sounds may include squeaking, whimpering, and low-pitched rasping. Territories are scent-marked by pressing the midventral gland against objects while using the front legs to pull the body along. The midventral gland extends from the sternum to near the pelvic region and its secretion stains the adjacent pelage a yellowish brown. Home ranges of adults overlap extensively, and the animals defend the territory immediately around their den site. Houses (middens) are large piles of sticks, twigs, and other items that are accumulated underneath rock ledges, at or in caves, or in rocky crevices. These serve as refuges from weather, places to store food, and sites to escape from predators. Latrines for defecation and urination are usually near houses on flat surfaces. Both sexes deposit scent from their midventral glands near latrines. Young are reared in nests made of bark, grasses, roots, and shredded wood. Nests are about 46 cm in diameter, with a cavity about 12 cm in diameter. The outside is made of coarse materials, and the inside is lined with fine materials. Nests are often in rocky crevices or on ledges in caves.

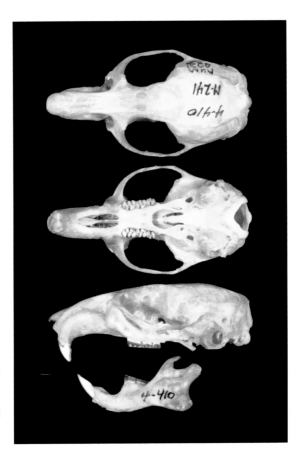

Dorsal, ventral, and lateral views of the cranium, and lateral view of the mandible of a male Allegheny woodrat. Greatest length of cranium is 50.9 mm.

LIFE HISTORY The midventral gland becomes active during the breeding season in both sexes. Breeding occurs all year. Allegheny woodrats may have 2–3 litters each year, and most litters are born from May to October. Following a gestation of 30–36 days, about 2 young are born (range is 1–4). Newborns are pink and hairless, their eyes and ears are closed, and they weigh 15–17 g. By 5 days old, young are covered with fine hairs, and by 2 weeks old, they are fully furred. By 2 weeks, eyes are sensitive to light, and by 3 weeks, eyes are fully open. Mother and young remain together until weight reaches at least 115 g. Sexual maturity is reached at 3–4 months old and females may have their first litter at 10 months old. Life span may be 3–5 years in the wild.

PARASITES AND DISEASES Ectoparasites and inhabitants of nests include mites (*Androlaelaps*, *Aplodontopus*, *Cheyletus*, *Chortoglyphus*,

Dermacarus, Echinonyssus, Eucheyletia, Eulaelaps, Euschoengastia, Eu-trombicula, Haemogamasus, Hypoaspis, Macrocheles, Myocoptes, Neotrombicula, Ornithonyssus, Proctolaelaps, Pygmephorus), ticks (*Dermacentor, Ixodes*), and fleas (*Epitedia, Orchopeas, Peromyscopsylla*). Endoparasites include protozoans (*Eimeria*) and nematodes. The ascarid nematode *Baylisascaris procyonis*, which causes fatal neurological disease in intermediate hosts such as woodrats, is believed to have contributed to declines in populations of Allegheny woodrats. The primary host of this parasite is the raccoon, which is common in habitats occupied by Allegheny woodrats. Allegheny woodrats may collect feces of raccoons that contain eggs of *B. procyonis*, deposit the feces on middens, and inadvertently expose themselves to this potentially deadly parasite.

CONSERVATION STATUS High conservation concern in Alabama and listed as under review by the United States Fish and Wildlife Service.

COMMENTS *Neotoma* is from the Greek *neos*, meaning "new," and *tomos*, meaning "a cut" or "slice." The specific name *magister* is Latin for "chief" or "magistrate."

REFERENCES Kinsey (1976), Cudmore (1986), Hayes and Richmond (1993), Best (2004*a*), Castleberry et al. (2006).

Golden Mouse
Ochrotomys nuttalli

IDENTIFICATION A small rodent with golden-orange pelage on the back and creamy white to yellowish hair on the feet and underside of the body. The tail is faintly bicolored; dark on top and pale beneath.

DENTAL FORMULA i 1/1, c 0/0, p 0/0, m 3/3, total = 16.

SIZE AND WEIGHT Average and range in size of 25 specimens from Alabama:

total length, 157 (140–190) mm / 6.3 (5.6–7.6) inches
tail length, 71 (63–84) mm / 2.8 (2.5–3.4) inches
hind foot length, 18 (16–19) mm / 0.7 (0.6–0.8) inch
ear length of 13 specimens, 16 (14–16) mm / 0.6 (0.6–0.6) inch
weight of 14 specimens, 18.4 (14.3–21.4) g / 0.6 (0.5–0.8) ounce

Although males average slightly larger, there is no significant sexual dimorphism.

DISTRIBUTION Statewide in Alabama. Geographic range of the species is from southern Illinois, Kentucky, and eastern Oklahoma across the

southeastern United States. Two subspecies occur in Alabama; *O. n. aure-*
olus in the north and *O. n. nuttalli* in southern Alabama.

Nest of a golden mouse constructed in vegetation above the ground.

ECOLOGY Golden mice occur in association with deciduous oak-hickory
(*Quercus-Carya*) and pine (*Pinus*) forests in a variety of habitats from
densely forested lowlands and floodplain communities to sandy-soiled
uplands with considerable undergrowth and vines. Habitats occupied
may include moist thickets, brushy areas, woods, canebrakes, swampy
woodlands, grassy fields, patches of briers (*Rubus*, *Smilax*), rocky hill-
sides, and rock-strewn ravines. Short-term flooding (up to 8 days) seems
to have little impact, but floods lasting 3 weeks may significantly decrease
populations. Diet includes invertebrates, acorns, and seeds, including
those of sumacs (*Rhus*), wild cherries (*Prunus*), dogwoods (*Cornus*), poison ivy (*Toxicodendron*), and blackberries (*Rubus*). Predators include
owls, hawks, gray and red foxes, long-tailed weasels, and bobcats.

BEHAVIOR This docile, primarily nocturnal, and semiarboreal rodent is
active on the ground and in trees and other elevated vegetation. Golden
mice have a prehensile-like tail that they use for balance when they move
through vegetation, and they may also wrap the tail around vegetation

After being disturbed by researchers, a golden mouse flees from her nest with a youngster still attached and nursing.

to keep the body stable while they are resting or foraging. Golden mice may construct nests and feeding shelters 1–10 m up in trees, in stumps, or beneath logs or rocks on the ground. The nest is a spherical mass of leaves, shredded bark, and grasses and is often supported on vegetation such as briers, honeysuckles (*Lonicera*), or grapevines (*Vitis*). Nests are usually about 150–200 mm long, 100–125 mm wide, and 100–200 mm tall, and they weigh 10–30 g. The inside is lined with finely shredded bark, feathers, fur, or similar materials. Sometimes, seeds are stored in nests and husks are often found there. At one end of the nest is an opening that is usually partially closed and about 25 mm in diameter. Feeding shelters are similar, but not as bulky as nests. Golden mice carry seeds to feeding shelters, where they open and eat them. When frightened, golden mice may remain motionless or climb the nearest object.

LIFE HISTORY Breeding season is March to early October, with peaks in late spring and early autumn. Following a gestation of 26–30 days, 1–4 young are born (average is 3). Females may become pregnant while still nursing a previous litter; this can result in the birth of a litter about 25 days after the birth of the previous litter. One study revealed that in captivity, 17 litters can be born within 18 months and females up to 6–7 years old may give birth. Newborns are reddish and have relatively smooth

Distribution of the golden mouse in Alabama and North America.

skin, their viscera are visible through the walls of the abdomen, and they can produce a loud, rasping squeak. Eyes and ears are closed, but vibrissae on the rostrum are well developed. Average and range in size (in mm) of newborns are: total length, 51 (38–58); tail length, 15 (13–18); hind foot length, 7 (6–8); weight, 2.7 g (1.8–3.6 g). By 2 days, the prehensile nature of the tail is noticeable. By 6 days, lower incisors erupt, and by 7–8 days, upper incisors erupt. By 7 days, the back is covered with a sleek, velvety, reddish-brown pelage. By 8–12 days, young first react to sound, and by 10–14 days, the eyes open. Weaning occurs at 17–21 days old. They are the same size as adults by 8–10 weeks, but there is slow, continual growth throughout life. Compared to other rodents in the family Cricetidae, golden mice have among the longest life spans in the wild (some live more than 7 years).

PARASITES AND DISEASES Ectoparasites include mites (*Androlaelaps, Bakerdania, Eulaelaps, Euschoengastia, Glycyphagus, Labidophorus, Laelaps, Lasioseius, Melichares, Myocoptes, Ornithonyssus, Radfordia*), ticks (*Dermacentor, Haemaphysalis, Ixodes*), lice (*Hoplopleura*), and fleas (*Ctenophthalmus, Doratopsylla, Epitedia, Orchopeas*). Occasionally, golden mice are parasitized by larval botflies (*Cuterebra*). Bacteria (*Escherichia, Grahamella*) have also been reported. Endoparasites include protozoans (*Eimeria, Retortomonas*), cestodes (*Raillietina, Taenia*), and nematodes (*Carolinensis, Gongylonema, Litomosoides, Longistriata, Physaloptera, Rictularia, Syphacia*).

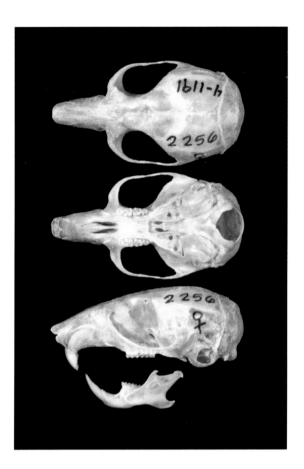

Dorsal, ventral, and lateral views of the cranium, and lateral view of the mandible of a female golden mouse. Greatest length of cranium is 25.9 mm.

CONSERVATION STATUS Lowest conservation concern in Alabama.

COMMENTS *Ochrotomys* is from the Greek *ōchra*, meaning "yellow-ochre," and *mys*, meaning "mouse"; *nuttalli* is a patronym honoring Thomas Nuttall (1786–1859), a British botanist and zoologist. The subspecific name *aureolus* is Latin for "golden" or "splendid."

REFERENCES Linzey and Packard (1977), Forrester (1992), Best (2004*a*), Whitaker et al. (2007), Beolens et al. (2009).

Cotton Deermouse
Peromyscus gossypinus

IDENTIFICATION A small rodent with hind feet that are more than 22 mm long from the heel to the tip of the longest toenail. Color is grayish in juveniles and dark golden brown in adults. Ears are prominent and eyes bulge slightly. Underside is creamy white. Tail has short dark hairs on top, short white hairs beneath, and a blending of these colors on the sides. Overall length of the skull is more than 28 mm.

DENTAL FORMULA i 1/1, c 0/0, p 0/0, m 3/3, total = 16.

SIZE AND WEIGHT Average and range in size of 44 specimens from Alabama:

 total length, 153 (137–199) mm / 6.1 (5.5–8.0) inches
 tail length, 74 (59–95) mm / 3.0 (2.4–3.8) inches
 hind foot length, 23 (22–24) mm / 0.9 (0.9–1.0) inch
 weight of 23 specimens, 29 (20–40) g / 1.0 (0.7–1.6) ounces
Males are larger than females.

DISTRIBUTION Statewide in Alabama, with the subspecies *P. g. gossypinus* in the south and *P. g. megacephalus* in the north. Geographic range of

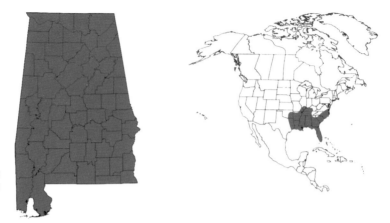

Distribution of the cotton deermouse in Alabama and North America.

the species is eastward from eastern Oklahoma and Texas to the Atlantic Ocean and southward from southern Illinois to the Gulf of Mexico.

ECOLOGY Habitats occupied include bottomland hardwood forests, wet and dry hammocks, and swamplands. Cotton deermice also occur along the edges of cleared fields and salt savannas and in fallow fields, palmetto (*Sabal*) thickets, dunes near beaches, pine and turkey oak (*Pinus–Quercus laevis*) forests, mixed pine-hardwood forests, and sandy pine scrublands. Cotton deermice may also inhabit rocky bluffs, ledges, caves, and buildings. Diet is omnivorous and includes invertebrates and plant materials that are encountered opportunistically. Predators include owls, striped skunks, gray and red foxes, coyotes, bobcats, and long-tailed weasels.

BEHAVIOR Although primarily nocturnal, cotton deermice may be active at dusk and dawn. They are excellent climbers and may place nests in vegetation about 0.8 m above the ground. They are also excellent swimmers, divers, and diggers. Nesting sites may be under logs, in stumps, under piles of brush, under palmettos, in moss on floating logs, or in burrows of gopher tortoises (*Gopherus polyphemus*). When available, nesting materials may include fibers from palms and palmettos and up to 450 g of cotton (*Gossypium*). When fleeing danger, cotton deermice may run across the ground, up trees, onto branches, or onto other types of vegetation. They have been captured 4.6 m above the ground in hardwood trees, and several were found 6.1 m above the ground in a live oak tree (*Quercus virginiana*), inhabiting an abandoned nest of an eastern gray squirrel. Cot-

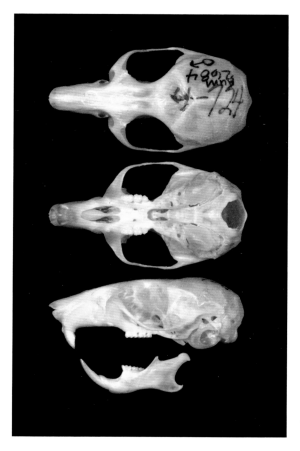

Dorsal, ventral, and lateral views of the cranium, and lateral view of the mandible of a male cotton deermouse. Greatest length of cranium is 28.9 mm.

ton deermice frequently inhabit swampy, wet areas and habitats that are prone to flooding.

LIFE HISTORY Breeding occurs throughout the year, but there is a decline in breeding activity during summer. Several litters may be produced during the year. A litter of 1–7 young (average is 4) is born following a gestation of 23–30 days. Giving birth takes about 55 minutes. Mothers lick newborns clean, consume the afterbirth, and respond to squeaking vocalizations of the babies. Newborns are hairless except for vibrissae on the rostrum, their eyes and ears are closed, and their incisors have not erupted. Average measurements (in mm) of 45 newborns were: total length, 47.2; tail length, 11.0; hind foot length, 6.5. Their average weight was 2.2 g. By 5 days, hairs cover the back, but the underside remains bare.

By 7 days, incisors have erupted. By 10 days, young are fully furred and they appear alert and respond to noises. Eyes open 12–14 days after birth. At 3–4 weeks, weaning occurs. By about 60 days of age, young are about the size of adults. At about 45 days old, males are reproductively mature, and females can mate by about 73 days of age. Life span is probably 4–5 months, with a maximum of about 2 years in the wild.

PARASITES AND DISEASES Ectoparasites include mites (*Androlaelaps, Dermatophagoides, Eulaelaps, Euschoengastia, Gahrliepia, Glycyphagus, Haemogamasus, Listrophorus, Ornithonyssus, Prolistrophorus, Pygmephorus, Walchia*), ticks (*Dermacentor, Ixodes*), lice (*Hoplopleura*), and fleas (*Ctenocephalides, Ctenophthalmus, Leptopsylla, Orchopeas, Polygenis, Stenoponia, Xenopsylla*). Cotton deermice may be parasitized by larval botflies (*Cuterebra*). Endoparasites include acanthocephalans (*Macracanthorhynchus, Moniliformis*), cestodes (*Hymenolepis, Spirometra, Taenia*), nematodes (*Aspicularis, Capillaria, Carolinensis, Gongylonema, Mastophorus, Physaloptera, Protospirura, Pterygodermatites, Trichostrongylus*), trematodes (*Fibricola, Scaphiostomum, Zonorchis*), and crustaceans (*Porocephalus*). Diseases include Black Creek Canal virus (*Hantavirus*), Cowbone Ridge virus (*Flavivirus*), leptospirosis (*Leptospira*), Lyme disease (*Borrelia*), Saint Louis encephalitis (*Flavivirus*), toxoplasmosis (*Toxoplasma*), and tularemia (*Francisella*).

CONSERVATION STATUS Lowest conservation concern in Alabama.

COMMENTS *Peromyscus* is from the Greek *peronē*, meaning "pointed," and *myskos*, meaning "mouse." The specific name *gossypinus* is Latin for "cotton." The subspecific name *megacephalus* is from the Greek *megas*, meaning "great," and *kephalē*, meaning "a head."

REFERENCES Whitaker (1968), Wolfe and Linzey (1977), Forrester (1992), Oliver et al. (1993), Durden (1995), Glass et al. (1998), Best (2004*a*), Whitaker et al. (2007).

White-footed Deermouse
Peromyscus leucopus

IDENTIFICATION A small rodent with hind feet that are up to 22 mm long from the heel to the tip of the longest toenail. Back is pale to dark brown in adults and grayish in juveniles. Underside is whitish, ears are dark, and tops of feet are white. Tail is dark on top and grades to whitish below.

DENTAL FORMULA i 1/1, c 0/0, p 0/0, m 3/3, total = 16.

SIZE AND WEIGHT Average and range in size of 30 specimens from Alabama:
>   total length, 145 (131–171) mm / 5.8 (5.2–6.8) inches
>   tail length, 60 (44–70) mm / 2.4 (1.8–2.8) inches
>   hind foot length, 18 (16–21) mm / 0.7 (0.6–0.8) inch
>   ear length, 16 (14–18) mm / 0.6 (0.6–0.7) inch
>   weight of 22 specimens, 18.9 (12.4–25.2 g) / 0.6 (0.4–0.8) ounce
There is no sexual dimorphism.

—

DISTRIBUTION Northern Alabama. Geographic range of the species is from southern Canada to southern Mexico and from central Arizona to the Atlantic Ocean. White-footed deermice are not present in Florida or in southern Alabama, Georgia, and South Carolina.

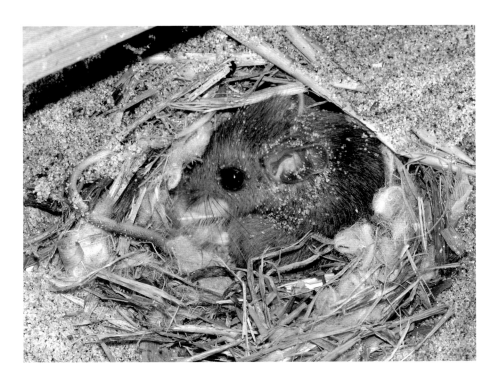

Nests constructed by white-footed deermice and by many other species of small rodents may be lined with feathers, fur, and shredded vegetation.

ECOLOGY White-footed deermice occur in the greatest numbers in a variety of habitats that include sufficient cover, such as brushy fields and deciduous forests. They may be present in lower densities in grassy fields and coniferous forests. Diet includes insects, fungi, and plant materials such as seeds, fruits, and green leaves. Predators include owls, snakes, striped skunks, gray and red foxes, coyotes, bobcats, and long-tailed weasels. Screech owls (*Otus asio*) are more likely to successfully prey on transient white-footed deermice than on resident mice that are more familiar with their home range.

BEHAVIOR White-footed deermice are primarily nocturnal but may be active during the daytime in winter. They do not hibernate and are semiarboreal. The tail is used for balance and as a prop while climbing. This mouse also spends considerable time on the ground. Nests are often aboveground and may be in piles of rocks, under logs or stumps, or in burrows. Nests are constructed from grasses, leaves, hair, feathers, floss from milkweeds, shredded bark, and mosses. When food is abundant, white-footed deermice may hoard it in burrows or cavities for con-

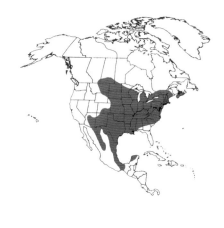

Distribution of the white-footed deermouse in Alabama and North America.

sumption later. During times of low temperatures, consumption of food increases and is especially high when females are nursing litters that are larger than average. Compared to males and nonlactating females, lactating females require about 25% more energy. In winter, white-footed deermice may become torpid in response to low temperatures. At other times of the year, they may have daily torpor of up to 3 hours in response to low temperatures or shortages of food. During winter, white-footed deermice occur in groups of 2–6 individuals that are either all in torpor or all active. They are excellent swimmers and may disperse across ice or swim distances of more than 200 m between islands.

LIFE HISTORY Breeding season is probably nearly year-round in Alabama, with a peak in warmer months. Each year, 4–5 litters may be born. Gestation in nonlactating females is 20–23 days but may be extended to 34–37 days in lactating females, possibly due to delayed implantation. Litters average 4–5 young (range is 1–8). Newborns weigh about 2 g and have closed eyes and ears, a pinkish body, and teeth that have not erupted. By 4–5 days, incisors have erupted; by 9–11 days, ears are open; and by 10–12 days, eyes are open. By 16–25 days, young may begin exploring the area surrounding their nest. When young are about 20–40 days old, their mother abandons them. Most juveniles disperse an average of about 100 m but may disperse as far as 1 km. Females become sexually mature at 33–45 days of age. At about 70 days old, young are fully grown. Life span is probably 4–6 months, but some individuals may live 12 months in the wild. Life span may be 7–8 years in captivity.

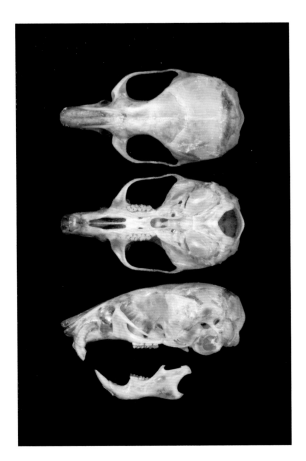

Dorsal, ventral, and lateral views of the cranium, and lateral view of the mandible of a male white-footed deermouse. Greatest length of cranium is 25.3 mm.

PARASITES AND DISEASES Ectoparasites include mites (*Androlaelaps, Blarinobia, Brevisterna, Cheladonta, Cheyletus, Dermacarus, Dermanyssus, Euschoengastia, Euschoengastoides, Eutrombicula, Fonsecia, Gahrliepia, Glycyphagus, Haemogamasus, Hirstionyssus, Laelaps, Leptotrombidium, Listrophorus, Microtrombicula, Myobia, Myocoptes, Neotrombicula, Notoedres, Odontacarus, Ornithonyssus, Pseudoschoengastia, Radfordia, Resinacarus, Trombicula*), lice (*Hoplopleura, Polyplax*), ticks (*Dermacentor, Ixodes*), and fleas (*Catallagia, Corrodopsylla, Ctenophthalmus, Doratopsylla, Echidnophaga, Epitedia, Hystrichopsylla, Megabothris, Monopsyllus, Nosopsyllus, Orchopeas, Peromyscopsylla, Stenoponia*). White-footed deermice may be parasitized by larval botflies (*Atrypoderma, Cuterebra*). Endoparasites include protozoans, cestodes (*Cladotaenia, Paruterina, Taenia*), nematodes (*Ascaris, Aspicularis, Capillaria, Mastophorus,*

Nematospiroides, Rictularia, Syphacia), and trematodes (*Alaria, Brachylaema, Entosiphonus, Postharmostomum*). Diseases include babesiosis (*Babesia*), hantaviruses (*Hantavirus*), human granulocytic anaplasmosis (*Anaplasma*), and Lyme disease (*Borrelia*).

CONSERVATION STATUS Lowest conservation concern in Alabama.

COMMENTS *Peromyscus* is from the Greek *peronē*, meaning "pointed," and *myskos*, meaning "mouse." The specific name *leucopus* is from the Greek *leukon*, meaning "white," and *pous*, meaning "foot."

REFERENCES Blair (1948), Metzgar (1967), Whitaker (1968), Lackey (1978), Lackey et al. (1985), Morzunov et al. (1998), Stafford et al. (1999), Best (2004a), Jaffe et al. (2005), Whitaker et al. (2007).

North American Deermouse

Peromyscus maniculatus

IDENTIFICATION Color is brown to brownish gray, with the middle region of the back slightly darker. Feet are white and the tail may be sharply bicolored. North American deermice and white-footed deermice are difficult to distinguish based on external characters, so it is possible that the two have been misidentified in northern Alabama where their ranges overlap. There are 2 morphological forms of North American deermice, both of which may be in Alabama. The form that is known to occur in Alabama is small and has a short, bicolored tail and small ears. The other form is larger and has a longer tail and larger ears.

DENTAL FORMULA i 1/1, c 0/0, p 0/0, m 3/3, total = 16.

SIZE AND WEIGHT Average and range in size of 18 specimens from Kentucky:

    total length, 178 (164–198) mm / 7.1 (6.6–7.9) inches
    tail length, 89 (77–97) mm / 3.6 (3.1–3.9) inches
    hind foot length, 20 (19–22) mm / 0.8 (0.8–0.9) inch
    weight, 16–24 g / 0.6–0.8 ounce
Females are larger than males.

Distribution of the North American deermouse in Alabama and North America.

DISTRIBUTION The first verified record of the North American deermouse in Alabama was a lactating female collected in an overgrown hayfield near Cox Sink Pond, 34°46'53.8"N, 86°18'20.0"W, Jackson County, on 1 October 2011. The species is probably widely distributed north of the Tennessee River in Alabama. Geographic range of the North American deermouse is from northwestern Canada into northern Alabama, from the Pacific Ocean to east of the Appalachian Mountains, and southward into southern Mexico.

ECOLOGY North American deermice are among the most abundant, widespread, and adaptable mammals in North America. They occur in all habitats within their range including tundra, coniferous and deciduous forests, grasslands, swamps, scrublands, and deserts. In Alabama, the North American deermouse has been recorded only from an overgrown hayfield within a mosaic of habitats that included hayfields, tree-lined ravines with dense understories, steep forested slopes with rocky outcrops, and mixed-hardwood forests with scattered pines (*Pinus*). The species is expected to inhabit all of these habitats, as well as human-made structures such as abandoned homesites and barns. Diet is mostly seeds and mast on the forest floor and in clearings, but North American deermice also eat blueberries (*Vaccinium*), raspberries (*Rubus*), insects, brown and green caterpillars, centipedes, and even small birds and mice. Because they are abundant in most of their geographic range, North American deermice are an important source of food for many vertebrates. Predators include a wide variety of snakes, owls, hawks, and mammals.

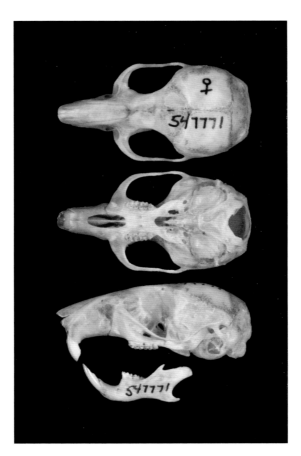

Dorsal, ventral, and lateral views of the cranium, and lateral view of the mandible of a female North American deermouse. Greatest length of cranium is 22.6 mm.

BEHAVIOR North American deermice are nocturnal, arboreal, and terrestrial. They may spend days in nests they construct themselves, in nests of squirrels, or in cavities well aboveground. In winter, they will huddle in communal nests. North American deermice use a typical walking gait when moving slowly, such as when foraging. When they move rapidly, rather than trotting like shrews and woodland voles, they leave 2 closely spaced pairs of tracks that are more like tiny rabbit tracks. The tail often leaves a mark in the snow or sand. North American deermice do not hibernate and they store nuts and seeds for use in winter.

LIFE HISTORY Breeding occurs from early spring into autumn but decreases slightly during warmer months. Gestation is 23–24 days; 27 days if the female becomes pregnant while still nursing a previous litter. There are 4–5 young/litter (range is 1–7). Young may be born in a stump, beneath a log, or elsewhere on the ground. Newborns are hairless, their eyes and ears are closed, and they weigh about 1.8 g. At 2–4 days, ears open; at 7–8 days, incisors erupt; and at 12–16 days, eyes open. By 3 weeks old, young are weaned, and they become sexually mature in about another 3 weeks. Life span is 5–13 months in the wild.

PARASITES AND DISEASES Ectoparasites include mites (numerous genera in the families Amerosiidae, Glycyphagidae, Laelapidae, Listrophoridae, Macrochelidae, Macronyssidae, Myobiidae, Myocoptidae, Pygmephoridae, Trombiculidae), ticks (*Dermacentor, Ixodes*), lice (*Hoplopleura, Polyplax*), and fleas (*Atyphlocerus, Catallagia, Ceratophyllus, Conorhinopsylla, Corrodopsylla, Ctenophthalmus, Doratopsylla, Echidnophaga, Epitedia, Hoplopsyllus, Hystrichopsylla, Megabothris, Nearctopsylla, Nosopsyllus, Opisodasys, Orchopeas, Peromyscopsylla, Rhadinopsylla, Stenoponia*). North American deermice may be parasitized by larval botflies (*Cuterebra*). Endoparasites include protozoans, cestodes (*Catenotaenia, Choanotaenia, Cladotaenia, Cysticercus, Hydatigera, Hymenolepis, Mesocestoides, Paruterina, Prochoanotaenia, Taenia*), nematodes (*Aspicularis, Capillaria, Gongylonema, Longistriata, Nematospiroides, Nippostrongylus, Protospirura, Rictularia, Syphacia, Trichuris*), and trematodes (*Brachylaema, Euryhelmis, Postharmostomum*). Diseases include Rocky Mountain spotted fever (*Rickettsia*) and hantaviruses (*Hantavirus*). The first documented cases of hantavirus in North America were linked to North American deermice in the western United States. To reduce the chances of contracting hantavirus, humans should wear a respirator in enclosed areas where North American deermice have urinated, defecated, or nested.

CONSERVATION STATUS Undetermined in Alabama.

COMMENTS *Peromyscus* is from the Greek *peronē*, meaning "pointed," and *myskos*, meaning "mouse." The specific name *maniculatus* is from the Latin *manicula*, meaning "a little hand," and *atus*, meaning "provided with."

REFERENCES Blair (1948), Whitaker (1968), Coultrip et al. (1973), Bowers and Smith (1979), Hall (1981), Whitaker and Hamilton (1998), Jaffe et al. (2005), Whitaker et al. (2007).

Oldfield Deermouse

Peromyscus polionotus

IDENTIFICATION This small rodent has hind feet that are usually less than 19 mm long from the heel to the tip of the longest toenail. This is the smallest deermouse in Alabama. It is similar in size to the eastern harvest mouse, but the oldfield deermouse does not have grooved upper incisors. Color of back varies with color of substrate, from pale in coastal areas to varying shades of brown in inland areas. Underside and feet are white. Tail is bicolored; brownish on top and white beneath.

DENTAL FORMULA i 1/1, c 0/0, p 0/0, m 3/3, total = 16.

SIZE AND WEIGHT Average and range in size of 25 specimens from Alabama:
　　total length, 118 (109–132) mm / 4.7 (4.4–5.3) inches
　　tail length, 45 (32–54) mm / 1.8 (1.3–2.2) inches
　　hind foot length, 17 (15–20) mm / 0.7 (0.6–0.8) inch
　　ear length, 14 (10–15) mm / 0.6 (0.4–0.6) inch
　　weight of 12 specimens, 13.0 (10.1–17.6) g / 0.5 (0.4–0.6) ounce
Females average larger than males.

This endangered subspecies of the oldfield deermouse is known as the Alabama beach mouse (*Peromyscus polionotus ammobates*). Note that the pale color closely matches the color of the sandy beach.

DISTRIBUTION Statewide, except for some areas in northwestern and southwestern Alabama. Geographic range of the species includes most of Alabama, Georgia, Florida, and South Carolina and adjacent areas of Mississippi, Tennessee, and North Carolina. The species is also present on some coastal islands.

ECOLOGY Oldfield deermice occupy sites that are disturbed and in various early stages of succession, such as beaches and fallow fields. As sites gain more vegetation and approach stabilization, populations of oldfield deermice gradually disappear. Sandy soils are an important component of habitat, and burrows are often placed within sparse vegetation on sandy slopes. In front of burrows are fan-shaped aprons of excavated sand. Burrows usually have a passage up to 1 m deep leading to 1 or more chambers. Often, an escape tunnel is present from the nesting chamber to just below the surface. Nests are made of dried grasses and other fibers. Entrance tunnels are plugged with sand several centimeters below the surface, presumably to reduce the chance of capture by a predator. Usually, there are 4–6 burrows within the home range, but there may be as many as 20. In the ever-changing frontal dunes of coastal habitats, this is often the only species of small mammal that is a permanent resident. Local densities may be 18–26 mice/hectare. Toward the center of its range, it

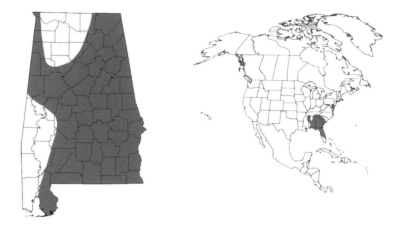

Distribution of the oldfield deermouse in Alabama and North America.

occurs in relatively isolated, small populations in mixed pine-hardwood forests where the canopy is open and the understory is sparse. Roadway embankments are also occupied. Oldfield deermice are rarely associated with buildings. Seeds are the most commonly consumed foods. Populations in coastal areas consume primarily wind-deposited seeds such as sea oats (*Uniola paniculata*) and bluestems (*Andropogon*). When available, acorns (*Quercus*) are a significant portion of the diet for both inland and coastal populations. Oldfield deermice also consume a variety of insects including beetles, leafhoppers, true bugs, and ants. Predators include various snakes such as coachwhips (*Masticophis flagellum*), pygmy rattlesnakes (*Sistrurus miliarius*), eastern diamondback rattlesnakes (*Crotalus adamanteus*), owls, great blue herons (*Ardea herodias*), gray and red foxes, long-tailed weasels, striped skunks, raccoons, and domestic dogs and cats.

BEHAVIOR Oldfield deermice are nocturnal and rarely active during the daytime. They are curious and active throughout the year. Nightly activity is directly related to weather conditions and the phase of the moon. These mice are less active on nights with bright moonlight and clear skies; they rarely leave their burrows during nights with a full moon. Generally, activity increases on warm and cloudy nights and when conditions change from clear to cloudy. While oldfield deermice are active throughout the night, peaks of activity occur shortly after dusk and again after midnight. They often move among burrows, but they move less frequently when litters are present. While oldfield deermice are capable of moving more than 5 km and often travel 0.5 km each night, most spend their life

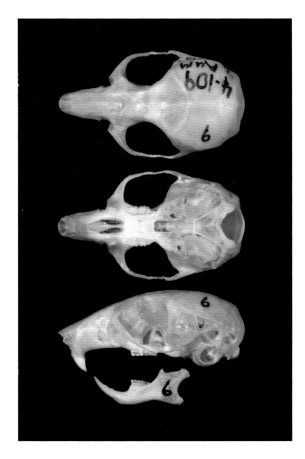

Dorsal, ventral, and lateral views of the cranium, and lateral view of the mandible of a female old-field deermouse. Greatest length of cranium is 22.5 mm.

within a few hundred meters of where they were born. Juveniles disperse an average of 160 m, which is about the width of a home range, away from where they were born.

LIFE HISTORY This is one of the few monogamous species of deermice. Breeding season is year-round but typically slows during summer and peaks during late autumn to early winter in correlation with availability of food. Pair-bonding is strong and there is parental cooperation in rearing young. Gestation averages 28 days; a postpartum estrus is common. An average litter has 4 young (range is 1–8). Beginning at 118 days of age, a captive female gave birth to 26 litters (139 young), with an average interval between litters of 30 days. Average life span is about 8–9 months in the wild, but some oldfield deermice live 1–2 years. Life span in captivity may be more than 4 years.

PARASITES AND DISEASES Ectoparasites include mites (*Androlaelaps, Cheyletus, Euschoengastia, Eulaelaps, Glycyphagus, Haemogamasus, Hypoaspis, Leptotrombidium, Ornithonyssus, Radfordia*), ticks (*Dermacentor, Ixodes*), lice (*Hoplopleura, Polyplax*), and fleas (*Ctenophthalmus, Echidnophaga, Peromyscopsylla, Polygenis, Xenopsylla*). Oldfield deermice may be parasitized by larval botflies (*Cuterebra*). Endoparasites include acanthocephalans (*Moniliformis*), nematodes (*Carolinensis, Gongylonema, Physaloptera, Pterygodermatites, Syphacia, Trichostrongylus*), and trematodes (*Zonorchis*). Diseases include toxoplasmosis (*Toxoplasma*) and tularemia (*Francisella*).

CONSERVATION STATUS Listed as moderate conservation concern in Alabama, except for the subspecies *P. p. ammobates* (Alabama beach mouse) and *P. p. trissylepsis* (Perdido Key beach mouse), which are listed as highest conservation concern in Alabama and as endangered by the United States Fish and Wildlife Service.

COMMENTS *Peromyscus* is from the Greek *peronē*, meaning "pointed," and *myskos*, meaning "mouse." The specific name *polionotus* is from the Greek *polios*, meaning "gray," and *ōtos*, meaning "ear." The subspecific name *ammobates* is from the Greek *ammos*, meaning "sand," and *batēs*, meaning "one that treads or haunts"; *trissylepsis* is possibly from the Greek *trissōs*, meaning "triple," and *lēpsis*, meaning "seizing" or "catching."

REFERENCES Whitaker (1968), Forrester (1992), Whitaker and Hamilton (1998), Best (2004*a*), Whitaker et al. (2007), Nims et al. (2008), M. C. Wooten (in litt.).

Eastern Harvest Mouse
Reithrodontomys humulis

IDENTIFICATION A small mouse that has grooved upper incisors, a grayish-brown and cinnamon back, paler sides, and a grayish, cinnamon-tinged underside. Tail is brownish on top and blends to whitish on the underside.

DENTAL FORMULA i 1/1, c 0/0, p 0/0, m 3/3, total = 16.

SIZE AND WEIGHT Average and range in size of 25 specimens from Alabama:

     total length, 113 (99–126) mm / 4.5 (4.0–5.0) inches

     tail length, 50 (41–59) mm / 2.0 (1.6–2.4) inches

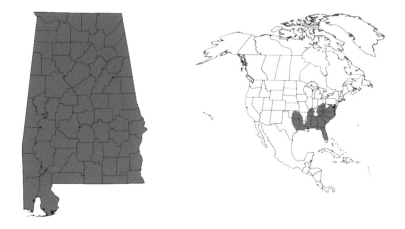

Distribution of the eastern harvest mouse in Alabama and North America.

hind foot length, 15 (13–19) mm / 0.6 (0.5–0.8) inch

ear length of 15 specimens, 11 (10–17) mm / 0.4 (0.4–0.7) inch

weight of 10 specimens, 6.9 (4.7–8.6) g / 0.2 (0.2–0.3) ounce

There is no apparent sexual dimorphism.

DISTRIBUTION Statewide in Alabama. Geographic range of the species is southward from southern Ohio to the Gulf of Mexico and from central Oklahoma and eastern Texas across most of the southeastern United States.

ECOLOGY Fallow fields are common habitats, but eastern harvest mice also occur along grass-covered roadside embankments and irrigation ditches, in grassy fields, in honeysuckle (*Lonicera*) thickets, and in wet meadows. Dense vegetation, regardless of species, is favored. The somewhat globular nests are usually on the edge of the home range and are constructed of finely shredded grasses. They vary from 5 cm in diameter in summer to 11 cm in winter. Nests may be placed in the middle of grassy clumps but are usually on the ground, hidden at the base of a grassy clump or in a shallow depression that the eastern harvest mouse excavates. One nest was in an abandoned nest of a hispid cotton rat. Up to 4 nests are used in the same week. Eastern harvest mice probably do not build runways; instead, they use runways made by hispid cotton rats. Diet is predominantly seeds, green vegetation, and insects, especially grasshoppers. Predators include barn owls (*Tyto alba*), other birds, snakes, red foxes, coyotes, bobcats, and long-tailed weasels.

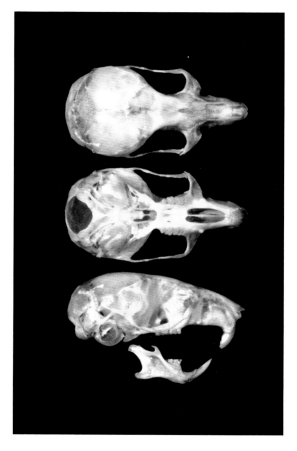

Dorsal, ventral, and lateral views of the cranium, and lateral view of the mandible of a female eastern harvest mouse. Greatest length of cranium is 19.5 mm.

BEHAVIOR Although primarily nocturnal, eastern harvest mice may be active during daylight, especially in winter. They do not hibernate. These are highly social animals and several may occupy the same nest, especially in winter. In captivity, mothers have killed and partially consumed their own young, and siblings may kill and consume their littermates. A mother will defend young by scooping them under her body and biting or pushing at the threat. She will retrieve young from outside the nest by grasping the baby in her mouth at any convenient place, including a foot, and returning the baby to the nest. She then thoroughly cleans the young. Occasionally, food is stored in nests, especially seeds of grasses and sedges.

LIFE HISTORY Breeding may occur throughout the year in Alabama, but peaks occur during spring and autumn. Following a gestation of 21–24 days, 2–4 young are born (range is 1–8). Eyes and ears are closed and the body is naked except for some unpigmented, nearly microscopic hairs. Average measurements of newborns are: total length, 37.3 mm; tail length, 10.7 mm; hind foot length, 5.3 mm; weight, 1.2 g. At 1 day, hair begins to darken, and by 5–8 days, hair is brownish on the back and grayish white on the underside. By 4 days, the semiprehensile tail responds to touch. By about 10 days, eyes and ears open. By 12 days of age, young can sit erect on their hind legs and groom in a manner similar to adults; they start at the snout, then move to the ears, and end with the tail from rump to tip. Weaning begins at about 9 days after birth and is completed by 23 days. Juveniles leave the nest when they are about 30 days old and weigh at least 5 g. Ages at first pregnancy in 2 captive females were 11 and 20 weeks; intervals between litters of 1 female were 24–49 days. Life span is probably 5–6 months in the wild. In captivity, eastern harvest mice have lived more than 2 years.

PARASITES AND DISEASES Ectoparasites include mites (*Cheyletus*) and fleas (*Polygenis*). Endoparasites include trematodes (*Zonorchis*). Reasons for eastern harvest mice having only a few species of parasites may include a diet composed mostly of seeds, a small body, usually noncolonial behavior, and small populations.

CONSERVATION STATUS Moderate conservation concern in Alabama.

COMMENTS *Reithrodontomys* is from the Greek *rheithron*, meaning "stream" or "channel," *odous*, meaning "tooth," and *mys*, meaning "mouse"; thus, "channel-toothed mouse," referring to the grooved upper incisors of this mouse; *humulis* is from the Latin *humilis*, meaning "small."

REFERENCES Forrester (1992), Stalling (1997), Best (2004*a*).

Marsh Oryzomys
Oryzomys palustris

IDENTIFICATION Once known as the marsh rice rat, this medium-sized rat is grayish brown on the back and much paler on the underside. Often, hairs along the midline of the back are much darker than elsewhere on the back and on the sides. Hairs on top of the tail are darker than on the underside.

DENTAL FORMULA i 1/1, c 0/0, p 0/0, m 3/3, total = 16.

SIZE AND WEIGHT Average and range in size of 25 specimens from Alabama:

> total length, 232 (200–253) mm / 9.3 (8.0–10.1) inches
> tail length, 114 (97–127) mm / 4.6 (3.9–5.1) inches
> hind foot length, 31 (27–32) mm / 1.2 (1.1–1.3) inches
> ear length of 5 specimens, 15 (13–17) mm / 0.6 (0.5–0.7) inch
> weight of 2 specimens, 58.2 (56.2–59.0) g / 2.0 (2.0–2.1) ounces

Weights to 80 g (2.8 ounces) have been recorded. Males average larger than females.

DISTRIBUTION Statewide in Alabama. Probable geographic range of the species is from northeastern Arkansas to the Atlantic Ocean and from

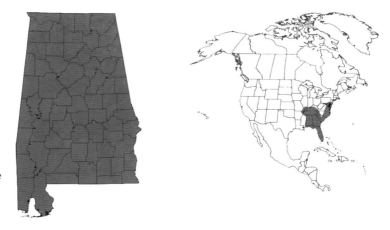

Distribution of the marsh oryzomys in Alabama and North America.

southern Illinois and central Kentucky southward to the Gulf of Mexico and across Florida. Examination of genetic variation indicates that the western subspecies *O. p. texensis* is probably specifically distinct from *O. p. palustris*, which occurs in Alabama and other southeastern states.

ECOLOGY This semiaquatic rodent occurs in wet marshes and meadows, hydric hammocks, and swamps. In Alabama, the marsh oryzomys is widely distributed in wetland habitats with dense vegetation, including marshes at higher elevations near Mount Cheaha. Diet is omnivorous and includes fungi (*Endogone*), succulent parts of plants, seeds, insects, blue (*Callinectes*) and fiddler (*Uca*) crabs, snails, fish, clams, baby turtles (*Graptemys*, *Chrysemys*), eggs of American alligators (*Alligator mississippiensis*), eggs and young of birds, and carrion of common muskrats, deermice, and birds. Owls, especially barn owls (*Tyto alba*), are significant predators. Other predators include barred owls (*Strix varia*), northern harriers (*Circus cyaneus*), cottonmouths (*Agkistrodon piscivorus*), raccoons, red foxes, American minks, and striped skunks.

BEHAVIOR Marsh oryzomys are nocturnal. They make runways, dig burrows, and are excellent divers and swimmers. Air trapped in the fur gives added buoyancy and provides insulation from cold water when they swim. They can swim underwater for up to 10 m and they readily swim 300 m to cross channels between islands. Marsh oryzomys groom repeatedly, which may be helpful in retaining the water-repellent quality of the pelage. Nests are about 13 cm in diameter and made of grasses and

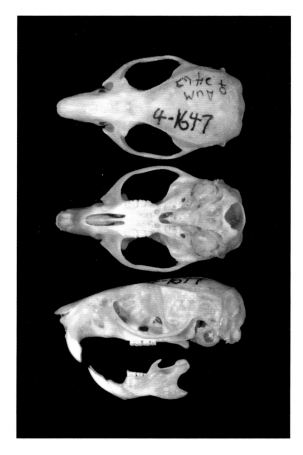

Dorsal, ventral, and lateral views of the cranium, and lateral view of the mandible of a female marsh oryzomys. Greatest length of cranium is 31.8 mm.

sedges. They are usually placed on the ground under debris, at the bases of shrubs, or at the ends of shallow burrows. They may also be attached to vegetation in areas that are flooded often. Marsh oryzomys also commandeer and modify nests of marsh wrens (*Cistothorus palustris*) and red-winged blackbirds (*Agelaius phoeniceus*), and they may use dens of common muskrats as nesting sites. High-pitched squeaks are emitted by newborns and juveniles in the nest, and by adults when fighting.

LIFE HISTORY Breeding season is probably throughout the year. Gestation is 21–28 days and is probably followed by a postpartum estrus. Each year, 5–6 litters can be produced. An average litter is 5 young (range is 4–6). Newborns have a sparse covering of fine hairs, some pigmentation on the back, closed eyes and ears, well-formed toes and claws, and well-developed vibrissae, and they weigh 3–4 g. At 1 day old, young emit high-pitched vocalizations. By 2 days old, young can crawl; by 6–7 days, incisors erupt; by 8 days, ears open; and by 8–11 days, young take solid food, their eyes open, and their pelage becomes sleek and shiny. Weaning occurs at 11–20 days. Range in weight (in g) at 10 days old was 8–17; at 20 days, 18–27; at 40 days, 27–40; at 60 days, 40–60; and at 120 days, 50–80. Both sexes reach sexual maturity at 50–60 days of age. Life span is probably 7–12 months in the wild.

PARASITES AND DISEASES Ectoparasites include mites (*Androlaelaps, Eutrombicula, Gigantolaelaps, Laelaps, Listrophorus, Ornithonyssus, Oryzomysia, Protolistrophorus, Radfordia, Trombicula*), ticks (*Dermacentor, Ixodes*), lice (*Hoplopleura*), and fleas (*Polygenis*). Endoparasties include protozoans (*Eimeria, Isospora, Toxoplasma*), cestodes (*Cladotaenia, Hymenolepis, Taenia*), nematodes (*Angiostrongylus, Capillaria, Hassalstrongylus, Litomosoides, Mastophorus, Monodontus, Parastrongylus, Physaloptera, Pterygodermatites, Skrjabinoclava, Strongyloides, Syphacia, Trichostrongylus*), trematodes (*Ascocotyle, Brachylaeme, Catatropis, Echinochasmus, Fibricola, Gynaecotyla, Levinseniella, Lyperosomum, Maritrema, Microphallus, Odhneria, Parvatrema, Probolocoryphe, Stictodora, Urotrema, Zonorchis*), and crustaceans (*Protocephalus*). Diseases and disorders include Bayou virus (*Hantavirus*), the bacterial pathogen *Bartonella*, bacterial periodontal disease, encephalomyocarditis (*Cardiovirus*), the genetic disorder kyphosis, which is characterized by a misshapen or hunchbacked vertebral column, Saint Louis encephalitis (*Flavivirus*), and Tamiami virus (*Arenavirus*).

CONSERVATION STATUS Lowest conservation concern in Alabama.

COMMENTS *Oryzomys* is from the Greek *oryza*, meaning "rice," and *mys*, meaning "mouse"; *palustris* is Latin for "marshy." The subspecific name *texensis* is Latin for "of Texas."

REFERENCES Esher et al. (1978), Wolfe (1982), NeSmith and Cox (1985), Webster (1987), Hunt and Ogden (1991), Forrester (1992), Forys and Dueser (1993), Schmidt and Engstrom (1994), Kosoy et al. (1996, 1997), Best (2004*a*), McIntyre et al. (2005), Whitaker et al. (2007), Hanson et al. (2010).

Hispid Cotton Rat
Sigmodon hispidus

IDENTIFICATION A medium-sized rat with pelage that has a mixture of dark brown to nearly black hairs with grayish and yellowish hairs intermixed. Sides and back are about the same color, but the underside is pale to dark grayish and may be yellowish along the sides. Tail is sparsely haired and dark brown on top, blending to pale grayish on the bottom.

DENTAL FORMULA i 1/1, c 0/0, p 0/0, m 3/3, total = 16.

SIZE AND WEIGHT Average and range in size of 40 specimens from Alabama:

> total length, 233 (135–300) mm / 9.3 (5.4–12.0) inches
> tail length, 97 (82–114) mm / 3.9 (3.3–4.6) inches
> hind foot length, 30 (26–35) mm / 1.2 (1.0–1.4) inches
> ear length of 18 specimens, 18 (15–21) mm / 0.7 (0.6–0.8) inch
> weight of 16 specimens, 95.9 (70.8–145.0) g / 3.4 (2.5–5.1) ounces

Males average larger than females in some characters, but there is no significant sexual dimorphism.

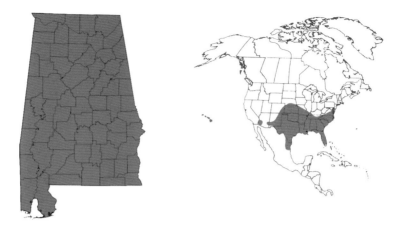

Distribution of the hispid cotton rat in Alabama and North America.

DISTRIBUTION Statewide in Alabama. Geographic distribution of the hispid cotton rat is from western Arizona to the Atlantic Ocean and from southern Nebraska to northeastern Mexico.

ECOLOGY Hispid cotton rats occur most frequently in grass-dominated habitats, including sites with scattered shrubs, trees, and brush. In Alabama, this species is abundant in grassy hay fields and in fallow fields and pastures with grasses and shrubby cover that are in the early stages of succession. Overgrown highway embankments and rights-of-way serve as suitable habitats and potential routes for dispersal. In wetland areas near ponds, streams, swamps, and marshes, hispid cotton rats often occur with marsh oryzomys, and in drier habitats, with oldfield deermice and eastern harvest mice. Hispid cotton rats do not hoard food. Populations fluctuate greatly within and among years, possibly related to temperature and precipitation, which may influence availability of food and the amount of vegetative cover. Diet includes grasses, fruits, seeds, leaves, green stems, and insects. Hispid cotton rats have been reported to destroy nests of northern bobwhite quail (*Colinus virginianus*). Predators include barn owls (*Tyto alba*), barred owls (*Strix varia*), red-tailed hawks (*Buteo jamaicensis*), snakes, coyotes, gray and red foxes, bobcats, striped skunks, and long-tailed weasels.

BEHAVIOR Hispid cotton rats do not hibernate and they are active at any time during the day or night. They may climb up onto vines of Japanese honeysuckles (*Lonicera japonica*). While swimming, they propel them-

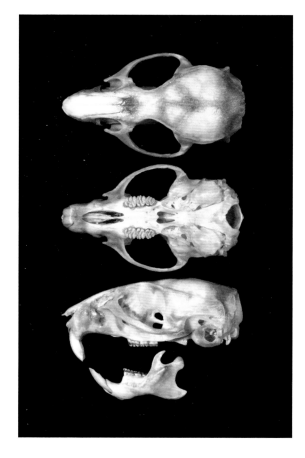

Dorsal, ventral, and lateral views of the cranium, and lateral view of the mandible of a male hispid cotton rat. Greatest length of cranium is 35.6 mm.

selves with their hind feet while holding their forefeet near their body; they do not swim underwater. Their fur becomes wet almost immediately upon immersion, giving them a tendency to sink. Nests on the surface of the ground and in burrows are made of woven grasses and range from cup-shaped to hollow ball-shaped structures with a single opening.

LIFE HISTORY Breeding probably occurs throughout the year, but peaks are in spring and autumn. Several litters may be produced each year. Females mate within 24 hours of giving birth. Gestation is about 27 days. An average litter has 3–9 young (range is 1–15). At birth, young are covered by a fine coat of pale, reddish-tinted hairs that are thickest around the head; while the skin on top of the head, back, and tail is slate gray, the underside is pink, and the eyes and ears are closed. Average weight of newborns is 6.8–7.2 g (range is 3.5–8.0 g). Usually by 18–36 hours, eyes are open. Weaning occurs 10–15 days after birth. By 3 months old, males are sexually mature. By 38–40 days old, some females become pregnant. Life span is probably less than 1 year in the wild.

PARASITES AND DISEASES Ectoparasites include mites (*Androlaelaps, Atricholaelaps, Eulaelaps, Eutrombicula, Gigantolaelaps, Haemogamasus, Laelaps, Macrocheles, Ornithonyssus, Prolistrophorus, Radfordia*), ticks (*Amblyomma, Dermacentor, Ixodes*), lice (*Hoplopleura*), and fleas (*Ctenocephalides, Echidnophaga, Epitedia, Hoplopsyllus, Leptopsylla, Polygenis, Xenopsylla*). Endoparasites include fungi, protozoans (*Toxoplasma*), cestodes (*Monoecocestus, Raillietina, Taenia*), nematodes (*Capillaria, Gongylonema, Hassalstrongylus, Litomosoides, Mastophorus, Monodontus, Physaloptera, Physocephalus, Pterygodermatites, Strongyloides, Syphacia, Trichostrongylus*), trematodes (*Gynaecotyla, Maritrema, Microphallus, Nudacotyle, Plagiorchis, Probolocoryphe, Zonorchis*), and crustaceans (*Porocephalus*). Diseases include Black Creek Canal virus (*Hantavirus*), Chagas disease (*Trypanosoma*), Cowbone Ridge virus (*Flavivirus*), diphtheria (*Corynebacterium*), eastern equine encephalomyelitis (*Alphavirus*), encephalomyocarditis (*Cardiovirus*), Highlands J virus (*Alphavirus*), leptospirosis (*Leptospira*), Lyme disease (*Borrelia*), poliomyelitis (*Enterovirus*), rabies (*Lyssavirus*), Saint Louis encephalitis (*Flavivirus*), tularemia (*Francisella*), and Venezuelan equine encephalomyelitis (*Alphavirus*). Hispid cotton rats are used in laboratory studies of these and other diseases.

CONSERVATION STATUS Lowest conservation concern in Alabama.

COMMENTS *Sigmodon* is from the Greek *sigma*, the Greek letter "Σ," and *odous*, meaning "tooth." This refers to the S-shaped cusp pattern on the last molar. The specific name *hispidus* is Latin for "rough" or "shaggy," referring to the fur.

REFERENCES Komarek (1939), Kinsella (1974), Cameron and Spencer (1981), Burgdorfer and Gage (1987), Forrester (1992), Rollin et al. (1995), Niewiesk and Prince (2002), Carroll et al. (2004), Best (2004*a*), Whitaker et al. (2007), Bradley et al. (2008).

Old World Rats and Mice

Family Muridae

Muridae is the most diverse family of mammals in the world, with 150 genera and 730 species. Members of this family are usually small, with total length ranging from about 110 to 800 mm. These rodents occupy myriad habitats and may be terrestrial, arboreal, fossorial, or semiaquatic. The native range includes Europe, Africa, Asia, Australia, and oceanic islands, but two genera (*Mus, Rattus*) have been introduced throughout the world and are commensal with humans. Three species have been introduced into Alabama (house mouse, brown rat, roof rat).

House Mouse
Mus musculus

IDENTIFICATION A small mouse with upper incisors that do not have grooves but do have a small notch on the back and near the tip. Tail appears hairless and wild animals are usually gray. Underside of body is usually slightly paler than the back and sides. Because of selective breeding by humans, this species, which is widely used as a pet and in scientific research, may be white, gray, brown, black, different shades of these or other colors, or a mixture of colors.

DENTAL FORMULA i 1/1, c 0/0, p 0/0, m 3/3, total = 16.

SIZE AND WEIGHT Average and range in size of 25 specimens from Alabama:

> total length, 159 (140–197) mm / 6.4 (5.6–7.9) inches
> tail length, 75 (62–102) mm / 3.0 (2.5–4.1) inches
> hind foot length, 17 (12–19) mm / 0.7 (0.5–0.8) inch
> ear length of 17 specimens, 13 (8–14) mm / 0.5 (0.3–0.6) inch
> weight of 7 specimens, 17 (13–25) g / 0.6 (0.5–1.0) ounce

There is no sexual dimorphism.

Distribution of the house mouse in Alabama and North America.

DISTRIBUTION Statewide in Alabama. Natural distribution is across nearly all of Europe, Africa, and Asia. House mice now occur throughout the world in close association with humans.

ECOLOGY In the wild, house mice live in a variety of habitats associated with humans including homes, farm buildings, especially where feed for pets and livestock is stored, fallow fields, and agricultural lands. Wild house mice, including those living in homes and farm buildings, are generally considered pests because of the damage they do, the foods they consume or contaminate, and the diseases and parasites they may transmit to humans and other animals. Nests are loose structures of cloth, paper, or other soft materials and are lined with finer shredded substances. They may be almost anywhere in buildings or other structures including grain bins, interior walls, attics, farm machinery, unused motor vehicles, piles of rags or clothing, boxes of newspapers, appliances, woodpiles, cabinets, or silverware drawers in the kitchen. Diet includes most foods consumed by humans, pets, and livestock, including grains, grasses, fruits, berries, and other plant materials. Predators are as varied and include snakes, owls, hawks, domestic dogs and cats, gray and red foxes, coyotes, raccoons, bobcats, skunks, long-tailed weasels, and humans. House mice are also reared as food for pets such as fish, lizards, birds, and snakes. House mice can be captured in traps baited with almost any type of food.

BEHAVIOR House mice are active explorers of their environment and are excellent jumpers, climbers, and swimmers. They do not hibernate. They

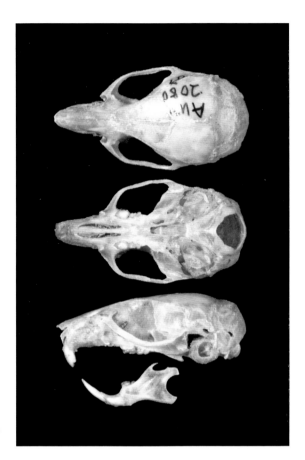

Dorsal, ventral, and lateral views of the cranium, and lateral view of the mandible of a male house mouse. Greatest length of cranium is 22.2 mm.

are most active at night but may move about in darkened homes or other buildings at any time of the day or night. Vocalizations include audible and ultrasonic high-pitched squeaks that are emitted by both young and adults.

LIFE HISTORY House mice breed year-round in captivity and during warmer months in the wild. During courtship, the male makes complex ultrasonic calls, and he sniffs and follows the female. A female may have 5–10 litters/year. Gestation is 19–21 days and litters average 6–8 young (range is 1–14). At birth, young are pink and hairless, their eyes and ears are closed, and they weigh about 0.5–1.5 g. By 3 days, fur begins to grow. By 7–14 days, eyes and ears open. Weaning occurs about 3 weeks after birth, when young weigh 10–12 g. Males reach sexual maturity at about 8 weeks and females at about 6 weeks, but breeding can take place as early as 5 weeks of age. Life span is probably 5–6 months in the wild, but house mice have lived 4–6 years in captivity.

PARASITES AND DISEASES Ectoparasites include mites (*Androlaelaps, Dermacarus, Echinolaelaps, Echinonyssus, Eulaelaps, Eutrombicula, Glycyphagus, Haemogamasus, Listrophorus, Myocoptes, Myobia, Ornithonyssus, Orycteroxenus, Pygmephorus, Radfordia, Xenoryctes*), ticks (*Dermacentor, Ixodes*), lice (*Hoplopleura, Polyplax*), and fleas (*Ctenophthalmus, Echidnophaga, Leptopsylla, Orchopeas, Xenopsylla*). Endoparasites include amoebas (*Endamoeba*), protozoans (*Trichomonas*), cestodes (*Catenotaenia, Hymenolepis, Rodentolepis, Taenia*), nematodes (*Aspiculuris, Calodium, Capillaria, Gongylonema, Heligmosomoides, Heterakis, Mastophorus, Protospirura, Syphacia, Trichuris*), and trematodes (*Brachylaima*). Diseases include leptospirosis (*Leptospira*), Lyme disease (*Borrelia*), lymphocytic choriomeningitis (*Arenavirus*), murine typhus (*Rickettsia*), plague (*Yersinia*), rickettsialpox (*Rickettsia*), and tularemia (*Francisella*). Because they are small, relatively inexpensive to maintain, quick to reproduce, and share many traits with humans, house mice are the most commonly used mammal in medical research.

CONSERVATION STATUS Exotic in Alabama.

COMMENTS *Mus* is Latin for "mouse"; *musculus* is Latin for "mouse."

REFERENCES Harkema (1936), Forrester (1992), Nowak (1999), Oliver et al. (1999), Milazzo et al. (2003), Best (2004*a*), Whitaker et al. (2007), Whitaker and Mumford (2009).

Brown Rat

Rattus norvegicus

IDENTIFICATION A large rat with coarse pelage. Tail appears hairless and is shorter than, or about as long as, head and body. Wild animals are usually grayish brown, but various patterns of color and mottling may occur. Underside of body is usually paler than back and sides. This species, which is widely used as a pet and in scientific research, may be white, gray, brown, black, different shades of these or other colors, or a mixture of colors.

DENTAL FORMULA i 1/1, c 0/0, p 0/0, m 3/3, total = 16.

SIZE AND WEIGHT Average and range in size of 15 specimens from Alabama:

    total length, 362 (290–434) mm / 14.5 (11.6–17.4) inches
    tail length, 168 (135–206) mm / 6.7 (5.4–8.2) inches
    hind foot length, 41 (35–56) mm / 1.6 (1.4–2.2) inches
    weight of 5 specimens, 189.3 (87.5–357.0) g / 6.6 (3.1–12.5)
        ounces
Males average larger than females.

DISTRIBUTION Statewide in Alabama. Original natural distribution was probably central Asia. Brown rats now occur throughout the world in close association with humans.

ECOLOGY Most habitats occupied by brown rats are closely associated with humans and their buildings, including homes and businesses, yards, alleys, barns, cellars, garbage dumps, and sewers. When away from humans, brown rats may burrow and live along damp riverbanks. Their original natural habitat in central Asia may have been forests and brushy areas. Diet is omnivorous and may include invertebrates, seeds, fruits, leaves, other plant materials, livestock feeds, birds and their eggs, fishes, carrion, and almost any type of food that is consumed by humans. They may also attack and bite humans. Besides humans, who may wage campaigns to eradicate local populations, predators include domestic dogs and cats, snakes, birds, and a variety of carnivores.

BEHAVIOR Brown rats are highly social: they exhibit social hierarchies, they may care for young other than their own, they groom each other, and they huddle and sleep together. Although primarily nocturnal, brown rats may be active at any time during the day or night. They are good swimmers and can swim on the surface or under water. Brown rats are excellent diggers and excavate extensive burrows that they use to escape from predators, to store food, to protect nests, and to avoid extreme temperatures. Nests may be made of leaves, twigs, trash discarded by humans, or other items. Brown rats have well-developed senses of

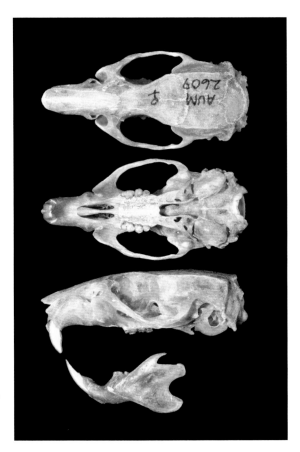

Dorsal, ventral, and lateral views of the cranium, and lateral view of the mandible of a female brown rat. Greatest length of cranium is 49.9 mm.

touch, hearing, and smell. These senses allow them to travel and locate resources in total darkness. Vocalizations and noises include teeth grinding, squeaks, chirps, and ultrasonic calls that are used to communicate between individuals during social interactions, such as during play, during mating, and between young and their mother. Another ultrasonic vocalization is a booming noise males produce when calm, grooming, or preparing to rest. Groups of brown rats may play among themselves by biting necks, jumping, chasing, and boxing.

LIFE HISTORY Breeding is year-round but usually peaks in warmer months. If a large portion of a population is exterminated, the reproductive rate of the remaining brown rats will increase until the original population has been restored. During a year, females may produce 5–7 litters. They may mate again about 18 hours after giving birth. Gestation is about

21 days and average litters have 6–8 young (range is 1–14). Newborns are pink and hairless, their eyes and ears are closed, and they weigh about 5 g. At 14–17 days, eyes open. Young leave the nest and weaning occurs at 21–28 days after birth. Males reach sexual maturity at 3 months and females at 4 months. Life span may reach 1–2 years in the wild but is usually 8–12 months. In captivity, brown rats may live 4 years.

PARASITES AND DISEASES Ectoparasites include mites (*Alliea, Androlaelaps, Cheyletus, Cosmolaelaps, Echinolaelaps, Echinonyssus, Eulaelaps, Glycyphagus, Gigantolaelaps, Haemogamasus, Hypoaspis, Laelaps, Macrocheles, Ornithonyssus, Pygmephorus, Radfordia, Zibethacarus*), ticks, lice (*Hoplopleura, Polyplax*), and fleas (*Ctenocephalides, Echidnophaga, Leptopsylla, Nosopsyllus, Polygenis, Xenopsylla*). Endoparasites include amoebas (*Endamoeba*), protozoans (*Chilomastix, Trichomonas, Trypanosoma*), cestodes (*Hymenolepis, Taenia*), nematodes (*Capillaria, Heterakis, Nippostrongylus, Trichosomoides*), and trematodes. Diseases include cryptosporidiosis (*Cryptosporidium*), hantavirus pulmonary syndrome (*Hantavirus*), plague (*Yersinia*), Q fever (*Coxiella*), rat-bite fever (*Spirillum, Streptobacillus*), toxoplasmosis (*Toxoplasma*), viral hemorrhagic fever, and Weil's disease (*Leptospira*). Because they are relatively inexpensive to maintain, quick to reproduce, and share many traits with humans, brown rats are used widely in medical research.

CONSERVATION STATUS Exotic in Alabama.

COMMENTS *Rattus* is Latin for "rat"; *norvegicus* is Latin for "of Norway."

REFERENCES Harkema (1936), Barnett (1963), Forrester (1992), Nowak (1999), Best (2004*a*), Whitaker et al. (2007).

Roof Rat
Rattus rattus

IDENTIFICATION A large rat with a tail that appears hairless and is usually longer than the head and body. Ears are long enough to cover the eyes when pushed forward. Color is usually grayish brown, slate gray, or nearly black, and underside of body is usually paler than back and sides.

DENTAL FORMULA i 1/1, c 0/0, p 0/0, m 3/3, total = 16.

SIZE AND WEIGHT Average and range in size of 7 specimens from Alabama:

     total length, 359 (332–406) mm / 14.4 (13.3–16.2) inches
     tail length, 185 (150–235) mm / 7.4 (6.0–9.4) inches
     hind foot length, 32 (24–36) mm / 1.3 (1.0–1.4) inches
     ear length of 3 specimens, 18 (10–22) mm / 0.7 (0.4–0.9) inch
     weight of 1 specimen, 103 g (3.6 ounces)

Males average larger than females.

DISTRIBUTION Statewide in Alabama. Original geographic distribution was southeastern Asia, but the species now occurs across Europe and much of Asia and North America, and there are populations in Africa,

Distribution of the roof rat in Alabama and North America.

Australia, and South America. Roof rats are closely associated with humans but are not as widely distributed as brown rats.

ECOLOGY Roof rats usually do not live in sewers or habitats where they would need to swim daily. However, they can swim several hundred meters during dispersal to islands. Roof rats may live in homes, feed stores, cotton gins, barns, corncribs, warehouses, upper floors of buildings, vine-covered fences, or trees in forested areas. Nests are usually in buildings, occasionally in burrows, and may be made of insulation from buildings, sticks, and leaves. Roof rats are omnivores. Diet includes seeds, fruits, other vegetative materials, and foods provided by humans to dogs, cats, poultry, swine, cattle, and other livestock. Roof rats also eat invertebrates, birds and their eggs, and lizards. They can become significant pests by consuming agricultural crops, especially within their original geographic range in southeastern Asia. They can also strip bark off trees, contaminate foods used by humans and livestock with urine and feces, and cause damage to buildings, electrical wiring, and other objects by chewing. In urban areas, domestic cats are significant predators. In less populated areas, predators may include snakes, owls, and various carnivorous mammals. Humans have eliminated or reduced populations of roof rats by trapping and poisoning them; shooting is a relatively ineffective method of controlling populations of these rodents.

BEHAVIOR Roof rats have well-developed senses of touch, smell, taste, and hearing. They are color-blind and their visual sense responds mostly

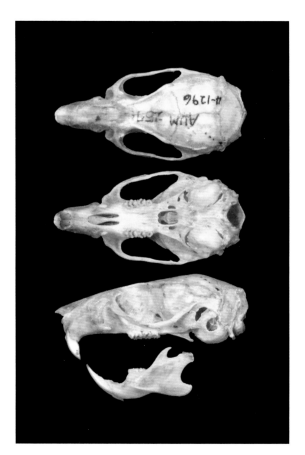

Dorsal, ventral, and lateral views of the cranium, and lateral view of the mandible of a male roof rat. Greatest length of cranium is 39.3 mm.

to degrees of light or dark. They are active at night, do not hibernate, and are good climbers. Roof rats may live on or near the ground, but in buildings, they often occupy attics and move among sites by running along rafters, pipes, or utility wires. Their long tail aids in balancing their body as they travel on narrow runways. Foraging usually begins about sunset and continues through the night, but some may occur during daylight hours. When roof rats encounter large food items, they usually carry them to a safe place to be consumed. Roof rats may store food for later consumption in attics, walls, and woodpiles, or behind shipping boxes and crates. Social groups of 50–60 individuals may occur; these have a dominant male and 2–3 dominant females. Because roof rats often live in warehouses and shipping docks, they may be transported as stowaways in shipping containers.

LIFE HISTORY Breeding is throughout the year, but in cooler climates peaks occur in warmer months. Each year, 3–6 litters of 5–7 young (range is 1–10) are born. Gestation is 21–29 days. Newborns are hairless and their eyes are closed. By 2 weeks after birth, pelage is developing, eyes open, and young can move about more easily. At 3–4 weeks old, young are weaned and become independent from their mother. However, young may stay in the nest until they are fully grown. Sexual maturity is attained at 3–5 months. Life span is about 12–15 months in the wild, but roof rats may live 4 years in captivity.

PARASITES AND DISEASES Ectoparasites include mites (*Androlaelaps, Cheyletus, Cosmolaelaps, Eutrombicula, Laelaps, Ornithonyssus, Radfordia*), ticks (*Amblyomma, Dermacentor, Haemaphysalis*), lice (*Hoplopleura, Polyplax*), and fleas (*Ctenocephalides, Echidnophaga, Leptopsylla, Nosopsyllus, Orchopeas, Polygenis, Xenopsylla*). Roof rats may also be parasitized by larval botflies (*Cuterebra, Dermatobia*). Endoparasites include protozoans, acanthocephalans, cestodes (*Hymenolepis, Raillietina, Taenia*), nematodes (*Calodium, Capillaria, Heterakis, Mastophorus, Nippostrongylus, Physaloptera, Syphacia, Trichinella*), and trematodes (*Zonorchis*). Diseases include bubonic plague (*Yersinia*), Chagas disease (*Trypanosoma*), hantavirus (*Hantavirus*), hepatitis E virus (*Hepevirus*), toxoplasmosis (*Toxoplasma*), typhus (*Rickettsia*), and Weil's disease (*Leptospira*).

CONSERVATION STATUS Exotic in Alabama.

COMMENTS *Rattus* is Latin for "rat."

REFERENCES Forrester (1992), Nowak (1999), Favorov et al. (2000), Milazzo et al. (2003), Best (2004*a*), Whitaker et al. (2007).

New World Porcupines

Family Erethizontidae

The 5 genera and 16 species in this family are medium to large in size and have hairs modified as spines with overlapping barbs. Total length usually ranges from 450 to 760 mm. Some genera have a relatively spineless and prehensile tail, while others have a relatively short tail with many spines. All species are at least partially arboreal. Members of this family occupy a wide variety of habitats from the High Arctic of North America through the tropics of Central and South America and into northern Argentina. One species (North American porcupine) occurred in what is now Alabama in recent times but is no longer present.

North American Porcupine
Erethizon dorsatum

IDENTIFICATION A large rodent with back, sides, and tail covered with long quills. Legs and tail are short. Muzzle is short and broad. Prominent incisors are dull yellow to deep orange. Pelage is composed of quills, hair, and underfur. Quills may be up to 75 mm long and 2 mm in diameter and may number more than 30,000. There are no quills on the underside of the body.

DENTAL FORMULA i 1/1, c 0/0, p 1/1, m 3/3, total = 20.

SIZE AND WEIGHT Range in size:
 total length, 790–1,330 mm / 31.6–53.2 inches
 tail length, 145–300 mm / 5.8–12.0 inches
 hind foot length, 75–91 mm / 3.0–3.6 inches
 ear length, 25–42 mm / 1.0–1.7 inches
 weight, 3.5–18 kg / 7.7–39.6 pounds
Males are larger than females.

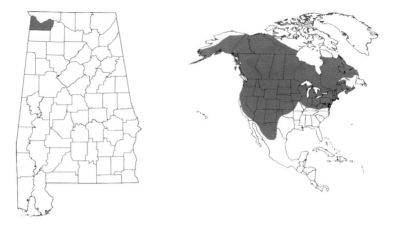

DISTRIBUTION Once occurred in what is now northern Alabama, as evidenced by the presence of 3 jaw bones of this species in an archaeological site at Little Bear Creek, Colbert County, which was dated to 3,000–5,000 years ago. Current geographic range of the species is from northern Alaska across nearly all of Canada and the contiguous United States (except coastal and southern California, southwestern Arizona, and the Southeast) and into northern Mexico.

ECOLOGY North American porcupines usually occur in habitats with mixed softwood and hardwood communities. However, they also occur in nonforested grassland and agricultural habitats with few trees. In these more open habitats, they may live and forage on grasses, crops, and shrubs and trees planted in yards and along roadways. Diet in winter is primarily inner bark of trees and needles of evergreen trees. In summer, diet is more varied and includes roots, stems, leaves, berries, catkins, seeds, flowers, nuts, aquatic plants, and grasses. North American porcupines may cause damage by gnawing trees in forests and orchards, ornamental trees, agricultural crops, rubber tires, plumbing, and woodpiles. Primary predators are humans and fishers, but other predators are great horned owls (*Bubo virginianus*), cougars, American minks, long-tailed weasels, red foxes, coyotes, bobcats, and domestic dogs. Fire, collisions with motor vehicles, and falls from trees also kill North American porcupines.

BEHAVIOR North American porcupines do not hibernate and are active throughout the year. Although they are primarily nocturnal, they may be active at any time during the day or night. They are solitary most of the

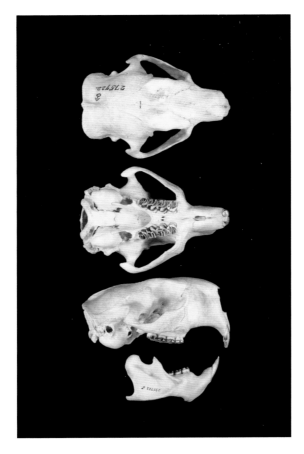

Dorsal, ventral, and lateral views of the cranium, and lateral view of the mandible of a male North American porcupine. Greatest length of cranium is 95.5 mm.

year, but several may share a den in winter. Dens may be in caves, logs, or hollow trees and are often marked by large piles of fecal pellets. They are not defended and may be used by several individuals on a rotating basis. If a den is not available, North American porcupines will spend the winter in a tree, which they will defend from intruders. They can swim to cross small bodies of water, but they tend to avoid water. North American porcupines are good climbers; they climb trees for food and protection and possibly to avoid insects. They tend to return to the same trees to forage, thus severely damaging some trees and leaving others unharmed. Vocalizations include moans, whines, grunts, coughs, squeaks, mews, snorts, barks, owl-like hoo-hoos, and shrill screeches. When disturbed, males chatter their teeth. Courtship is elaborate and includes much vocalization.

LIFE HISTORY North American porcupines mate during autumn or early winter. Copulation is by the usual rear-mount method of rodents. Gestation is about 210 days (range is 205–217). During April–June, 1 young is born, but there are a few records of multiple births. Newborns weigh 340–640 g; their eyes are open; their incisors, premolars, and molars are erupted; and their quills are soft, well formed, and hardened within an hour after birth. Soon after birth, young can climb and assume normal defensive posture. At 1–2 weeks, young can survive on vegetation, but they continue to nurse and remain close to their mother. Females become sexually mature at about 1.5 years old. Life span is probably 5–6 years in the wild, but some have lived more than 10 years in the wild.

PARASITES AND DISEASES Ectoparasites include mites (*Sarcoptes*), ticks (*Dermacentor, Ixodes*), lice (*Trichodectes*), and fleas (*Ceratophyllus*). Endoparasites include cestodes (*Monoecocestus, Cittotaenia*), nematodes (*Dipetalonema, Dirofilaria, Molinema, Wellcomia*), and crustaceans (*Porocephalus*). Diseases and disorders include encephalitis, toxoplasmosis (*Toxoplasma*), hepatic lepidosis, tick fever (*Coltivirus*), and tularemia (*Francisella*).

CONSERVATION STATUS There is no evidence that North American porcupines currently occur in Alabama.

COMMENTS This species is included here because it is an extant species, it occurred in what is now Alabama in relatively recent times, and it still occurs widely in North America. *Erethizon* is from the Greek *ērithizon*, meaning "to provoke"; *dorsatum* is from the Latin *dorsum*, meaning "back."

REFERENCES Barkalow (1961), Best and Kennedy (1972), Woods (1973), Fitzgerald et al. (1991), Barigye et al. (2007).

Coypus

Family Myocastoridae

There is 1 genus and 1 species in this family. This is a medium-sized mammal (about 8 kg) with a stocky body, webbed toes, and a round, scaly, sparsely haired tail. Native range is from Brazil and Paraguay southward to the tip of South America. Coypus have been widely introduced in the world, including the southeastern United States, where wild populations have become established. The coypu has been introduced into Alabama.

Coypu
Myocastor coypus

IDENTIFICATION Also known as the nutria, this large rodent has a well-arched back and short legs. Tail is long, round, slightly tapered toward the end, and scantily haired. Incisors are broad and orange. Lips close behind the incisors to allow gnawing while under water. The yellowish-brown to dark-brown pelage contains soft, dense underfur and long, coarse guard hairs. There are white hairs on the chin and rostrum.

DENTAL FORMULA i 1/1, c 0/0, p 1/1, m 3/3, total = 20.

SIZE AND WEIGHT Average and range in size:
head and body length, 521 (472–575) mm / 20.8 (18.9–23.0) inches
tail length, 375 (340–405) mm / 15.0 (13.6–16.2) inches
hind foot length, 135 (120–150) mm / 5.4 (4.8–6.0) inches
ear length, 27 (25–30) mm / 1.1 (1.0–1.2) inches
Weights of 2 males from Alabama were 5.2 and 5.6 kg (11.4 and 12.3 pounds). Males average larger than females.

DISTRIBUTION Coypus possibly occur statewide in Alabama, but recent

sightings are from southern counties. They have been introduced into Europe, Asia, Africa, and North America. Native geographic range of the species is southern South America.

Coypus can stay submerged more than 10 minutes when foraging or avoiding predators.

ECOLOGY Coypus live in a variety of aquatic habitats including waterways, rivers, lakes, and marshes, especially in areas with succulent or emerging vegetation along the banks. Burrows 1–6 m long are often constructed on banks with 45–90° slopes, and they have several openings. Platform nests may be on floating logs or other objects or constructed on vegetation. Coypus can damage drainage systems, many kinds of agricultural crops, natural plant communities, sugarcane fields, fruit and nut trees, conifer trees, and deciduous trees. Diet includes a great variety of plants, emergent and succulent vegetation, alfalfa, rice, ryegrass, and seedlings of bald cypress (*Taxodium distichum*). Coypus have many predators including gars (*Lepisosteus*), American alligators (*Alligator mississippiensis*), turtles, large snakes, red-shouldered hawks (*Buteo lineatus*), northern harriers (*Circus cyaneus*), domestic dogs, coyotes, red foxes, red wolves, cougars, and humans.

BEHAVIOR Coypus are nocturnal but become more active in the daytime when nighttime temperatures are cold. They are highly gregarious. During predawn hours on cold nights, they often huddle together to con-

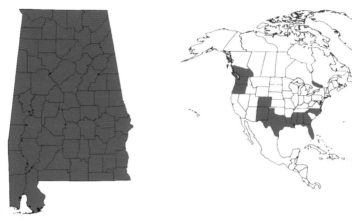

Distribution of the coypu in Alabama and North America.

serve energy. Groups usually number 2–13 animals and consist of related adult females, their offspring, and an adult male. Except during mating, females are dominant over males. Males vigorously defend a nest after a litter is born. Most of the active period is spent feeding, grooming, swimming, sunning, and sleeping. Coypus use their forefeet to hold and manipulate food. They defecate throughout their activity period, expelling most feces into the water, but coprophagy occurs when animals are at the nest. During dives, coypus can remain submerged for more than 10 minutes. Senses of hearing and touch are well developed. Females have 4–5 pairs of mammary glands located along their sides. This arrangement allows young to nurse while swimming and also allows the mother to sit in an alert position on the nest when young are nursing.

LIFE HISTORY Mating occurs throughout the year. Gestation is 127–139 days. Litters are usually born in open nests at the edge of water or in large nest chambers deep inside burrows. Litters average 3–6 young (range is 1–12). At birth, young are precocial and covered with soft, downy hair, and they weigh about 225 g. They are weaned at about 2 months old. Females may produce 3 litters and an average of 8–15 young each year under optimal conditions, but reproductive potential depends on type and availability of food, weather, predators, and disease. There is a postpartum estrus within 2 days of giving birth to a litter. At 6–15 months old, females give birth to their first litter. Life span is 5–7 years.

PARASITES AND DISEASES Ectoparasites include ticks (*Dermacentor, Ixodes*), lice (*Pitrufquenia*), and fleas (*Ceratophyllus*). Endoparasites include

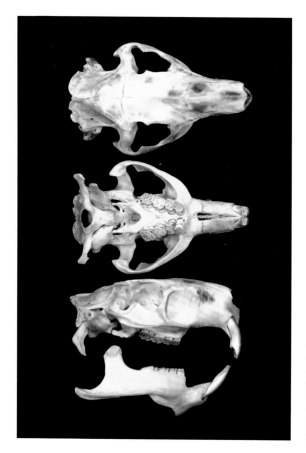

Dorsal, ventral, and lateral views of the cranium, and lateral view of the mandible of a female coypu. Greatest length of cranium is 117.2 mm.

cestodes (*Andrya, Anoplocephala, Coenurus, Cysticercus, Echinococcus, Hymenolepis, Multiceps, Polycephalus, Rodentolepis, Schizotaenia, Taenia*), nematodes (*Baylisascaris, Boehmiella, Capillaria, Dipetalonema, Filaria, Graphidioides, Heligmosomoides, Longistriata, Oxyuris, Strongyloides, Subulura, Toxocara, Trichocephalus, Trichostrongylus, Trichuris, Viannaia*), and trematodes (*Chiostichorchis, Dicrocoelium, Fasciola, Hippocrepis, Notocotylus, Paramphistomum, Stichorchis, Schistosoma*). Diseases and disorders include coccidiosis (coccidian protozoans), equine encephalomyelitis (*Alphavirus*), hepatitis, leptospirosis (*Leptospira*), papillomatosis, toxoplasmosis (*Toxoplasma*), paratyphoid fever (*Salmonella*), rabies (*Lyssavirus*), salmonellosis (*Salmonella*), sarcosporidiosis (*Sarcocystis*), and various rickettsial infections (*Rickettsia*).

CONSERVATION STATUS Exotic in Alabama.

COMMENTS *Myocastor* is from the Greek *myos* and *kastōr*, meaning "mouse beaver"; *coypus* is the South American Araucanian Indian word for coypu.

REFERENCES Babero and Lee (1961), Woods et al. (1992), Bounds et al. (2003), Best (2004*a*).

Hares, Pikas, and Rabbits

Order Lagomorpha

The 3 families, 13 genera, and 92 species of Lagomorpha are distinguished from Rodentia and other mammals by the presence of a second set of upper incisors located directly behind the first. The upper incisors of this second set are small and peg-like and lack a cutting edge. A third pair of upper incisors is lost soon after birth. Native populations occur nearly worldwide, except southern South America, Antarctica, Madagascar, the Philippines, Australia, New Zealand, most of Indonesia, and many oceanic islands including islands in the Caribbean. One family (Leporidae) occurs in Alabama.

Rabbits and Hares

Family Leporidae

The 11 genera and 61 species in this family have ears that are longer than they are wide, with a lower margin that opens well above the skull. Hind limbs are usually much longer than forelimbs, the tail is short and recurved, and the maxilla has numerous fenestrae. Total length ranges from 250 to 700 mm. Native populations of leporids range across nearly all of North America (except interior Greenland), South America (except the southern cone), Europe, Asia (except the southern Arabian Peninsula), and Africa (except Madagascar). Four species of leporids (swamp rabbit, eastern cottontail, Appalachian cottontail, marsh rabbit) occur in Alabama.

Swamp Rabbit
Sylvilagus aquaticus

IDENTIFICATION A large rabbit with a dark, rusty-brown to nearly black head and back. Underside of body, throat, and tail are white. This is the largest rabbit in Alabama.

DENTAL FORMULA i 2/1, c 0/0, p 3/2, m 3/3, total = 28. A second pair of upper incisors lies directly behind the front pair.

SIZE AND WEIGHT Average and range in size:
> total length, 501 (452–552) mm / 20.0 (18.1–22.1) inches
> tail length, 59 (50–74) mm / 2.4 (2.0–3.0) inches
> hind foot length, 101 (90–113) mm / 4.0 (3.6–4.5) inches
> ear length, 70 (60–80) mm / 2.8 (2.4–3.2) inches
> weight of 144 males from Alabama, 2,059 (1,525–2,722) g / 4.5
>   (3.3–6.0) pounds
> weight of 180 females from Alabama, 2,035 (1,389–2,722) g / 4.5
>   (3.0–6.0) pounds

There is no sexual dimorphism in size.

Distribution of the swamp rabbit in Alabama and North America.

DISTRIBUTION Except for some southern areas of Alabama, swamp rabbits occur statewide. Two subspecies are present in the state: *S. a. aquaticus* occurs north of Baldwin and Mobile counties and *S. a. littoralis* is present in Baldwin and Mobile counties. Geographic range of the species is from southern Illinois to the Gulf of Mexico and from central Oklahoma to western South Carolina.

ECOLOGY Swamp rabbits occur primarily in swamps, river bottoms, marshes, and lowland areas throughout their range. They may construct shelter forms in vegetation or use holes in the ground or in trees. Their shelter forms and runways may be in tangled marshy vegetation and among briers (*Smilax*). Shelter forms may also be on top of vegetation-covered stumps, in low crotches of trees, on fallen logs, in tangles of Japanese honeysuckles (*Lonicera japonica*), in canebrakes, and in open grassy areas in floodplains. Diet is a variety of plants including crossvines (*Bignonia capreolata*), sedges, poison ivy (*Rhus radicans*), grasses, briers (*Smilax*), seedlings, swamp grass (*Carex lupulina*), blackberries (*Rubus*), hazelnuts (*Corylus*), deciduous holly (*Ilex decidua*), and spicebrush (*Lindera benzoin*). Predators are humans, domestic dogs, American alligators (*Alligator mississippiensis*), bobcats, and coyotes.

BEHAVIOR Swamp rabbits are territorial and vocal, and males exhibit chinning, a pheromone-marking display. They are excellent swimmers and will readily take to water to avoid predators. When pursued by a dog, one swamp rabbit entered the water, swam under the bank, and remained submerged except for its nose and eyes. Swamp rabbits are usually active

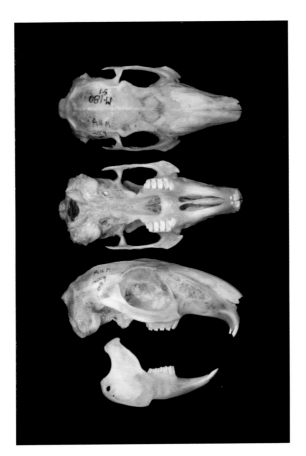

Dorsal, ventral, and lateral views of the cranium, and lateral view of the mandible of a male swamp rabbit. Greatest length of cranium is 91.0 mm.

late in the afternoon, at night, and early in the morning. Primarily during the day, swamp rabbits are coprophagous. They often use elevated sites for defecation. These latrines can provide evidence of swamp rabbits at specific locations. Nests may be against or under fences or the bases of trees, boards, or similar structures. Sometimes nests are built a few days before giving birth, but sometimes they are built the same night. Complete with side openings, they are about 4–7 cm deep, 15 cm wide, and 18 cm tall and may be lined with abdominal fur from the mother. Young may be scattered during birthing, and they begin nursing following birth of the litter. Most nursing is at dawn and dusk. Males do not participate in caring for young.

LIFE HISTORY Breeding may be year-round in some parts of the geographic range. In Alabama, peak breeding season is January to February.

First litters are born in mid-March and second litters in August. In captivity, females have averaged 3.5 litters/year (range is 1–5). Gestation is 35–40 days, but 36–37 days is most common. Size of litters varies seasonally and geographically but averages 3–4 young (range is 1–6). At birth, young are fully furred, their eyes are closed, and their average weight is about 60 g. At 5–8 days, eyes open; at 12–15 days, young leave the nest; and by 10 months, young are fully grown. Until weaning, young continue to nurse, even after leaving the nest. Males are sexually mature at 10–12 months. At 23–30 weeks, females reach sexual maturity, and they usually have their first litter during the year following their birth. Average life span is about 2 years.

PARASITES AND DISEASES Ectoparasites include mites (*Psoroptes, Trombicula*), ticks (*Haemaphysalis*), and fleas (*Hoplopsyllus*). Endoparasites include protozoans (*Enteromonas, Giardia, Trypanosoma*), cestodes (*Cysticercus, Cittotaenia, Multiceps, Raillietina*), nematodes (*Graphidium, Nematodirus, Obeliscoides, Parasalurus, Trichostrongylus, Trichuris*), and trematodes (*Hasstilesia*). Diseases include tularemia (*Francisella*) and Rocky Mountain spotted fever (*Rickettsia*).

CONSERVATION STATUS Low conservation concern in Alabama.

COMMENTS *Sylvilagus* is from the Latin *sylva*, meaning "forest," and the Greek *lagos*, meaning "hare"; *aquaticus* is Latin for "water dweller." The subspecific name *littoralis* is New Latin for "belonging to the sea shore."

REFERENCES Ward (1934), Hill (1967), Chapman and Feldhamer (1981), Schwartz and Schwartz (2001), Chapman and Litvaitis (2003), Best (2004*a*).

Eastern Cottontail
Sylvilagus floridanus

IDENTIFICATION This rabbit has long, dense, brownish to grayish pelage on the back and it is white on the underside of the body and tail.

DENTAL FORMULA i 2/1, c 0/0, p 3/2, m 3/3, total = 28. A second pair of upper incisors lies directly behind the front pair.

SIZE AND WEIGHT Average and range in size of 16 specimens from Alabama:

    total length, 440 (376–598) mm / 17.6 (15.0–23.9) inches
    tail length, 48 (25–94) mm / 1.9 (1.0–3.8) inches
    hind foot length, 89 (41–106) mm / 3.6 (1.6–4.2) inches
    ear length of 10 specimens, 62 (57–64) mm / 2.5 (2.3–2.6) inches
    weight of 9 specimens, 1,234 (965–1,542) g / 2.7 (2.1–3.4) pounds
Females are slightly larger than males.

DISTRIBUTION Statewide in Alabama. Geographic range of the eastern cottontail is from southern Canada into northwestern South America. In the United States, eastern cottontails occur from western Arizona to the Atlantic Ocean.

Eastern cottontails are born helpless, nearly hairless, and with their eyes and ears closed. They grow rapidly and remain in the nest until about 2 weeks old.

Ecology Eastern cottontails occupy widely diverse habitats including urban areas, agricultural fields, hedgerows, woodlands, deserts, grasslands, scrublands, swamps, hardwood forests, boreal forests, and rainforests. Nests are slanting holes in the ground and average 125–180 mm long, 104–126 mm wide, and 91–119 mm deep. They are often made of leaves, stems, and grasses and lined with fur from the mother. Diet varies seasonally and contains a wide range of plant material including woody species such as apples (*Malus*), sumacs (*Rhus*), maples (*Acer*), and raspberries (*Rubus*), and herbaceous species such as bluegrass (*Poa*), timothy (*Phleum*), orchard grass (*Dactylis*), clover (*Trifolium*), and wild carrot (*Daucus*). As in other leporids, coprophagy occurs in eastern cottontails; 2–3 soft fecal pellets are eaten directly from the anus before they touch the ground. Eastern cottontails are an important food item for many animals including hawks, owls, crows, raccoons, ringtails, fishers, long-tailed weasels, gray and red foxes, coyotes, bobcats, and domestic cats. Humans also hunt them for food.

Behavior Activity peaks in morning and late afternoon, but eastern cottontails may be active at any time of the day or night. They may use brush piles, hedges, fencerows, and dense stands of vegetation for refuge sites. They make shelter forms by clearing sites under vegetation, especially under grassy clumps; they use these forms for resting, shelter, and

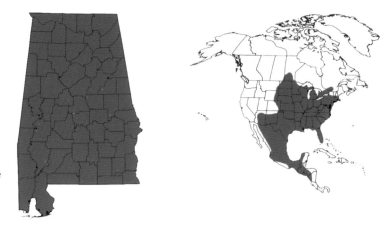

Distribution of the eastern cottontail in Alabama and North America.

hiding from predators. In desert areas or when temperatures are high, eastern cottontails will excavate a shallow layer of soil from the bottom of the shelter form and press the underside of their bodies onto the freshly exposed soil, presumably to help eliminate excess heat. Grooming may include licking the paws and moving them across the head from eyes to muzzle, licking the body and legs, scratching with the hind legs, and biting and cleaning the feet. There are 2 general types of avoidance behavior: flushing, which is a rapid and often zigzagging run at maximum speed to an established travel lane for escape; and slinking, where the animal remains close to the ground with ears back and moves more slowly to cover. Vocalizations include distress cries, which are high-pitched screams emitted by frightened or injured animals that alert other animals; squeals, which are made during copulation by either sex; and grunts, which are made by females when their nest is approached by an intruder.

LIFE HISTORY Breeding season varies greatly across the geographic range, but in Alabama, breeding begins in early January and may continue into autumn. Gestation is 28–29 days (range is 25–35 days). During a year, 5–7 litters may be born; average is 3–6 young/litter. At birth, young are covered with fine hair, their eyes are closed, and they can crawl about in the nest. Their total length is 90–110 mm, hind foot length is 21–23 mm, and weight is 35–45 g. At 4–5 days, their eyes open, and at 14–16 days, they are able to leave the nest. Some young breed during the year they are born, but most begin breeding the following year. Average life span is 15 months, but potential life span is at least 10 years in the wild.

Dorsal, ventral, and lateral views of the cranium, and lateral view of the mandible of a male eastern cottontail. Greatest length of cranium is 73.3 mm.

PARASITES AND DISEASES Ectoparasites include mites (*Androlaelaps, Cheyletiella, Marsupialichus, Neotrombicula, Ornithonyssus, Pygmephorus*), ticks (*Amblyomma, Dermacentor, Haemaphysalis, Ixodes*), and fleas (*Cediopsylla, Ceratophyllus, Ctenocephalides, Ctenophthalmus, Hoplopsyllus, Odontopsyllus*). Eastern cottontails may be parasitized by larval botflies (*Cuterebra*). Endoparasites include protozoans (*Eimeria, Enteromonas*), cestodes (*Cittotaenia, Ctenotaenia, Cysticercus, Hydatigera, Mosgovoyia, Multiceps, Raillietina, Taenia*), nematodes (*Ascaris, Baylisascraris, Capillaria, Dermatoxys, Dirofilaria, Graphidium, Heterakis, Longistriata, Nematodirus, Obeliscoides, Passalurus, Physaloptera, Protostrongylus, Trichostrongylus, Trichuris*), and trematodes (*Fasciola, Hasstilesia*). Diseases include tularemia (*Francisella*), Tyzzer's disease (*Clostridium*), and viral infections (*Herpesvirus*).

Conservation Status Lowest conservation concern in Alabama.

Comments *Sylvilagus* is from the Latin *sylva*, meaning "a forest," and the Greek *lagōs*, meaning "hare"; *floridanus* is Latin for "of Florida," referring to the origin of the type specimen.

References Harkema (1936), Erickson (1947), Chapman et al. (1980), Schwartz and Schwartz (2001), Best (2004*a*), Whitaker et al. (2007).

Appalachian Cottontail

Sylvilagus obscurus

IDENTIFICATION Compared to the similar eastern cottontail, the Appalachian cottontail is slightly smaller, it has shorter ears, there are more black hairs on the back, there is a narrow black patch on top of the head between the ears, and it does not have the reddish or rusty patch on the nape that is present on the eastern cottontail.

DENTAL FORMULA i 2/1, c 0/0, p 3/2, m 3/3, total = 28. A second pair of upper incisors lies directly behind the front pair.

SIZE AND WEIGHT Average and range in size of males and females, respectively:

>total length, 405 (386–415), 411 (387–430) mm / 16.2 (15.4–16.6), 16.4 (15.5–17.2) inches
>
>tail length, 43 (22–57), 47 (30–65) mm / 1.7 (0.9–2.3), 1.9 (1.2–2.6) inches
>
>hind foot length, 93 (89–97), 93 (87–96) mm / 3.7 (3.6–3.9), 3.7 (3.5–3.9) inches

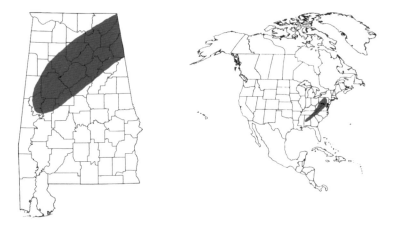

Distribution of the Appalachian cottontail in Alabama and North America.

ear length, 57 (54–59), 58 (54–63) mm / 2.3 (2.2–2.4), 2.3 (2.2–2.5) inches

weight, 1,000 (750–1,300), 1,100 (850–1,350) g / 2.2 (1.6–3.3), 2.8 (2.1–3.4) pounds

Females average larger than males.

DISTRIBUTION Northern Alabama. Geographic range of the species is from northern Pennsylvania into northern Alabama.

ECOLOGY The Appalachian cottontail inhabits primarily a mosaic of highland habitats that include dense stands of deciduous and coniferous forests. In the central Appalachian Mountains, it typically lives at elevations above 570 m in coniferous forest, mixed oak (*Quercus*) forest, or 6–9-year-old clear-cuts with cover including mountain laurels (*Kalmia latifolia*), rhododendrons (*Rhododendron maximum*), and blueberries (*Vaccinium*). Appalachian cottontails occupy microhabitats with over 80% vegetative cover to a height of 1.5 m above the substrate. Large, unfragmented (or at least connected) areas of suitable habitat are required for maintenance of viable populations. Appalachian cottontails are coprophagous. Diet in summer includes grasses, clovers, shrubby plants, fruits, seeds, twigs, and bulbs. Diet in autumn and winter includes shrubby and herbaceous plants, such as common rush (*Juncus effusus*), serviceberry (*Amelanchier*), honeysuckle (*Lonicera*), and chokeberry (*Photinia*). Predators include snakes, hawks, owls, bobcats, gray and red foxes, and coyotes.

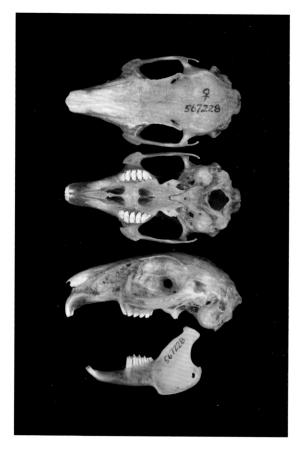

Dorsal, ventral, and lateral views of the cranium, and lateral view of the mandible of a female Appalachian cottontail. Greatest length of cranium is 72.6 mm.

BEHAVIOR Appalachian cottontails are secretive and rarely venture into the open. They may be active at any time during the day or night, but most activity is in late afternoon, at night, and in early morning. Holes are used to escape from predators. Home range is 6–13 hectares from May to September (when leaves are on vegetation) and 2–9 hectares from October to April (when leaves are off vegetation). For males, home ranges are larger from May to September than from October to April, but there are no seasonal differences in home ranges of females. Nests may be in brush, woods, hayfields, and grasslands. They are similar to those of the eastern cottontail and consist of depressions in the ground about 10 cm deep and 13 cm wide. One nest was in a hole 46 cm deep. About 3 days before litters are born, Appalachian cottontails construct nests at night, cover them with leaves and twigs, and line them with grass and fur.

LIFE HISTORY Breeding season is from March to September. Gestation is about 28 days. Litters average 4 young (range is 3–8). Possibly, females produce 5–6 litters each year. Eyes are closed at birth and open at 7–10 days. Average measurements (in mm) at birth, 7 days old, and 16 days old, respectively, are: length from crown of head to rump, 74, 94, 133; tail length, 14, 17, 23; hind foot length, 11, 34, 50; ear length, 10, 18, 31. At about 16 days old, young are weaned and independent of the nest, and they can hop and run quickly. Some females reach sexual maturity and breed their first year, but most mate for the first time during the spring after their birth. Males are potentially reproductively active during the breeding season following the one in which they were born, but not before. Average life span is probably 12–15 months in the wild.

PARASITES AND DISEASES Ectoparasites include ticks (*Haemaphysalis*) and fleas (*Ctenocephalides*). Larval botflies (*Cuterebra*) occasionally parasitize Appalachian cottontails. Endoparasites include protozoans (*Eimeria*), cestodes (*Cittotaenia, Cysticercus*), and nematodes (*Dirofilaria, Obeliscoides, Passalurus*).

CONSERVATION STATUS High conservation concern in Alabama, listed as under review by the United States Fish and Wildlife Service, and listed on the International Union for Conservation of Nature and Natural Resources Red List of Threatened Species as near threatened.

COMMENTS *Sylvilagus* is from the Latin *sylva*, meaning "a forest," and the Greek *lagōs*, meaning "hare"; *obscurus* is Latin for "obscure" or "dusky."

REFERENCES J. A. Chapman (1975), Chapman and Stauffer (1981), Chapman et al. (1992), Chapman and Litvaitis (2003), Best (2004*a*), Boyce and Barry (2007), Hartman and Barry (2010).

Marsh Rabbit

Sylvilagus palustris

IDENTIFICATION This rabbit has a dark brown body and small, slender, dark reddish to buffy feet. Tail is small, ears are short and broad, and underside is reddish brown to pale brown with white on the abdomen. The marsh rabbit is similar in color to the swamp rabbit but is smaller, the underside of its tail is not white, and its feet are smaller and more slender.

DENTAL FORMULA i 2/1, c 0/0, p 3/2, m 3/3, total = 28. A second pair of upper incisors lies directly behind the front pair.

SIZE AND WEIGHT Average size:
> total length, 431 mm / 17.2 inches
> tail length, 36 mm / 1.4 inches
> hind foot length, 89 mm / 3.6 inches
> ear length from notch (dry), 49 mm / 2.0 inches

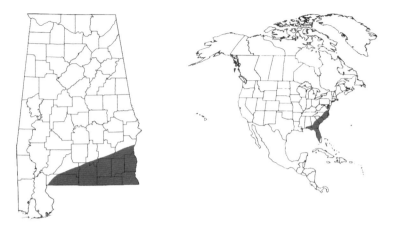

Distribution of the marsh rabbit in Alabama and North America.

DISTRIBUTION Only in southeastern counties in Alabama. Geographic range of the species is from southeastern Alabama through most of Florida and northeastward to southeastern Virginia.

ECOLOGY The most important factor limiting the distribution of marsh rabbits is water. These rabbits occur only in damp or swampy lowlands and marshes. They are most common in brackish-water areas, although they were once associated with freshwater marshes. In Florida, a small population was in a hammock habitat that contained trees such as magnolias (*Magnolia grandiflora*), tupelos (*Nyssa sylvatica*), and sweetgums (*Liquidambar styraciflua*), and shrubs such as blackberries (*Rubus betulifolius*) and dewberries (*Rubus continentalis*). A larger population was among cattails (*Typha*) that grew along the edge of a pond. Diet is a variety of plant material such as vines of cantellas (*Cantella repanda*) and briers (*Smilax*), leaves of sweetgum trees, and seeds of tupelo trees. Predators include American alligators (*Alligator mississippiensis*), great horned owls (*Bubo virginianus*), barred owls (*Strix varia*), barn owls (*Tyto alba*), northern harriers (*Circus cyaneus*), red-tailed hawks (*Buteo jamaicensis*), bobcats, gray foxes, coyotes, long-tailed weasels, and American minks. Cottonmouths (*Agkistrodon piscivorus*) and eastern diamondback rattlesnakes (*Crotalus adamanteus*) feed on young marsh rabbits. Primary causes of death of marsh rabbits are human-related, such as fire and domestic dogs. When chased by dogs, marsh rabbits often make turns and double back to avoid capture, but they are easily tired.

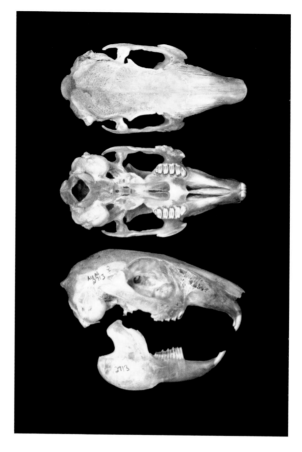

Dorsal, ventral, and lateral views of the cranium, and lateral view of the mandible of a female marsh rabbit. Greatest length of cranium is 79.6 mm.

BEHAVIOR Marsh rabbits are active primarily at dusk and during the night. They do not live in burrows, but they do use their long toenails to dig holes that are sloped about 60° and are about 30 cm deep. Days are spent in shelter forms, which are often well hidden in the middle of brier patches. Shelter forms are well-screened sites where leaves and twigs have been scraped away, exposing the bare soil. They may be next to logs, among fallen branches, in patches of grass, or protected by the large fans of cabbage palms (*Sabal palmetto*). Marsh rabbits have excellent hearing and can readily detect motion. When approached, they crouch in the shelter form and remain motionless, but if forced to leave, they will run slowly for 7–10 m and stop. When they must choose between land and water as an escape route, they usually choose land. However, marsh rabbits are excellent swimmers, and with alternate paddling of their appendages, they seem to enjoy the water. After swimming about 30 m, they may tire, stop swimming, and float with only their head and rump above water. Marsh rabbits construct nests that are about 36 cm across and about 20 cm deep and are made of soft grasses and rabbit fur, with a floor about 2.5 cm thick and made of the same material. One nest was under burned sedges about 9 m from the edge of water. Young in nests can exit or enter from any side. Vocalizations include whines and squeals. They often make a squeaking noise just before they run from a threat. Fecal pellets, which average 9 mm long and 8 mm wide and weigh 1.2 mg, are deposited in piles, which are larger along active runways, or they are scattered on logs or stumps, in mud along edges of ponds, or on floating vegetation. Instead of hopping, marsh rabbits usually walk, particularly when moving across soft mud. Their tracks are distinguished easily in the mud by the exceptionally long toenails.

LIFE HISTORY Breeding is year-round. Gestation is about 30–37 days. Litters average 3–5 young. Number of litters is 6–7/year. Life span is unknown but is possibly 12–15 months.

PARASITES AND DISEASES Ectoparasites include mites (*Androlaelaps, Cheyletiella, Ornithonyssus*), ticks (*Dermacentor, Haemaphysalis, Ixodes*), and fleas (*Cediopsylla, Ctenocephalides, Hoplopsyllus*). Larval botflies (*Cuterebra*) may also occur in the skin of the neck and between the forelegs. Endoparasites include protozoans, cestodes (*Cittotaenia, Cysticercus, Multiceps, Taenia*), nematodes (*Dermatoxys, Dirofilaria, Longistriata, Obeliscoides*), and trematodes (*Hasstilesia*). Diseases and disorders include respiratory infections, toxoplasmosis (*Toxoplasma*), and tularemia (*Francisella*).

CONSERVATION STATUS High conservation concern in Alabama.

COMMENTS *Sylvilagus* is from the Latin *sylva*, meaning "a forest," and the Greek *lagōs*, meaning "hare"; *palustris* is Latin for "marshy."

REFERENCES Nelson (1909), Blair (1936), Erickson (1947), Chapman and Willner (1981), Forrester (1992), Best (2004*a*), Whitaker et al. (2007).

Shrews, Moles, Desmans, and Relatives

Order Soricomorpha

Soricomorpha includes 4 families, 45 genera, and 428 species. Members usually have a long and pointed rostrum, a low braincase that does not rise much above the dorsal surface of the rostrum, and eyes that are small relative to the size of the head. Geographic range of the order is worldwide, except Antarctica, the Australian region, and southern South America. Two families (Soricidae, Talpidae) are widely distributed in Alabama.

Shrews

Family Soricidae

Containing 26 genera and 376 species, this family occurs throughout most of North America, all of Europe, Asia, and Africa (including Madagascar), and northwestern South America. The skull is long and narrow, with no zygomatic arch, postorbital process, or auditory bullae. Rostrum is long and pointed, eyes are small, and ears usually have pinnae. Shrews may occupy tundra, forest, marsh, grassland, and desert habitats. In Alabama, there are 6 species of shrews (northern short-tailed shrew, southern short-tailed shrew, North American least shrew, smoky shrew, American pygmy shrew, southeastern shrew) represented by 3 genera (*Blarina*, *Cryptotis*, *Sorex*). In Alabama and elsewhere, large numbers of shrews die each year in bottles thrown onto roadsides. Shrews enter the bottles in search of food or shelter and frequently cannot escape, especially from bottles with the opening oriented uphill, and they die. Littering is not only dirty and unsightly but can be deadly for some of our native wildlife.

Northern Short-tailed Shrew
Blarina brevicauda

IDENTIFICATION This relatively large shrew has a short tail and a nearly uniform silvery, slate-gray, or black pelage, sometimes with brownish-tipped hairs. Underside is slightly paler than the back. Eyes are minute and external ears are inconspicuous and concealed by the soft pelage. Rostrum is elongate, pointed, and somewhat proboscis-like. This is the largest short-tailed shrew in Alabama. Northern short-tailed shrews can be distinguished from southern short-tailed shrews by the diploid number of chromosomes (48–50 for the northern, 31–46 for the southern).

DENTAL FORMULA i 4/2, c 1/0, p 2/1, m 3/3, total = 32. Although there is consensus on the total number of teeth, there is disagreement about the number of each kind of tooth. Actually, each upper tooth row has 1 falciform incisor, 5 unicuspids, the fourth premolar, and 3 molars, and each lower tooth row has 1 precumbent incisor, 1 unicuspid, the fourth premolar, and 3 molars. Teeth have dark chestnut-brown tips.

SIZE AND WEIGHT Average and range in size of 22 specimens from Alabama:

     total length, 109 (101–125) mm / 4.4 (4.0–5.0) inches
     tail length, 24 (20–28) mm / 1.0 (0.8–1.1) inches
     hind foot length, 14 (13–15) mm / 0.6 (0.5–0.6) inch
     weight of 13 specimens, 16.3 (11.7–19.7) g / 0.6 (0.4–0.7) ounce
There is no significant sexual dimorphism.

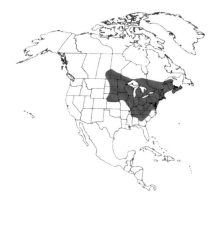

Distribution of the northern short-tailed shrew in Alabama.

DISTRIBUTION Northern Alabama. Geographic range of the species is from the Great Plains of southern Canada to central Georgia and from western Nebraska to the Atlantic Ocean.

ECOLOGY Northern short-tailed shrews occupy a wide range of habitats including grasslands, agricultural areas, fallow fields, coniferous-deciduous forests, and marshes. Dense understory, deep leaf litter, and favorable temperature and precipitation have been suggested as important components of habitat. Saliva of northern short-tailed shrews is poisonous to humans and other animals. Poison is produced by submaxillary glands and exits through ducts at the base of the lower incisors. When small mammals are bitten, this toxin may cause them to die of respiratory failure and other complications. In addition to incapacitating or killing prey, the toxin may assist in digestion of proteins. Diet includes fungi (*Endogone*), plant materials, larval and adult insects, slugs and snails, earthworms, millipedes, centipedes, spiders, mollusks, and a variety of small vertebrates. Predators include fish (*Salvelinus*, *Lepomis*), snakes (*Agkistrodon*, *Nerodia*, *Pituophis*), owls (*Aegolius*, *Asio*, *Bubo*, *Otus*, *Strix*), hawks (*Buteo*, *Circus*, *Falco*), bobcats, coyotes, gray and red foxes, long-tailed weasels, striped skunks, American minks, raccoons, Virginia opossums, and domestic dogs and cats.

BEHAVIOR Northern short-tailed shrews are generally solitary, with females and young being more sociable than males and older individuals. They are semifossorial, with tunnels usually within the top 10 cm of soil,

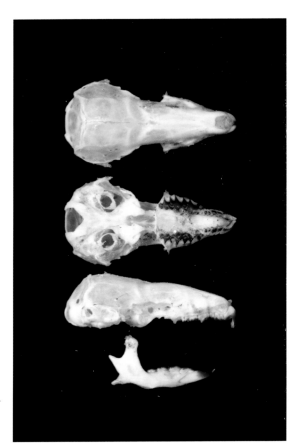

Dorsal, ventral, and lateral views of the cranium, and lateral view of the mandible of a female northern short-tailed shrew. Greatest length of cranium is 23.6 mm.

but with some as deep as 50 cm below the surface. Northern short-tailed shrews dig along and through rotten logs and often use the runways and tunnels of voles and moles. They spend relatively little time on the surface of the ground and have been reported to climb trees. Their spherical nests are underground and may be lined with vegetation and the fur of voles. Feces, which are about 2.5 cm long, dark green, and twisted into a corkscrew shape, are rarely found in nests. Instead, they are usually deposited neatly along the side of a tunnel or outside the opening of a nest. Northern short-tailed shrews are most active in the morning and on cloudy days, but movement is characterized by about 4–5 minutes of activity followed by intervening periods of inactivity. This pattern is repeated almost continuously throughout the day. Total time of activity is about 16% of the day. To compensate for their increased need for energy in cold weather, they consume about 43% more food in winter than in summer. Northern short-tailed shrews use echolocation to aid in exploring their environment. Ultrasonic clicks allow them to detect passageways and objects.

LIFE HISTORY Breeding season is usually from February to September, with peaks during spring and from late summer to early autumn. Gestation is 21–22 days. Litters average 4–5 young (range is 1–7). Newborns are pink and hairless except for short vibrissae, and their eyes and ears are closed. At 2, 8, and 13 days, respectively, measurements (in mm) are: total length, 31, 61, and 73; tail length, 4, 9.5, and 12; hind foot length, 4.5, 9, and 16; weight, 1.3, 6.2, and 9 g. At 25 days, weaning occurs. Young grow rapidly and are about the size of adults when they leave the nest. Sexual maturity may be reached during the year in which they are born, but first litters are probably born the following spring. Life span may be 10–22 months in the wild, but some have lived 30–33 months in captivity.

PARASITES AND DISEASES Ectoparasites include mites (*Amorphacarus*, *Androlaelaps*, *Blarinobia*, *Echinonyssus*, *Eulaelaps*, *Glycyphagus*, *Haemogamasus*, *Laelaps*, *Listrophorus*, *Myonyssus*, *Olistrophorus*, *Orycteroxenus*, *Protomyobia*, *Pygmephorus*, *Xenoryctes*), fleas (*Ctenophthalmus*, *Doratopsylla*), and beetles (*Leptinus*). Northern short-tailed shrews may also be parasitized by larval botflies (*Cuterebra*). Endoparasites include acanthocephalans (*Centrorhynchus*), cestodes (*Hymenolepis*, *Protogynella*, *Pseudodiorchis*, *Oochoristica*), nematodes (*Angiostrongylus*, *Capillaria*, *Longistriata*, *Parastrongyloides*, *Physaloptera*, *Porrocaecum*, *Spirura*), and trematodes (*Brachylaima*, *Ectosiphonus*, *Entosiphonus*, *Panopistus*).

CONSERVATION STATUS Moderate conservation concern in Alabama.

COMMENTS *Blarina* is a coined name (a made-up word); *brevicauda* is from the Latin *brevis* and *cauda*, meaning "short-tailed."

REFERENCES Blair (1948), Rausch and Tiner (1949), Chandler and Melvin (1951), Oswald (1958), Lumsden and Zischke (1962), Wittrock and Hendrickson (1979), Whitaker and French (1984), George et al. (1986), Genoways and Choate (1998), Best (2004*a*), Whitaker et al. (2007).

Southern Short-tailed Shrew
Blarina carolinensis

IDENTIFICATION This relatively large shrew has a short tail, short legs, and a nearly uniform dark slate-gray pelage. Underside is slightly paler than the back. Eyes are minute and external ears are inconspicuous and concealed by the soft pelage. Rostrum is elongate, pointed, and somewhat proboscis-like. This is the smallest short-tailed shrew in Alabama. Southern short-tailed shrews can be distinguished from northern short-tailed shrews by the diploid number of chromosomes (31–46 for the southern, 48–50 for the northern).

DENTAL FORMULA i 4/2, c 1/0, p 2/1, m 3/3, total = 32. Although there is consensus on the total number of teeth, there is disagreement about the number of each kind of tooth. Each upper tooth row has 1 falciform incisor, 5 unicuspids, the fourth premolar, and 3 molars, and each lower tooth row has 1 precumbent incisor, 1 unicuspid, the fourth premolar, and 3 molars. Teeth have dark chestnut-brown tips.

SIZE AND WEIGHT Average and range in size of 24 specimens from Alabama:

> total length, 93 (83–99) mm / 3.7 (3.3–4.0) inches
> tail length, 19 (17–22) mm / 0.8 (0.7–0.9) inch

Distribution of the southern short-tailed shrew in Alabama and North America.

hind foot length, 11 (10–12) mm / 0.4 (0.4–0.5) inch

weight of 1 specimen, 9 g (0.3 ounce)

There is no significant sexual dimorphism.

DISTRIBUTION Statewide, except for parts of northern and eastern Alabama. Geographic range of the species is from southern Illinois to the Gulf of Mexico and from eastern Texas to the Atlantic Ocean.

ECOLOGY Southern short-tailed shrews live in diverse habitats. These include natural and human-made forests of pines (*Pinus*) in all stages of succession, but these shrews are more common in older forests than in newly regenerated forests. They occur in disturbed habitats such as strip-mined areas in various stages of reclamation, fallow agricultural fields, roadsides, and areas with large-scale blowdowns of trees caused by hurricanes or tornadoes. Natural habitats vary from dry to wet and include grasslands, brushy areas, cane bottomlands, swampy areas, tidal marshes, hardwood forests, mixed pine-hardwood forests, and dense thickets with briers (*Smilax*) and honeysuckles (*Lonicera*). Southern short-tailed shrews are believed to be one of the most abundant species of small mammals within their geographic range. There are strong fluctuations in abundance through the year, with peaks during late spring and autumn following peaks in the breeding season. Diet includes vegetative matter, slugs and snails, fungi (*Endogone*), earthworms, adult and larval beetles, adult and larval moths, flies and midges, centipedes, spiders, ants, and eggs of turtles. Predators include snakes (*Agkistrodon, Elaphe,*

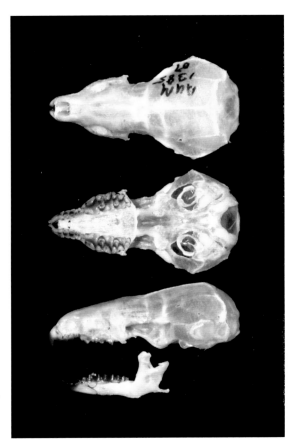

Dorsal, ventral, and lateral views of the cranium, and lateral view of the mandible of a male southern short-tailed shrew. Greatest length of cranium is 20.0 mm.

Masticophis), owls (*Asio, Otus, Tyto*), hawks (*Accipiter, Buteo*), coyotes, red foxes, and domestic dogs and cats. One short-tailed shrew was recovered from the stomach of a green sunfish (*Lepomis cyanellus*).

BEHAVIOR Southern short-tailed shrews travel along surface and subsurface runways. Surface runways are almost invariably under some type of ground cover such as leaves, pine needles, or grass. Two nests, one containing 5 young and the other containing 6, were in tunnels about 30 cm underground. These 2 nests were constructed of root fibers and blades of various kinds of grasses. Another nest was made of leaves and was inside a rotten log. Home range is estimated at 1–11 hectares, depending on where the study was conducted and the method of analysis. Average minimum distance traveled is about 40 m, but southern short-tailed shrews have traveled more than 600 m.

LIFE HISTORY Breeding occurs from March to November, with peaks from April to June and from September to November. In southwestern Alabama, a lactating female was observed in late December, indicating that the breeding season is longer in southern latitudes. Litters average 4–5 young (range is 1–6). Possibly, 3–4 litters can be produced by an adult female each year. Young are probably born hairless and with eyes and ears closed, but there is no information on the development or life span of southern short-tailed shrews.

PARASITES AND DISEASES Ectoparasites include mites (*Androlaelaps, Asiochirus, Bakerdania, Blarinobia, Comatacarus, Cyrtolaelaps, Echinonyssus, Eucheyletia, Eulaelaps, Euryparasitus, Euschoengastia, Glycyphagus, Haemogamasus, Histiostoma, Hypoaspis, Myonyssus, Olistrophorus, Oryctoxenus, Proctolaelaps, Protomyobia, Prowichmannia, Pygmephorus, Scutacarus, Xenoryctes*), ticks (*Dermacentor*), fleas (*Ctenophthalmus, Doratopsylla, Stenoponia*), and beetles (*Leptinus*). Endoparasites include acanthocephalans (*Centrorhynchus*), cestodes (*Cryptocotylepis*), nematodes (*Capillaria, Longistriata, Physaloptera, Porrocaecum*), and trematodes (*Brachylaima, Brachylecithum, Ectosiphonus, Panopistus*).

CONSERVATION STATUS Moderate conservation concern in Alabama.

COMMENTS *Blarina* is a coined name (a made-up word); *carolinensis* means "of Carolina," a reference to the origin of the type specimen.

REFERENCES Forrester (1992), Genoways and Choate (1998), McCay (2001), Best (2004a), Whitaker et al. (2007).

North American Least Shrew
Cryptotis parva

IDENTIFICATION A small, short-tailed, brownish shrew with inconspicuous ears, minute black eyes, and a long, pointed rostrum. Fur is fine, dense, short, and velvety. The species is distinguished from other shrews in Alabama by the presence of 4 unicuspid teeth on either side of the upper jaw.

DENTAL FORMULA i 1/1, u 4/1, p 1/1, m 3/3, total = 30. There is consensus on the total number of teeth, but there is disagreement about the number of each kind of tooth (u = unicuspid). Teeth have chestnut-colored cusps.

SIZE AND WEIGHT Average and range in size of 24 specimens from Alabama:

     total length, 73 (56–82) mm / 2.9 (2.2–3.3) inches
     tail length, 17 (13–23) mm / 0.7 (0.5–0.9) inch
     hind foot length, 10 (8–11) mm / 0.4 (0.3–0.4) inch
     weight of 13 specimens, 3.8 (2.8–5.2) g / 0.1 (0.1–0.2) ounce
There is no sexual dimorphism.

DISTRIBUTION Statewide in Alabama. Geographic range of the North

Distribution of the North American least shrew in Alabama and North America.

American least shrew is from eastern New Mexico to the Atlantic Ocean and from southern Minnesota into northern Panama.

ECOLOGY North American least shrews often inhabit grassy, weedy, and brushy fields, but they may be present in marshes, arid regions with mesquites (*Prosopis*), agaves (*Agave*), and yuccas (*Yucca*), and in woodland habitats, including oak-maple (*Quercus-Acer*) forests, cabbage palm (*Sabal palmetto*) forests, and oak-pine (*Quercus-Pinus*) forests. In Central America, North American least shrews occur in pine-oak forests, in dense, humid, tropical forests, and in cloud forests. Nests may be in burrows or under flat objects such as rocks, logs, boards, or sheet metal. North American least shrews have been maintained in captivity for weeks with an adequate water supply and a diet of baby mice and insects. Each day, North American least shrews consume nearly their own weight in food, e.g., a 4.7-g animal consumed about 3.6 g of food/day. Food passes through their digestive system in 95–114 minutes. Diet includes fungi (*Endogone*), some plant material, earthworms, centipedes, snails and slugs, spiders, and insects, including crickets, grasshoppers, beetles (Scarabaeidae, Carabidae, larval beetles), larval lepidopterans, chinch bugs (Lygaeidae), bees, and aphids. Predators include snakes (*Heterodon, Lampropeltis*), long-eared owls (*Asio otus*), great horned owls (*Bubo virginianus*), barn owls (*Tyto alba*), short-eared owls (*Asio flammeus*), screech owls (*Megascops asio*), rough-legged hawks (*Buteo lagopus*), eastern spotted skunks, and domestic dogs and cats.

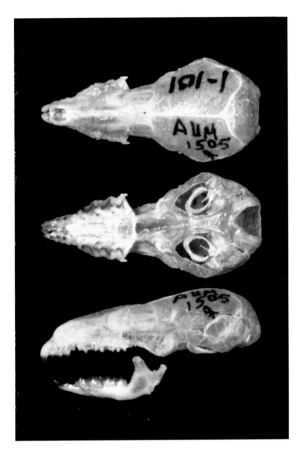

Dorsal, ventral, and lateral views of the cranium, and lateral view of the mandible of a female North American least shrew. Greatest length of cranium is 17.1 mm.

BEHAVIOR North American least shrews may be active anytime during the day or night, but most activity is at night. They are good swimmers, do not hibernate, and may emit birdlike chirping sounds and other vocalizations. Individuals have not been observed to fight, but a female may become aggressive when her babies are threatened. North American least shrews seem somewhat colonial and often huddle together in a nest. On one occasion, about 25 were discovered in a nest of leaves under a log. At another location, 31 were in a 100 x 150-mm diameter nest that was made of leaves and grass and was in a 50-mm-deep depression in the soil, under a log. Adjacent to this nest was a 100 x 150-mm pile of fecal droppings that was 12 mm deep, indicating that North American least shrews may occupy nests for a long time and that they have latrines associated with their nests. In a nest made of rootlets and grass, 5 North American least shrews were observed huddled together under a heavy slab of rock. Another nest about 10 cm below the surface of the ground was made of dried grass and contained 3 individuals. North American least shrews may use burrows constructed by other animals or may construct their own burrows by rooting with their snout and using their feet to remove soil. They make tiny runways or use those of other small mammals, such as voles, cotton rats (*Sigmodon*), and rice rats (*Oryzomys*). North American least shrews clean themselves by licking, and they comb their fur using their front feet. They sleep soundly and awake slowly and sluggishly, with trembling or convulsive movements similar to those of an animal waking from hibernation.

LIFE HISTORY Breeding occurs year-round, especially in southern populations, but peaks during March to November in northern populations. Gestation is 21–23 days. Their deciduous dentition is shed prior to birth. Litters average 4–5 young (range is 1–7). Females may mate again on the same day a litter is born, thus producing a litter about 21–23 days after the previous litter. Newborns are naked except for short vibrissae, their eyes are closed, and their toes have tiny claws. Approximate size is: total length, 22 mm; tail length, 3 mm; hind foot length, 2.5 mm; weight, 3 g. By day 3, the back becomes darker and crawling begins; by day 4, pigmentation appears on the nose and chin; by day 6, hair is visible; by days 10–11, ears open; and by day 14, eyes are open, young are fully haired, and they look like small versions of adults, except for their silvery pelage. Weaning occurs 21–22 days after birth. Young closely follow their mother until about 22–23 days old. By 25 days after birth, young weigh about the same as adults. Life span is probably 8–10 months in the wild, but 1 individual lived 21 months in captivity.

PARASITES AND DISEASES Ectoparasites include mites (*Androlaelaps, Blarinobia, Dermacarus, Echinonyssus, Eulaelaps, Euschoengastia, Eutrombicula, Glycyphagus, Haemogamasus, Hirstionyssus, Macrocheles, Myonyssus, Neotrombicula, Olistrophorus, Orycteroxenus, Protomyobia, Pseudoschoengastia, Pygmephorus*), ticks (*Ixodes*), and fleas (*Epitedia, Corrodopsylla, Ctenocephalides, Ctenophthalmus, Doratopsylla, Peromyscopsylla, Xenopsylla*). Endoparasites are uncommon but include protozoans (*Eimeria*). Diseases include leptospirosis (*Leptospira*).

CONSERVATION STATUS Moderate conservation concern in Alabama.

COMMENTS *Cryptotis* is Greek for "hidden ear"; *parva* is Latin for "small."

REFERENCES Whitaker (1974), Hoditschek et al. (1985), McAllister and Upton (1989), Forrester (1992), Best (2004*a*), Whitaker et al. (2007).

Smoky Shrew
Sorex fumeus

IDENTIFICATION The back of this shrew may be slate gray to blackish to dark brown, depending on the season. Underside of body is paler than the back, and the tail is bicolored; dark above and pale below.

DENTAL FORMULA i 3/1, c 1/1, p 3/1, m 3/3, total = 32.

SIZE AND WEIGHT Average and range in size:
   total length, 111 (101–122) mm / 4.4 (4.0–4.9) inches
   tail length, 44 (41–48) mm / 1.8 (1.6–1.9) inches
   hind foot length, 13 (12–14) mm / 0.5 (0.5–0.6) inch
   weight, 7.6 (6.5–9.9 g) / 0.3 (0.2–0.4) ounce
   weight of 29 males, 8.9 g (0.3 ounce) and of 24 females, 7.4 g (0.3 ounce)
Males average larger than females.

DISTRIBUTION In Alabama, the smoky shrew is known only from northern Jackson County, but it probably has a wider distribution in the state. Geographic range of the species is from southeastern Canada to northeastern Alabama and from western Kentucky to east of the Appalachian Mountains.

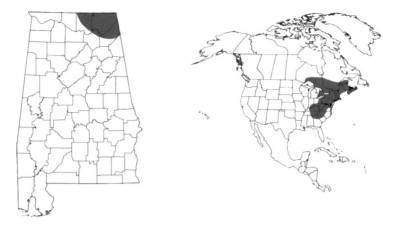

Distribution of
the smoky shrew
in Alabama and
North America.

ECOLOGY Habitat is primarily the cool, shaded floors of coniferous and deciduous forests, but smoky shrews may also live in bogs, swamps, grassy areas, talus slopes, and next to streams. Microhabitats often include thick leaf mold on top of soft soils, damp, moss-covered rocks, and decaying logs and brush. In Alabama, habitat includes steep slopes that drain toward the Tennessee River and is characterized by clear-cuts, selectively harvested forests, and deciduous forests dominated by black oaks (*Quercus velutina*), northern red oaks (*Q. rubra*), white oaks (*Q. alba*), and chestnut oaks (*Q. prinus*), with fewer hickories (*Carya*), sugar maples (*Acer saccharum*), and yellow poplars (*Liriodendron tulipifera*). Common species in the understory are flowering dogwoods (*Cornus florida*), eastern redbuds (*Cercis canadensis*), and sourwoods (*Oxydendrum arboreum*). Nests may be 10–23 cm underground, lined with fur and plant material, nearly spherical, and about 23 cm in circumference, They may be in hollow logs, within tunnels, or beneath debris on the forest floor. Diet consists of some plant material, fungi (*Endogone*), and a wide variety of invertebrates such as centipedes, earthworms, adult lepidopterans and beetles (Cantheridae, Scarabaeidae), spiders, sowbugs (Isopoda), flies (Chironomidae), and a variety of larval insects. Predators include owls (Strigidae, Tytonidae), hawks (Accipitridae, Falconidae), bobcats, gray and red foxes, long-tailed weasels, and northern short-tailed shrews.

BEHAVIOR Smoky shrews are active throughout the year and at any time during the day or night. They are very tolerant of cold conditions and may be active on top of snow in temperatures well below freezing. They

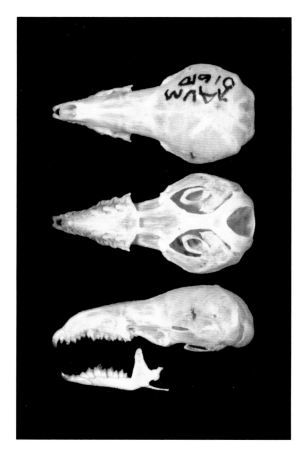

Dorsal, ventral, and lateral views of the cranium, and lateral view of the mandible of a female smoky shrew. Greatest length of cranium is 19.8 mm.

may not repair or construct their own runways and tunnels but will use those constructed by other small mammals. Smoky shrews have a weak sense of smell. They may use vision to some extent but often use their sense of touch, mediated through the vibrissae, to assess their environment. When alarmed, smoky shrews utter a high-pitched grating sound. If greatly disturbed, a smoky shrew will throw itself onto its back, and with legs spread and waving, repeatedly utter pulses of squeaks. As they forage with their twitching nose and vibrissae held aloft, smoky shrews emit a low, nearly continuous, twittering vocalization that is audible to humans. In captivity, smoky shrews consumed about one-half their body weight in food each day (foods provided were snails, beetles, earthworms, mouse flesh, centipedes, spiders, and sowbugs). When eating large food items, smoky shrews bite off small pieces as they hold the item down with their forepaws. They hold down small plethodontid salamanders and kill them with a bite severing the spinal cord. After eating the viscera, feet, and head, they usually abandon the rest.

LIFE HISTORY Breeding season is late March to early October. Gestation is 3 weeks or less and 2–3 litters may be produced each year. Litters average 5–6 young (range is 1–8). Females may become pregnant while nursing a litter. Following the mating season, most adults die before winter. By weaning at 1 month of age, young weigh about 4 g; they grow slowly and weigh 4–7 g during the first 7–10 months of life. Smoky shrews reach sexual maturity the year following their birth. Life span may be 10–12 months, possibly up to 14–17 months.

PARASITES AND DISEASES Ectoparasites include mites (*Amorphacarus, Androlaelaps, Bakerdania, Cheladonta, Echinonyssus, Eucheyletia, Euhaemogamasus, Euschoengastia, Glycyphagus, Haemogamasus, Haemolaelaps, Hirstionyssus, Macrocheles, Myobia, Myonyssus, Neotrombicula, Ornithonyssus, Orycteroxenus, Protomyobia, Pygmephorus, Xenoryctes*), ticks (*Ixodes*), and fleas (*Corrodopsylla, Ctenophthalmus, Doratopsylla, Epitedia, Nearctopsylla*). Endoparasites include nematodes (*Capillaria, Porrocaecum*) that have been found in the liver, coiled among muscles, and in and on the viscera.

CONSERVATION STATUS Undetermined in Alabama.

COMMENTS *Sorex* is Latin for "shrew-mouse"; *fumeus* is from the Latin *fumus*, meaning "smoke."

REFERENCES Hamilton (1940), Owen (1984), Whitaker and French (1984), Whitaker et al. (2007), Felix et al. (2009).

American Pygmy Shrew
Sorex hoyi

IDENTIFICATION This shrew is reddish brown or grayish brown on the back, paler on the sides, and whitish, grayish, or rusty gray on the underside. It has minute eyes and small ears; soft, silky fur; and an elongated rostrum that is constantly in motion. Scent glands along the sides of the body emit a strong musky odor. The American pygmy shrew can be distinguished from other species of *Sorex* by its unique unicuspid teeth; the third unicuspid in the upper jaw is so small that it cannot be seen without magnification.

DENTAL FORMULA i 3/1, c 1/1, p 3/1, m 3/3, total = 32.

SIZE AND WEIGHT This is one of the smallest mammals on earth; American pygmy shrews weigh about the same as a penny. Measurements of 4 males and 1 female, respectively, from Jackson County, Alabama, were:
    total length, 66, 73, 77, 65, 61 mm / 2.6, 2.9, 3.1, 2.6, 2.4 inches
    tail length, 25, 27, 26, 25, 23 mm / 1.0, 1.1, 1.0, 1.0, 0.9 inch
    hind foot length, 9, 9, 9, 9, 9 mm / 0.4, 0.4, 0.4, 0.4, 0.4 inch
    weight, 2.1–3 g / 0.1 ounce
There is no sexual dimorphism.

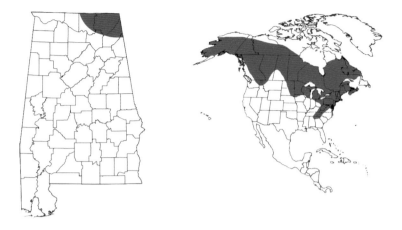

Distribution of the American pygmy shrew in Alabama and North America

DISTRIBUTION In Alabama, the American pygmy shrew is known only from northern Jackson County, but it probably has a wider distribution in the state. Geographic range of the species is from western Alaska to the Atlantic Ocean and from northern Canada into northern Alabama and Georgia.

ECOLOGY Habitat of American pygmy shrews includes nearly all types of forested communities, successional stages, and moisture regimes from subarctic forests to forested areas with grassy clearings, marshes, and swamps. They also occur in grasslands and nonforested habitats including agricultural lands. In the Southeast, American pygmy shrews may be in fallow fields, in clear-cuts dominated by early successional grasses, pines, and mixed pines and hardwoods, in mixed upland and bottomland mixed-hardwood forests, and in saturated forested wetlands. In Alabama, American pygmy shrews occur in moderately rolling terrain dominated by red maples (*Acer rubrum*), chestnut oaks (*Quercus montana*), red oaks (*Q. rubra*), white oaks (*Q. alba*), yellow poplars (*Liriodendron tulipifera*), hickories (*Carya*), sourwoods (*Oxydendrum arboreum*), sassafrasses (*Sassafras albidum*), virburnums (*Virburnum*), and vacciniums (*Vaccinium*). Burrows may be under leaf litter, under logs, or in roots of decaying stumps where there may be numerous small runways. Diet includes spiders; larval and adult beetles, dipterans, and lepidopterans; ants; springtails (Collembola); and many other kinds of insects and invertebrates. Predators include garter snakes (*Thamnophis*), hawks (*Buteo*), and domestic cats.

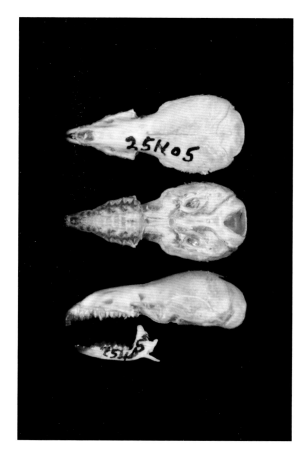

Dorsal, ventral, and lateral views of the cranium, and lateral view of the mandible of an American pygmy shrew of unknown sex. Greatest length of cranium is 15.1 mm.

BEHAVIOR American pygmy shrews are active during the day or night throughout the year. In winter, they may be active on top of snow and in runways within snow. From early June to early September, American pygmy shrews are not active on the surface; apparently, they spend most of this time in burrows. They can swim and climb. When sleeping, American pygmy shrews keep their legs under their body and their tail curled alongside. When awake, they are continually active; they hold their tail straight out from their body and it curves slightly upward. Because they are probably inefficient at constructing burrows, they may rely on burrows constructed by other animals. When threatened, American pygmy shrews quiver intensely and run to cover. During foraging, they sniff and move their bodies and snouts continuously. When consuming larger items, American pygmy shrews sit on their hind legs, similar to a tree squirrel, and hold the food with their forefeet. Vocalizations include high-pitched squeaks, purrs, and whistles.

LIFE HISTORY Breeding may occur year-round but is primarily from November to March, with a peak in January. Gestation is unknown but may be about 20–21 days, as in some other members of the genus. Litters average 4–6 young (range is 1–8). Although it has been suggested that American pygmy shrews have more than 1 litter each year, this has not been documented. At about 6 weeks old, they become active outside the burrow. By winter, young are almost fully grown. Average life span may be 6–8 months, but some have lived 16–17 months in the wild.

PARASITES AND DISEASES Ectoparasites include mites (*Amorphacarus*, *Echinonyssus*, *Haemogamasus*, *Orycteroxenus*, *Protomyobia*, *Pygmephorus*, *Xenoryctes*), ticks (*Ixodes*), and fleas (*Stenoponia*). Endoparasites include cestodes. No diseases are known.

CONSERVATION STATUS High conservation concern in Alabama.

COMMENTS *Sorex* is Latin for "shrew-mouse"; *hoyi* is a patronym honoring Philip R. Hoy (1816–1892), an American physician, explorer, and naturalist.

REFERENCES Long (1974), Whitaker and French (1984), Ryan (1986), Feldhamer et al. (1993), Laerm et al. (1996), Whitaker and Hamilton (1998), Best (2004*a*), Whitaker et al. (2007), Beolens et al. (2009).

Southeastern Shrew
Sorex longirostris

IDENTIFICATION This shrew is brownish on the back and yellowish to reddish brown on the underside. Southeastern shrews have an elongated rostrum, minute eyes, and small ears. Usually, southeastern shrews can be distinguished from other members of this genus in the southeastern United States by the third upper unicuspid being smaller than the fourth.

DENTAL FORMULA i 3/1, c 1/1, p 3/1, m 3/3, total = 32.

SIZE AND WEIGHT Average and range in size of 23 males and 23 females, respectively, from Alabama:

    total length, 81 (73–88), 83 (71–90) mm / 3.2 (2.9–3.5), 3.3 (2.8–3.6) inches

    tail length, 30 (27–33), 31 (26–34) mm / 1.2 (1.1–1.3), 1.2 (1.0–1.4) inches

    hind foot length, 10 (9–11), 10 (9–11) mm / 0.4 (0.4–0.4), 0.4 (0.4–0.4) inch

    weight, 3.2 (2.7–4.0), 3.7 (2.6–5.6) g / 0.1 (0.1–0.1), 0.1 (0.1–0.2) ounce

Females average larger than males.

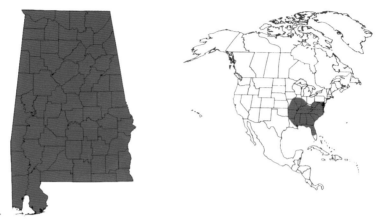

Distribution of the southeastern shrew in Alabama and North America.

DISTRIBUTION Statewide in Alabama. Geographic range of the southeastern shrew is from northern Illinois to central Florida and from western Arkansas to the Atlantic Ocean.

ECOLOGY Until the 1970s and 1980s, the southeastern shrew was believed to be very rare. Pitfall traps, which are open-ended buckets, food cans, or beverage cups buried in soil up to their rims and placed next to logs, rocks, or drift fences, were used to discover that this species was relatively common in many parts of the southeastern United States. Common habitats occupied by southeastern shrews are moist to wet areas, usually bordering marshes, swamps, or rivers. They may also be common in fallow fields, planted fields, dry upland hardwood forests, planted pines (*Pinus*), and dry sandy areas. In all habitats, southeastern shrews are frequently associated with dense ground cover of decaying leaves, grasses, sedges, rushes, blackberries (*Rubus*), kudzu (*Pueraria lobata*), Virginia creepers (*Parthenocissus quinquefolia*), poison ivy (*Rhus radicans*), or Japanese honeysuckles (*Lonicera japonica*). Diet includes vegetation, slugs, snails, spiders (including harvestmen), centipedes, larval lepidopterans, crickets, and adult beetles. Predators include cottonmouths (*Agkistrodon piscivorus*), barn owls (*Tyto alba*), barred owls (*Strix varia*), hooded mergansers (*Lophodytes cucullatus*), Virginia opossums, and domestic dogs and cats. When domestic cats kill southeastern shrews (and other kinds of shrews), they usually do not eat or mutilate them, possibly because of distasteful glands in the skin.

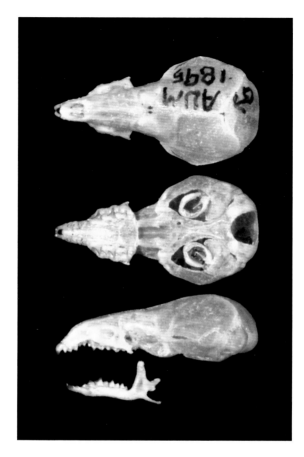

Dorsal, ventral, and lateral views of the cranium, and lateral view of the mandible of a male southeastern shrew. Greatest length of cranium is 15.0 mm.

BEHAVIOR Southeastern shrews are active day or night throughout the year. They are good climbers and move quickly. They construct small, shallow tunnels under dead leaves and in well-drained loamy soils near streams with permanent water. Leaf-lined nests may be in decaying logs. One nest, containing an adult and 4 young, was in a decaying log that lay across a small stream in a shaded ravine. The nest was deep within the log and about 0.7 m above the water. Another nest was in a partially rotten pine log at the crest of a well-drained hill. Another nest, consisting of a mass of cut leaves and containing 2 young southeastern shrews, was beneath the exfoliating bark on top of the log. There were holes leading into other parts of the log. In late November, a bulky nest composed of fine grasses was discovered under a piece of cardboard near a trash dump. Several southeastern shrews were in the nest, possibly a mother and her young or a winter aggregation of adults. Vocalizations include a series of soft, birdlike chirps. When shrews make these calls, they appear as if they are sniffing the air; they hold their head high and rapidly pop their nose up and down.

LIFE HISTORY Breeding season is from March to October, with a peak in April. Gestation is unknown but may be 13–21 days, as in other species of this genus. Litters average 3–4 young (range is 1–10). Two or more litters may be produced in a year. Young stay in the nest until almost fully grown. Two nestlings were 71 and 72 mm in total length, and tail lengths were 24 and 27 mm, respectively. Three young that had recently left the nest but were still dependent on their mother were 76, 78, and 78 mm in total length and 27, 28, and 29 mm in tail length, respectively. Most males and females reach sexual maturity during the year they are born. Some females produce litters during their first year of life. Most individuals probably live through only 1 winter. Life span is about 12 months, but some have lived more than 19 months in the wild.

PARASITES AND DISEASES Ectoparasites include mites (*Amorphacarus, Androlaelaps, Bakerdania, Cyrtolaelaps, Dermacarus, Echinonyssus, Glycyphagus, Haemogamasus, Hypoaspis, Neotrombicula, Ornithonyssus, Orycteroxenus, Parasitus, Proctolaelaps, Protomyobia, Pygmephorus, Xenoryctes*) and ticks (*Dermacentor*). Endoparasites include protozoans (*Eimeria*), cestodes, and nematodes (*Capillaria, Porrocaecum*).

CONSERVATION STATUS Moderate conservation concern in Alabama.

COMMENTS *Sorex* is Latin for "shrew-mouse"; *longirostris* is from the Latin *longus* and *rostrum*, meaning "long snout."

REFERENCES Cook (1942), Goodpaster and Hoffmeister (1952), Dusi (1959), Negus and Dundee (1965), French (1980*a*, 1980*b*), Rose (1980), Best (2004*a*), Whitaker et al. (2007).

Moles

Family Talpidae

The 17 genera and 39 species of moles are small, mouse-sized to rat-sized animals that are usually fossorial and have an elongated rostrum, small eyes, reduced or no pinnae, and forelimbs that are greatly enlarged and modified for digging or, in some, for swimming. The clavicle is unusually short and broad; it is frequently as broad as long. Geographic distribution is the Pacific coastal region and eastern North America, most of Europe, and across central Asia into Japan and southward into Malaysia. One species (eastern mole) occurs in Alabama.

Eastern Mole
Scalopus aquaticus

IDENTIFICATION Eastern moles have large, paddle-like front feet, an elongated rostrum, soft, velvety fur, a short tail, webbed toes on all feet, and no obvious eyes or ears. Color is silverish to nearly black.

DENTAL FORMULA i 3/2, c 1/0, p 3/3, m 3/3, total = 36.

SIZE AND WEIGHT Average and range in size of 22 specimens from Alabama:

total length, 132 (115–158) mm / 5.3 (4.6–6.3) inches
tail length, 20 (11–28) mm / 0.8 (0.4–1.1) inches
hind foot length, 18 (14–26) mm / 0.7 (0.6–1.0) inch
weight of 15 specimens, 49.8 (26.2–59.5) g / 1.7 (0.9–2.1) ounces
Males average larger than females.

DISTRIBUTION Statewide in Alabama. Geographic range of the eastern mole is from northern Michigan to northeastern Mexico and from southeastern Wyoming to the Atlantic Ocean.

In sandy-loam soils, surface ridges constructed by eastern moles can be large and conspicuous.

ECOLOGY Habitat includes moist, loamy or sandy soils; eastern moles are rare or absent in soils with dense clay, stones, or gravel. Woodlands and floodplains along rivers and streams are frequently occupied, but eastern moles may also occur in open, grassy habitats and cultivated fields. Their appetite appears insatiable. Each day, eastern moles consume food equal to 25–100% of their weight. Diet includes earthworms, adult and larval insects, and vegetation. Predators include snakes, barred owls (*Strix varia*), red-shouldered hawks (*Buteo lineatus*), red-tailed hawks (*Buteo jamaicensis*), bobcats, coyotes, gray and red foxes, and long-tailed weasels. Humans may kill moles that invade lawns, golf courses, or other places where they are unwelcome. Although not obtained from eastern moles, millions of mole skins were once used in caps, purses, tobacco pouches, garment trimmings, and powder puffs.

BEHAVIOR Eastern moles are active any time of the day or night throughout the year but reduce their activity during extremely cold weather in winter. At that time, they remain deep within their burrows, but they do not hibernate. Nests are 8–46 cm or more below the surface, inside a hollow dome 20 cm in diameter and 12 cm deep. They may contain dry leaves, grasses, or nothing at all. Tunnels may be constructed and

The stream-
lined body and
paddle-like front
feet of the east-
ern mole allow
it to move
easily through
sandy-loam soils.

occupied by 2–3 individuals. Tunnels are of two types: surface tunnels or ridges, which are used for collecting food, and deep passages, which are more permanent and used as living quarters and as passageways to foraging areas. When digging surface tunnels and ridges, eastern moles use their forefeet to push the soil aside. The snout is apparently active in searching for a suitable place for the next digging sidestroke of its fore-feet. Then, the animal twists the front half of its body, turning on its side, and pushes the soil upward with its forefeet. Although surface tunnels and ridges may average about 90 m long, they remain consistently at the same depth below the surface of the ground. Surface tunnels may be con-structed at a speed of about 4 m/hour. Tunnels in protected places, such as along fencerows, may be more than 1,000 m long. Excavations of deep passageways do not produce ridges on the surface, but their approximate location can be discerned by small mounds of soil. Deep tunnels are 3–60 cm deep (average is 10–35 cm). Usually, the deepest parts of tunnels are not directly under the mounds of soil but between them. There may be up to 20 mounds of soil above 1 deep passageway. Eastern moles have tremendous strength for the size of their body. They can lift a mass 26–32

Distribution of
the eastern mole
in Alabama and
North America.

times their own body weight, which is equivalent to a 67.5-kg human lifting 2,160 kg (150-pound human lifting 4,800 pounds). Eastern moles are very flexible; to turn around in a narrow tunnel, they tuck their head under their abdomen and pull themselves under their body. They are also good swimmers. Eastern moles are usually subterranean, but they may travel briefly on the surface searching for a mate or food, especially in areas where burrows are not connected or during times of drought and food shortages. Activity on the surface is verified by human observers and by the presence of eastern moles in the diets of nondigging predators such as hawks and owls.

LIFE HISTORY Breeding season is late spring, with a peak from March to April. Length of gestation is unknown, but estimates are 28–45 days. One litter, or possibly two, of 2–5 young is produced each year. Life span is unknown but may be about 3 years.

PARASITES AND DISEASES Ectoparasites may cause an eastern mole to stop feeding to scratch itself, run frantically about, or rub itself against some object. Ectoparasites include mites (*Androlaelaps*, *Bakerdania*, *Echinonyssus*, *Eulaelaps*, *Haemogamasus*, *Macrocheles*, *Ornithonyssus*, *Pygmephorus*, *Scalopacarus*, *Xenoryctes*), lice (*Haematopinoides*), fleas (*Ctenophthalmus*), and beetles (*Leptinus*). Endoparasites include nematodes (*Filaria*, *Gongylonema*, *Wuchereria*).

CONSERVATION STATUS Low conservation concern in Alabama.

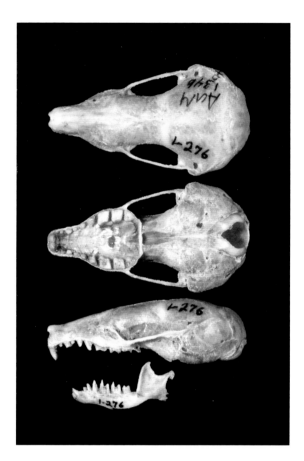

Dorsal, ventral, and lateral views of the cranium, and lateral view of the mandible of a female eastern mole. Greatest length of cranium is 34.6 mm.

COMMENTS *Scalopus* is from the Greek *skallo*, meaning "to dig," and *pous*, meaning "foot," referring to the digging feet; *aquaticus* is Latin for "found in the water." Linnaeus named the animal from a preserved specimen and knew nothing of its habits; he assumed that because it had webbed feet, the eastern mole was aquatic.

REFERENCES Arlton (1936), Yates and Schmidly (1978), Anonymous (1994), Best (2004*a*), Whitaker et al. (2007).

Bats

Order Chiroptera

Chiroptera is among the most diverse orders of mammals, with 18 families, 202 genera, and 1,116 species. This order is second in diversity only to Rodentia. Bats are the only true flying mammals and usually have a wing membrane extending from the sides of the body and legs and supported by elongated fingers on the forelimbs. There is usually an interfemoral membrane between the legs and tail. Most bats have a thumb that is not incorporated into the wing membrane. Echolocation is well developed in the suborder Microchiroptera (common bats) and rare in the suborder Megachiroptera (flying foxes and relatives). Bats occur worldwide, except in polar regions and on some oceanic islands. There are 2 families of microchiropteran bats present in Alabama (Molossidae, Vespertilionidae).

Free-tailed Bats

Family Molossidae

This family contains 16 genera and 100 species that occur in temperate and equatorial regions around the world. Although this is not the only family of bats that have members with a tail that protrudes beyond the free edge of the uropatagium (tail membrane), molossids are typically characterized by a tail that projects far beyond the free edge. Generally, these insectivorous bats are rapid, long-distance fliers; wings are often narrow and some species fly long distances during annual migrations. One species (Brazilian free-tailed bat) is present in Alabama.

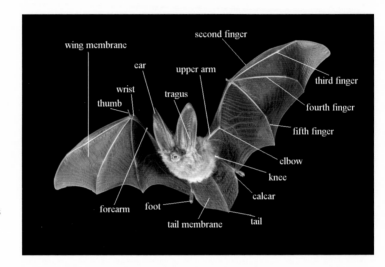

Anatomy of a bat (Townsend's big-eared bat, *Corynorhinus townsendii*).

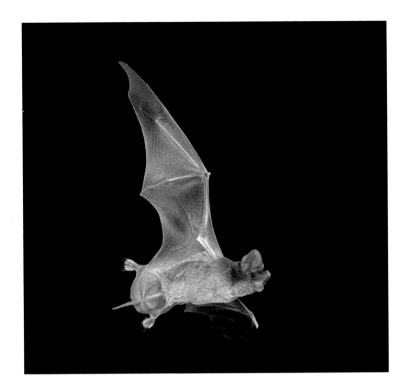

Brazilian Free-tailed Bat
Tadarida brasiliensis

IDENTIFICATION This is the only bat in Alabama that has a tail extending well beyond the edge of the interfemoral membrane. Pelage on the back is uniformly brownish to blackish and slightly paler on the underside. Long, bristly hairs extend beyond the toes.

DENTAL FORMULA i 1/2–3, c 1/1, p 2/2, m 3/3, total = 30–32.

SIZE AND WEIGHT Average and range in size of 78 males and 60 females, respectively, from Alabama:

> total length, 102 (92–115), 100 (91–108) mm / 4.1 (3.7–4.6), 4.0 (3.6–4.3) inches
>
> tail length, 35 (28–40), 35 (28–40) mm / 1.4 (1.1–1.6), 1.4 (1.1–1.6) inches
>
> foot length, 10 (7–13), 10 (7–12) mm / 0.4 (0.3–0.5), 0.4 (0.3–0.5) inch

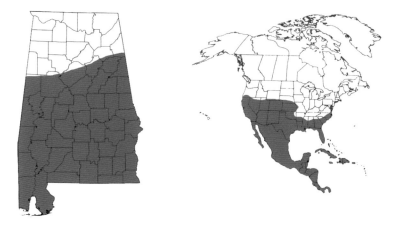

Distribution of the Brazilian free-tailed bat in Alabama and North America.

ear length, 19 (17–21), 19 (16–21) mm / 0.8 (0.7–0.8), 0.8 (0.6–0.8) inch

weight of 77 males, 12.5 (10.5–16.5) g / 0.4 (0.4–0.6) ounce

weight of 60 females, 12.7 (8.0–16.5) g / 0.5 (0.3–0.6) ounce

forearm length of 23 males from Louisiana, 42 (40–45) mm / 1.7 (1.6–1.8) inches

forearm length of 46 females from Louisiana, 43 (38–46) mm / 1.7 (1.5–1.8) inches

Males are slightly larger than females in some characters.

DISTRIBUTION Southern Alabama. Geographic range of the species is from southwestern Oregon to eastern North Carolina and from southern Nebraska to southern South America, except for the Amazon Basin and some regions east of the Andes, but including most islands in the Caribbean.

ECOLOGY In the southeastern United States, roosts are in human-made structures such as homes, churches, warehouses, sports stadiums, and government buildings. Before humans removed large hollow trees such as red mangroves (*Rhizophora*), black mangroves (*Avicennia*), white mangroves (*Laguncularia*), and cypresses (*Taxodium*), these were probably the natural roosts of Brazilian free-tailed bats. Elsewhere in the United States, Brazilian free-tailed bats usually roost in caves, but they may also occupy human-made structures. Roosts have a characteristic odor, which is similar to that of taco shells, tortillas, or corn chips and is produced by the chemical 2-aminoacetophenone. Diet includes moths, beetles, bugs,

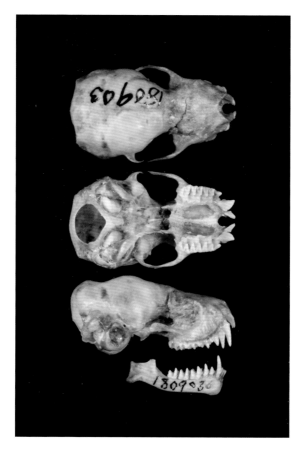

Dorsal, ventral, and lateral views of the cranium, and lateral view of the mandible of a male Brazilian free-tailed bat. Greatest length of cranium is 17.4 mm.

flies, ants, lacewings, termites, and spiders. Predators include black rat snakes (*Elaphe obsoleta*), coachwhips (*Masticophis flagellum*), copperheads (*Agkistrodon contortrix*), American kestrels (*Falco sparverius*), red-tailed hawks (*Buteo jamaicensis*), great horned owls (*Bubo virginianus*), barn owls (*Tyto alba*), Virginia opossums, striped skunks, and raccoons. Other causes of death are pesticides, collisions with transmission towers and associated guy wires, and entanglement in cacti, grass burs, and other vegetation.

BEHAVIOR Brazilian free-tailed bats are swift flyers and may cover more than 100 km in each night of foraging. They cannot initiate flight from a flat surface, so they must crawl to an elevated structure (a wall, boulder, or tree), climb, and launch themselves. Usually, they simply drop into flight from the ceiling of the roosting site. Although most foraging probably takes place 6–15 m above the ground, military radar and weather balloons have tracked Brazilian free-tailed bats foraging at altitudes of 3,000 m above the ground. A gland on the throat of both sexes becomes conspicuous during or just before the mating season and may be important in sexual activity, social dominance, or territoriality. Brazilian free-tailed bats do not hibernate. Populations in the southeastern and northwestern United States do not migrate, but they move among colonies during the year. Populations in southwestern North America may migrate more than 1,800 km to Mexico and beyond in autumn, but some individuals and populations overwinter in the United States. Maternity colonies consisting almost entirely of females are formed in spring. The largest aggregation of mammals in the world is the group of about 20,000,000 Brazilian free-tailed bats that reside in Bracken Cave, Texas, during summer.

LIFE HISTORY In the southeastern United States, ovulation and mating occur in March, gestation is about 11 weeks, and 1–2 young (occasionally triplets) are born in June. While giving birth, females hang with their head downward. Following birth, neonates take about 15 minutes to find a nipple and begin nursing. Their skin is smooth and nearly hairless except for hairs near the toes and thumbs, and neonates are about two-thirds the length of adults. Mothers are able to find and nurse their own young twice nightly because they leave their babies in nurseries within large colonies. Females usually do not carry their young, but when disturbed or in response to severe weather, they will carry their baby to a new roost. In transit, the baby is strongly attached to a nipple and is carried horizontally across her chest. Young begin flying at about 35 days old and can fly well by 38 days. Females reach sexual maturity by about 9 months old and males by about 12 months old. In the western United States, few adult males migrate northward; mating usually occurs in overwintering colonies in Mexico or farther south, and young are born in June or July. Life span is probably 8–10 years but may be up to 15 years.

PARASITES AND DISEASES Ectoparasites include mites (*Androlaelaps, Chiroptonyssus, Dentocarpus, Ewingana, Microtrombicula, Nycteriglyphus, Olabidocarpus, Steatonyssus*), ticks (*Ornithodoros*), fleas (*Sternopsylla*), and bat flies (*Basilia*). Endoparasites include cestodes (*Hymenolepis*), nematodes (*Anoplostrongylus, Molinostrongylus, Parallintoshius, Physaloptera, Rictularia, Seuratum, Tricholeiperia*), and trematodes (*Acanthatrium, Allassogonoporus, Conspicuum, Dicrocoelium, Limatu-*

lum, Ochoterenatrema, Paralecithodendrium, Plagiorchis, Platynosumum, Tremajoannes, Urotrema). Diseases include Chagas disease (*Trypanosoma*), eastern equine encephalitis (*Alphavirus*), histoplasmosis (*Histoplasma*), Japanese B encephalitis (*Flavivirus*), rabies (*Lyssavirus*), Rio Bravo virus (*Flavivirus*), Saint Louis encephalitis (*Flavivirus*), and western equine encephalitis (*Alphavirus*). Bats were not known to carry rabies until the 1950s; several bats of this species were among the first to test positive for this virus. In Alabama, 20 of 165 (12.1%) Brazilian free-tailed bats tested positive for rabies.

CONSERVATION STATUS High conservation concern in Alabama.

COMMENTS Guano was once a significant source of fertilizer for farms in California (it is still used in many parts of the world). In the early 1900s, mining operations in the western United States shipped large quantities of guano by train from caves to farms. Guano was also an important ingredient in gunpowder during the Civil War. During World War II, plans were made to attach incendiary bombs to Brazilian free-tailed bats and drop the bats with bombs attached onto cities in Japan, where the bats would seek shelter in homes and other structures and the bombs would then ignite and burn the buildings. The plan was abandoned, possibly because some of the bats ignited buildings on the military installation where tests were being conducted. *Tadarida* is a made-up word; *brasiliensis* is Latin for "of Brazil," referring to the type locality.

REFERENCES Crawford and Baker (1981), Wilkins (1989), Henry et al. (2000), Best and Geluso (2003), Best (2004*a*), López-González and Best (2006), McWilliams (2005), Nielsen et al. (2006), Hester et al. (2007), Harvey et al. (2011), Whitaker et al. (2007).

Common Bats

Family Vespertilionidae

The 48 genera and 407 species of common bats vary in size from small to moderately large, with forearm length ranging from 24 to 90 mm. Most are insectivores, but a few species consume fish and birds. Numerous members of this family are cave dwellers, but many live under exfoliating bark, in cracks and crevices of rocks and cliff faces, or in structures built by humans (e.g., houses, barns, wells, mines, birdhouses), and some roost in trees or under leaf litter. Most have 1 young each year, but some may have 3–4. Life span is probably 10–15 years, but 1 individual (little brown myotis) was known to survive in the wild for more than 30 years. In Alabama, 14 species represented by 7 genera (*Eptesicus, Lasiurus, Nycticeius, Perimyotis, Corynorhinus, Lasionycteris, Myotis*) have been verified to occur in the state, and another species is probably present (eastern small-footed myotis).

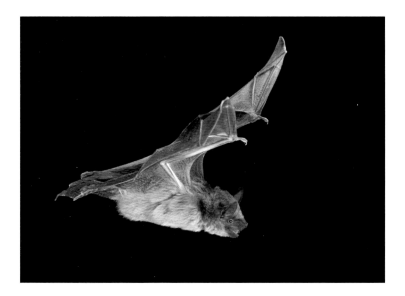

Big Brown Bat
Eptesicus fuscus

IDENTIFICATION This is one of the largest bats in Alabama. It is pale to dark brown, and the face, ears, wings, and most of the uropatagium are blackish. The tip of the tail extends about 3 mm beyond the uropatagium. This bat has 2 incisors and 1 premolar on each side of the upper jaw. The tragus is broad and tapers toward the tip, and the calcar is keeled.

DENTAL FORMULA i 2/3, c 1/1, p 1/2, m 3/3, total = 32.

SIZE AND WEIGHT Average and range in size of 47 males and 95 females, respectively, from Alabama:

> total length, 114 (101–124), 119 (87–132) mm / 4.6 (4.0–5.0), 4.8 (3.5–5.3) inches
>
> tail length, 42 (36–51), 45 (34–54) mm / 1.7 (1.4–2.0), 1.8 (1.4–2.2) inches
>
> foot length, 10 (8–14), 11 (8–13) mm / 0.4 (0.3–0.6), 0.4 (0.3–0.5) inch
>
> ear length, 17 (15–19), 17 (14–22) mm / 0.7 (0.6–0.8), 0.7 (0.6–0.9) inch
>
> forearm length of 2 males, 49 (48–49) mm / 2.0 (1.9–2.0) inches

forearm length of 5 females, 49 (46–51) mm / 2.0 (1.8–2.0) inches
weight of 47 males, 14.9 (10.0–25.0) g / 0.5 (0.4–0.9) ounce
weight of 94 females, 18.0 (10.5–27.5) g / 0.6 (0.4–1.0) ounce
Females are larger than males.

DISTRIBUTION Statewide in Alabama. Geographic range of the species is from central Canada into northwestern South America, and from the Pacific to the Atlantic Ocean and islands of the Caribbean, except some coastal regions of Central America.

ECOLOGY Big brown bats are more numerous in eastern deciduous forests than in western coniferous forests, but they occur in nearly all types of habitats within their range. In urban roosts, such as homes, schools, businesses, and churches, the size of access openings and height above the ground are important in determining suitable roosting sites. Although urban areas provide abundant roosting sites, big brown bats usually leave these areas to forage over less developed and more rural areas at night. They show no preference in foraging habitat. Small beetles (e.g., Scarabaeidae) are the most common prey, but big brown bats may also eat moths, true bugs, flies, ants, wasps, grasshoppers, and lacewings. Predators include various species of owls, common grackles (*Quiscalus quiscula*), American kestrels (*Falco sparverius*), long-tailed weasels, raccoons, rodents, and domestic cats.

One of these back-to-back bat houses was occupied by big brown bats and the other by Brazilian free-tailed bats.

BEHAVIOR Foraging occurs throughout the night but peaks about 1–2 hours after sunset. Distance to foraging areas averages 1–2 km. Although away from the roost most of the night, big brown bats have a total flying time of about 100 minutes; they spend most of the night in roosts away from the colony. Big brown bats are good swimmers. While drinking water or after collisions with objects, these bats may fall into the water, whereupon they promptly swim to shore, climb up onto vegetation, and fly away. Distance from summer to winter roosts is rarely more than 80 km. Males enter hibernation before females. Big brown bats may enter or leave hibernation at intervals throughout the winter. They hibernate singly or in clusters of about 2–20 individuals in buildings, mines, caves, under rocks, and in rock crevices. In subtropical habitats, big brown bats do not hibernate, but they do enter torpor when temperatures are low. When they are aroused from torpor, their heart rate increases from 12 to 800 beats/minute. After hibernation, adult females form maternity colonies.

Oily stains around attic vents are evidence that big brown bats or other species of bats are roosting there. Bats can be evicted when no young are present by placing loose netting over openings.

Males are usually solitary in summer, but they may roost with females or in groups of other males. In eastern North America, maternity colonies usually contain 25–75 adults (range is 5–700) and are in human-made structures such as attics, barns, and churches, but they may also be in hollow oak (*Quercus*) and beech (*Fagus*) trees. In western North America, maternity colonies may be in buildings, rocky crevices, or dead trees.

LIFE HISTORY Mating occurs from August to March, ovulation and fertilization is delayed until after hibernation, and young are born from May to July. In western North America, 1 young is born, but in eastern North America and Cuba, 2 are born. Lactation lasts 32–40 days. Newborns are hairless and almost immobile and weigh about 3.3 g. Their forearm length is about 16.8 mm, and their eyes and ears open within a few hours. Mothers and young communicate with ultrasonic calls. Juveniles begin to fly when 18–35 days old. Mothers and offspring spend considerable time together foraging and roosting, which may facilitate learning by the

Distribution of the big brown bat in Alabama and North America.

young. Males become sexually mature during their first autumn, but not all females mate at the end of their first year. Life span is probably 12–15 years, but some live more than 19 years in the wild.

PARASITES AND DISEASES Ectoparasites include mites (*Acanthophthirius, Alabidocarpus, Cheletonella, Chiroptonyssus, Cryptonyssus, Euschoengastia, Leptotrombidium, Macronyssus, Neospeleognathopsis, Neotrombicula, Nycteriglyphitis, Nycteriglyphus, Olabidocarpus, Parasecia, Perissopalla, Pteracarus, Spinturnix, Steatonyssus*), ticks (*Ornithodoros*), fleas (*Myodopsylla, Nycteridopsylla*), bat flies (*Basilia*), and bat bugs (*Cimex*). Endoparasites include protozoans (*Trypanosoma*), cestodes (*Hymenolepis, Vampirolepis*), nematodes (*Allintoshius, Capillaria, Cyrnea, Litomosoides, Maseria, Physocephalus, Rictularia, Seuratum*), and trematodes (*Acanthatrium, Allassogonoporus, Anenterotrema, Dicrocoelium, Glyptoporus, Limatulum, Ochoterenatrema, Paralecithodendrium, Plagiorchis, Postorchigenes, Prosthodendrium, Urotrema*). Rabies (*Lyssavirus*) occurs at a relatively low rate across the range of the species; in Alabama, 22 of 372 (5.9%) individuals tested positive for the rabies virus. Other diseases include Saint Louis encephalitis (*Flavivirus*) and histoplasmosis (*Histoplasma*). Because big brown bats hibernate in caves, they are susceptible to infection with white-nose syndrome (*Geomyces destructans*). However, the incidence of infection is low, possibly because big brown bats roost in drier parts of the hibernaculum where they are seldom covered with condensation during hibernation.

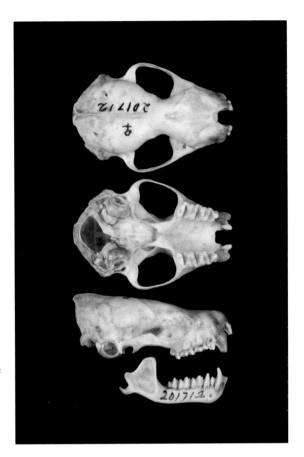

Dorsal, ventral, and lateral views of the cranium, and lateral view of the mandible of a female big brown bat. Greatest length of cranium is 20.6 mm.

CONSERVATION STATUS Lowest conservation concern in Alabama.

COMMENTS In Alabama, big brown bats may roost in attics of homes or other places where they are not welcome. It is important to know how to properly and legally evict nuisance bats. Do not kill or harm the bats—they will be replaced by other bats, their dead and decomposing bodies in attics and walls are smelly, and their rotting remains may be dangerous to humans and pets. Bats should be evicted during early spring (February–April) before young are born or during autumn (August–October) when young are able to fly and before bats go into hibernation. Do not try to evict bats during colder months or when young bats may be present. Locate all openings into the part of the building that is occupied by bats. Once the young bats are able to fly (August), attach netting or cloth above *all* access openings on the outside of the building. Netting should lie against the outside of the building, and it should extend about 0.5 m beyond the sides of any openings and hang about 1 m below the openings. Only the top of the netting should be attached to the building; this will allow bats to push out of the roost and fly away unharmed but will also keep them from re-entering the structure. After 5–7 days, all bats should be out of the building. Fill or patch *all* access openings, then remove the netting. *Eptesicus* is from the Greek *epiēn*, meaning "to fly," and *oikos*, meaning "house"; big brown bats often live in houses; *fuscus* is Latin for "brown."

REFERENCES Blankespoor and Ulmer (1970), Jones et al. (1973), Kurta and Baker (1990), Lotz and Font (1991), Henry et al. (2000), M. A. Menzel et al. (2001), Agosta and Morton (2003), Best (2004*a*), Hester et al. (2007), Neubaum et al. (2007), Whitaker et al. (2007), Cryan et al. (2010), Harvey et al. (2011).

Eastern Red Bat
Lasiurus borealis

IDENTIFICATION This bat is brick red to rusty red on the back, not mahogany colored. Underside is paler and there is a white patch on each shoulder. Interfemoral membrane is fully furred. Ears are short, broad, and rounded, and tragus is blunt and triangular.

DENTAL FORMULA i 1/3, c 1/1, p 2/2, m 3/3, total = 32.

SIZE AND WEIGHT Average and range in size of 10 males and 9 females, respectively, from Alabama:

total length, 104 (98–109), 106 (94–109) mm / 4.2 (3.9–4.4), 4.2 (3.8–4.4) inches

During the day, this mother eastern red bat and her 2 babies cling together as they hang from a branch in a tree.

tail length, 45 (40–50), 49 (36–55) mm / 1.8 (1.6–2.0), 2.0 (1.4–2.2) inches

foot length, 9 (6–10), 7 (6–9) mm / 0.4 (0.2–0.4), 0.3 (0.2–0.4) inch

ear length, 10 (7–12), 9 (7–12) mm / 0.4 (0.3–0.5), 0.4 (0.3–0.5) inch

forearm length of 2 males, 40 mm (40–41 mm) mm / 1.6 (1.6–1.6) inches

weight of 7 males, 8.9 (7.5–12.0) g / 0.3 (0.3–0.4) ounce

weight of 8 females, 8.1 (8.0–10.0) g / 0.3 (0.3–0.4) ounce

Males are more brightly colored than females.

DISTRIBUTION Statewide in Alabama. Geographic range of the species is from southern Canada to northern Mexico and from western Montana to the Atlantic Ocean.

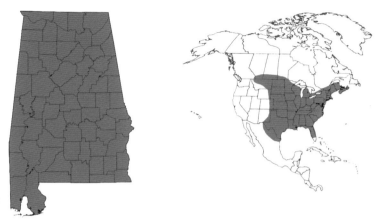

Distribution of the eastern red bat in Alabama and North America.

ECOLOGY Habitats include pine (*Pinus*) and mixed-hardwood communities and bottomland hardwood swamps, but eastern red bats also occur in pasturelands bordered by forests, along roadways, and in suburban and urban areas. Roosts generally provide dense cover and shade above but are open below. They are 1–15 m above the ground in woodpecker cavities, in clumps of Spanish moss (*Tillandsia usneoides*), in dense grass, under leaves of sunflowers (*Helianthus*), in leaf litter, under shingles of houses, and in many kinds of trees including sweetgums (*Liquidambar styraciflua*), black gums (*Nyssa sylvatica*), American elms (*Ulmus americana*), box elders (*Acer negundo*), wild plums (*Prunus americana*), sugar maples (*Acer saccharum*), wild black cherries (*Prunus serotina*), and at least 6 species of oaks (*Quercus*). In winter, eastern red bats may switch from roosting in trees to roosting in leaf litter when temperatures approach or drop below freezing. Diet includes moths, true bugs, beetles, ants and wasps, flies and mosquitoes, and ground-dwelling crickets. Predators include blue jays (*Cyanocitta cristata*), sharp-shinned hawks (*Accipiter striatus*), American kestrels (*Falco sparverius*), great horned owls (*Bubo virginianus*), Virginia opossums, and domestic cats. Many eastern red bats are killed by wind turbines during migration. Because eastern red bats may spend considerable time in leaf litter in winter, they are also susceptible to fires. During fires, eastern red bats have been observed in a state of torpor in burning leaf litter on the ground, attempting to crawl or fly after being overwhelmed by smoke or heat from the flames.

BEHAVIOR Eastern red bats are solitary, foliage-roosting bats. They usually roost in trees and shrubs along the edges of streams or fields, and

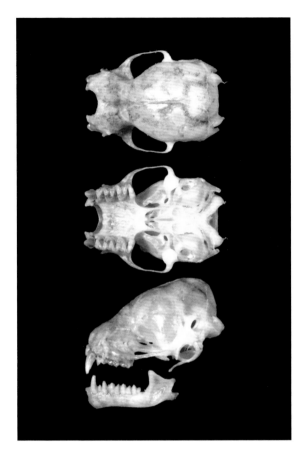

Dorsal, ventral, and lateral views of the cranium, and lateral view of the mandible of a male eastern red bat. Greatest length of cranium is 13.3 mm.

in urban areas. When they emerge to begin foraging, eastern red bats fly high using slow, fluttering, erratic flight. After 15–30 minutes, they descend and forage from within a few meters of the ground to treetop level. At this time, they fly straight or in wide circles that they interrupt only to chase and capture insects. Eastern red bats also appear to forage regularly over the same area. They migrate to southern states in autumn and may swarm at the mouths of caves, but they spend winter in leaf litter or in foliage. On warm afternoons in winter, eastern red bats may emerge from day roosts and fly about in daylight, presumably in search of food or water. Many eastern red bats that enter caves die, presumably because they become disoriented and cannot find their way out. They may migrate in groups. In Alabama, males and females seem to migrate separately, with males migrating last.

LIFE HISTORY Mating occurs in flight during August or September, and some coupled pairs fall to the ground. Fertilization occurs in spring and young are born in June. Lactation lasts about 38 days. Litters average 2–3 young (range is 1–5). Neonates are hairless, their eyes are closed, their forearm length is 12.7–16.6 mm, and their weight is about 0.5 g. Very young eastern red bats cling to their mother with their teeth, thumbs, and feet. As they grow older, young will hang by 1 or both feet from a leaf or twig and clasp the mother with their wings. A mother will transport young to new roosting sites about every 1–2 days, but she usually leaves them in the roost when she forages in the evening. By 3–6 weeks old, young can fly, and they are weaned by about 4–6 weeks of age. Life span is unknown but is probably 8–10 years.

PARASITES AND DISEASES Ectoparasites include mites (*Acanthophthirius, Pteracarus, Steatonyssus*), fleas (*Nycteridopsylla*), and bat bugs (*Cimex*). Endoparasites include protozoans (*Distoma*), cestodes (*Cycloskrjabinia, Oochoristica, Taenia*), nematodes (*Longibucca*), and trematodes (*Acanthatrium, Lecithodendrium, Prosthodendrium, Urotrema*). In Alabama, the incidence of rabies (*Lyssavirus*) detected in eastern red bats was 17% (51 of 300 tested).

CONSERVATION STATUS Lowest conservation concern in Alabama.

COMMENTS *Lasiurus* is from the Greek *lasios*, meaning "hairy," and *oura*, meaning "tail," a reference to the furred uropatagium; *borealis* is Latin for "northern."

REFERENCES Macy (1933), Blankespoor and Ulmer (1970), Shump and Shump (1982*a*), Saugey et al. (1989), M. A. Menzel et al. (1998), Laerm et al. (1999), Mager and Nelson (2001), Best (2004*a*), Hester et al. (2007), Mormann and Robbins (2007), Whitaker et al. (2007), Cryan (2008), Winhold et al. (2008).

Hoary Bat
Lasiurus cinereus

IDENTIFICATION This large, colorful bat is easily distinguished from all other bats in Alabama. Color is mixed dark brownish and grayish, with white-tipped hairs that give it a frosted or hoary appearance. Wrist and shoulder patches are whitish and the throat is distinctly yellowish. The interfemoral membrane is well furred; the ears are short, rounded, and edged with black; and the tragus is short and broad.

DENTAL FORMULA i 1/3, c 1/1, p 2/2, m 3/3, total = 32.

SIZE AND WEIGHT Average and range in size of 4 specimens from Alabama:

    total length, 124 (119–130) mm / 5.0 (4.8–5.2) inches
    tail length, 53 (49–57) mm / 2.1 (2.0–2.3) inches

Hoary bats and other tree-roosting bats are difficult to see among leaves and branches in their daytime roosts.

foot length, 11 (10–11) mm / 0.4 (0.4–0.4) inch

ear length, 13 (10–16) mm / 0.5 (0.4–0.6) inch

average forearm length for males and females, 52.6 and 54.2 mm (2.1 and 2.2 inches), respectively

weight, 20–35 g / 0.7–1.2 ounces

Females are larger than males.

DISTRIBUTION Statewide in Alabama. Geographic range of the hoary bat is the largest of any bat in the Americas and stretches from the Arctic of Canada into southern South America, except for a few regions in Central America and the Amazon Basin. This is the only native terrestrial mammal in Hawaii, and there are records from Iceland, Bermuda, Hispaniola, the Galapagos Islands, and the Orkney Islands in Scotland.

ECOLOGY Hoary bats are uncommon in the northern Rocky Mountains and eastern United States but common in the prairie states and the Pacific Northwest. In Alabama, most observations have been in forested habitats in northern counties during cooler months, but hoary bats occur as far south as the Mobile-Tensaw Delta during December. There are also a few

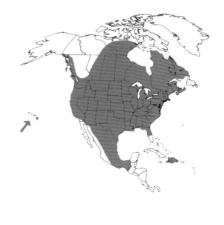

Distribution of the hoary bat in Alabama and North America.

scattered records throughout the state in warmer months. An observation of a young-of-the-year on 1 August in Baldwin County may indicate that a breeding population occurs in Alabama, but this has not been verified. Diet includes moths, beetles, flies, grasshoppers, termites, dragonflies, and wasps. Hoary bats have attacked tri-colored bats, possibly as a source of food or while defending a territory. Predators include red rat snakes (*Elaphe guttata*), owls, hawks, and American kestrels (*Falco sparverius*). There are several reports of hoary bats being impaled on barbed wire. Eastern red bats and hoary bats are killed in greater numbers by wind turbines than are any other species of bat in North America.

BEHAVIOR During warmer months, males and females tend to be separated geographically. Except during mating season and when mothers are with their young, hoary bats are solitary and usually roost in foliage. By day, hoary bats roost 3–5 m above the ground in trees such as elms (*Ulmus*), black cherries and plums (*Prunus*), box elders (*Acer negundo*), and Osage oranges (*Maclura pomifera*). Unusual roosting sites include woodpecker cavities, sites beneath planks of driftwood, caves, nests of eastern gray squirrels, and sides of buildings. In their roosts, hoary bats are usually well hidden from above but visible from below. They tend to roost near the edges of clearings and emerge in the evening to forage. During warm afternoons in winter, hoary bats may arouse from torpor and fly. Although they consume other invertebrates, they seem to prefer moths; a bat will approach a flying moth from behind, engulf the abdomen-thorax in its mouth, bite down, and allow the head and wings to drop to the ground. Hoary bats are migratory, but some overwinter in cold regions.

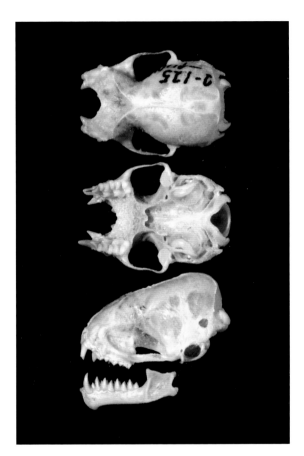

Dorsal, ventral, and lateral views of the cranium, and lateral view of the mandible of a male hoary bat. Greatest length of cranium is 17.9 mm.

They are fast and straight flyers, traits useful in long migrations. Hoary bats may aggregate into large groups during autumn and spring migrations. They may fly high above the ground; possibly during a migratory flight in October, a hoary bat was sucked into a jet engine at an altitude of 2,438 m (8,000 feet). These bats are also vulnerable to collisions with wind turbines, especially during autumn migration.

LIFE HISTORY Mating probably occurs in autumn during migration or at overwintering sites. Implantation of the embryo is delayed until spring. Young are born from mid-May to early July. Usually, 2 young are born (range is 1–4). Immediately following birth, the mother cradles the young in her wings and grooms them until they are clean and dry. At this time, young begin to groom themselves. Newborns have fine, silvery-gray hairs and are naked on their underside. Their eyes and ears are closed, and their forearm length is 16–20 mm. During the daytime, young cling to their mother, but she usually leaves them hanging from a twig or leaf when she goes foraging at night. By 3 days, ears open; by 12 days, eyes open; and by 33 days, young can fly. Life span is unknown but may be 8–10 years.

PARASITES AND DISEASES Ectoparasites include mites (*Chiroptonyssus, Acanthophthirius, Pteracarus, Steatonyssus*). Endoparasites include protozoans (*Distoma*), cestodes (*Oochoristica*), nematodes (*Longibucca, Physocephalus*), and trematodes (*Plagiorchis*). In some areas of Canada and the United States, the incidence of rabies (*Lyssavirus*) in hoary bats is 10–67%. However, of 18 hoary bats tested in Alabama, none had evidence of the rabies virus.

CONSERVATION STATUS Moderate conservation concern in Alabama.

COMMENTS *Lasiurus* is from the Greek *lasios*, meaning "hairy," and *oura*, meaning "tail," a reference to the furred uropatagium; *cinereus* is Latin for "ash-colored."

REFERENCES Peterson (1966), Blankespoor and Ulmer (1970), Shump and Shump (1982*b*), Peurach (2003), Best (2004*a*), Hester (2007), Whitaker et al. (2007), Cryan (2008), Hirt (2008), Kilgore (2008), Reimer et al. (2010).

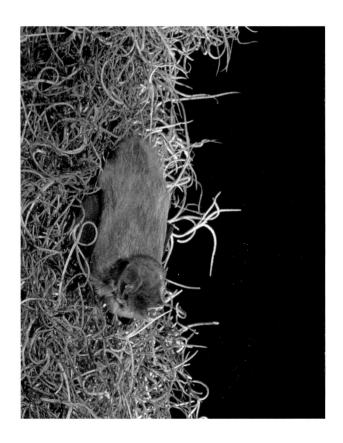

Northern Yellow Bat

Lasiurus intermedius

IDENTIFICATION This yellowish to yellowish-orange bat is one of the largest species of bats in Alabama (wingspan is about 375 mm). The uropatagium is furred on top from the body to about midway to the tip of the tail. There are no white patches on wrists or shoulders. The tragus is rounded and points forward. The northern yellow bat has 1 upper premolar on each side of the upper jaw, not 2 as in other members of this genus in Alabama.

DENTAL FORMULA i 1/3, c 1/1, p 1/2, m 3/3, total = 30.

SIZE AND WEIGHT Average and range in size of 14 males from Florida: total length, 127 (121–132) mm / 5.1 (4.8–5.3) inches

Distribution of the northern yellow bat in Alabama and North America.

tail length, 54 (51–60) mm / 2.2 (2.0–2.4) inches
foot length, 10 (8–11) mm / 0.4 (0.3–0.4) inch
ear length from notch, 16 (15–17) mm / 0.6 (0.6–0.7) inch
tragus length, 8 (5–11) mm / 0.3 (0.2–0.4) inch
forearm length, 52 (51–53) mm / 2.1 (2.0–2.1) inches
weight, 17 (14–20) g / 0.6 (0.5–0.7) ounce

Females average larger than males.

DISTRIBUTION In Alabama, the northern yellow bat is known only from the southernmost counties. Geographic range of the species is from coastal South Carolina southward around the Gulf of Mexico, and southward from the west coast of Mexico to Nicaragua, including Cuba and adjacent islands in the Caribbean.

ECOLOGY Habitats are usually wooded areas near permanent water, but northern yellow bats may be active over residential areas, especially in coastal communities. In the southeastern United States, northern yellow bats inhabit coniferous and deciduous forests, where they roost in clumps of Spanish moss (*Tillandsia usneoides*) hanging from oak trees (*Quercus*). Elsewhere, they may live in palm groves or pine-oak (*Pinus-Quercus*) woodlands, and they have been observed in Mexico hanging from the side of a large, open farm building among dried stalks of corn. In Georgia, the home range of a northern yellow bat in a pine-oak community was about 10 hectares and the foraging area was about 110 m away from

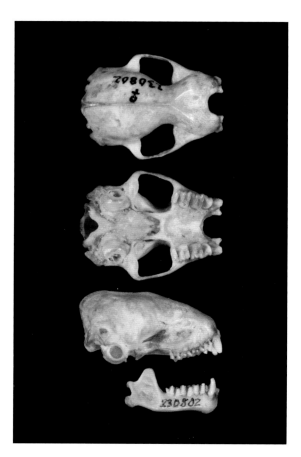

Dorsal, ventral, and lateral views of the cranium, and lateral view of the mandible of a female northern yellow bat. Greatest length of cranium is 18.6 mm.

the roosting area. This home range is much smaller than that of eastern red bats (453 hectares) and Seminole bats (429 hectares) but may not be representative of home ranges of northern yellow bats in general. Diet includes Hemiptera (true bugs), Odonata (damselflies and dragonflies), Diptera (Anthomyiidae—flies, Culicidae—mosquitoes), Coleoptera (Dytiscidae—predaceous diving beetles, Curculionidae—bark beetles), and Hymenoptera (Myrmicinae—ants). Predators include barn owls (*Tyto alba*), and probably snakes, hawks, and other types of owls. Northern yellow bats may also be killed by collisions with wind turbines, television towers, or supporting guy wires. At a television tower in Florida, 8 were killed (1 in March and 7 from August to October).

BEHAVIOR Northern yellow bats are believed to be solitary, but a day roost in eastern Mexico contained about 45 individuals (adult males, adult females, and young-of-the-year). They typically begin foraging well before dark. Northern yellow bats are high flying and forage 4–6 m above the ground. They seem to prefer foraging over open areas including pasturelands, golf courses, and airports. In Florida, they have been observed foraging in city parks and over berms and dunes in beachfront habitats. After young begin to fly, northern yellow bats form aggregations of up to 100 individuals, which are mostly females. They emerge from roosts and forage on warm nights throughout winter. However, when exposed to cool temperatures, northern yellow bats may become torpid.

LIFE HISTORY Breeding occurs in autumn and winter. Sperm are stored until spring. Gestation is unknown but is probably about 20–25 days after fertilization of the ovum. Young are born in May or June. Litters average 2–4 young (range is 1–4). Newborns weigh about 3 g and forearm length is about 16 mm. Young are flying by late June to early July. Life span is unknown but may be about 10–12 years.

PARASITES AND DISEASES Ectoparasites include mites (*Steatonyssus*). Endoparasites include trematodes (*Lecithodendrium*). Northern yellow bats have tested positive for rabies (*Lyssavirus*) elsewhere, but neither of the 2 tested from Alabama had evidence of the virus.

CONSERVATION STATUS High conservation concern in Alabama.

COMMENTS *Lasiurus* is from the Greek *lasios*, meaning "hairy," and *oura*, meaning "tail," a reference to the furred uropatagium; *intermedius* is Latin for "between" or "middle."

REFERENCES Rageot (1955), Baker and Dickerman (1956), La Val (1967), Linzey and Linzey (1969), Webster et al. (1980), Crawford and Baker (1981), Forrester (1992), M. A. Menzel et al. (1999), J. M. Menzel et al. (2003), M. A. Menzel et al. (2003), Best (2004a), Hester et al. (2007).

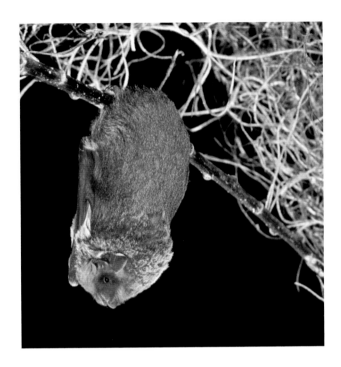

Seminole Bat

Lasiurus seminolus

IDENTIFICATION This bat has a dark mahogany pelage with white-tipped hairs that give it a frosted appearance. The underside is paler than the back. There are patches of whitish hairs on the thumbs, neck, and shoulders. The tragus tapers to a round tip that turns forward. The uropatagium is fully furred.

DENTAL FORMULA i 1/3, c 1/1, p 2/2, m 3/3, total = 32.

SIZE AND WEIGHT Average size of 12 males and 12 females, respectively:
    total length, 98, 104 mm / 3.9, 4.2 inches
    tail length, 40, 46 mm / 1.6, 1.8 inches
    foot length, 8, 8 mm / 0.3, 0.3 inch
    ear length, 9, 11 mm / 0.4, 0.4 inch
    forearm length, 40, 41 mm / 1.6, 1.6 inches
    weight, about 9 g, about 11 g / 0.3, 0.4 ounce
Females average larger than males.

Distribution of the Seminole bat in Alabama and North America.

DISTRIBUTION Statewide in Alabama. Geographic range of the species is from eastern Texas to the Atlantic Ocean and from eastern Virginia southward through Florida. There are isolated records from eastern Mexico and from as far north as New York.

ECOLOGY General distribution corresponds to the distribution of Spanish moss (*Tillandsia usneoides*), in which Seminole bats may roost. Seminole bats occur in forested habitats that frequently contain blackgums (*Nyssa sylvatica*), cypresses (*Taxodium*), hickories (*Carya*), maples (*Acer*), loblolly bays (*Gordonia lasianthus*), red bays (*Persea borbonia*), sweet bays (*Magnolia virginiana*), dogwoods (*Cornus*), loblolly pines (*Pinus taeda*), slash pines (*P. elliottii*), longleaf pines (*P. palustris*), water oaks (*Quercus niger*), and live oaks (*Q. virginiana*). Roosts are often in lower parts of the canopy, where Seminole bats hang from small limbs in or near clumps of pine needles at the tips of branches. The bats are nearly indistinguishable from pine cones when observed from the ground. Foraging areas are over and within these woodlands, especially pine forests, as well as over streams, ponds, grassy pastures, swamps, and marshes. Diet may be mostly beetles but also includes leafhoppers; long-legged, green, and house flies (Dolichopodidae, Muscidae, Tabanidae); damselflies; and flying ants. In addition to consuming a variety of night-flying insects, Seminole bats may glean prey such as flightless crickets (*Gryllus assimilis*) from leaves or branches. One was observed feeding on insects attracted to the flower spikes of a cabbage palm (*Sabal palmetto*). It continuously circled through the habitat, but each time it passed the palm, it landed for a few seconds on the horizontal frond and quickly captured an insect.

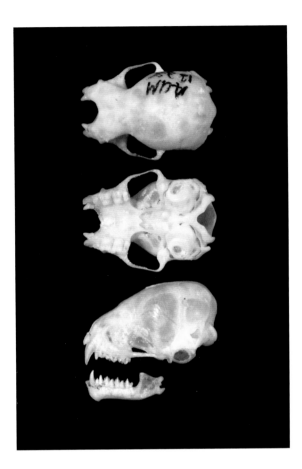

Dorsal, ventral, and lateral views of the cranium, and lateral view of the mandible of a male Seminole bat. Greatest length of cranium is 14.0 mm.

Predators include blue jays (*Cyanocitta cristata*) and probably snakes and owls. In addition, commercial harvesting of Spanish moss may be a significant cause of death in Seminole bats.

BEHAVIOR Although Seminole bats have been captured outside mines and caves in autumn, this species rarely, if ever, enters caves or mines. In summer, Seminole bats frequently select roosting sites in Spanish moss, but they also roost beneath loose bark and in clumps of foliage within the canopies of trees, especially pines, where they hang from leaves or small branches. In winter, roosts include the canopies of hardwood trees, hanging vines, clusters of pine needles suspended from understory vegetation, and leaf litter on the forest floor. Spanish moss seems to be used for roosting in cooler months, and pine trees appear to be used most often as roosts in warmer months, which is the time of parturition and lactation. Roosts average 16.3 m above the ground, and Seminole bats occupy them for about 1–2 days before moving to the next roosting site. Apparently, they do not enter deep hibernation that lasts through the winter but arouse and forage, especially in the southern part of their range, during warm evenings in winter. In cold weather, Seminole bats may go into torpor, and they usually do not fly when the temperature is below 18°C (64°F).

LIFE HISTORY Almost nothing is known about the life history of Seminole bats. They almost certainly mate in autumn or winter, females probably store sperm through the winter, and fertilization likely occurs in spring. Young are born in May or June. Litters average 3–4 young (range is 1–4). Life span is unknown but is probably at least 6–8 years.

PARASITES AND DISEASES No parasite is known from this species, but 23 of 201 (11.4%) Seminole bats tested positive for rabies (*Lyssavirus*) in Alabama. Data on the incidence of rabies in bats in Alabama were obtained from bats that may have been sick or dead when encountered and submitted for testing. Usually, the incidence of rabies in bats is less than 2%.

CONSERVATION STATUS Lowest conservation concern in Alabama.

COMMENTS *Lasiurus* is from the Greek *lasios*, meaning "hairy," and *oura*, meaning "tail," a reference to the furred uropatagium; *seminolus* refers to the Seminole Indians, who lived in the region where the type specimen was obtained.

REFERENCES Wilkins (1987), M. A. Menzel et al. (1998), M. A. Menzel et al. (1999), Laerm et al. (1999), Best (2004*a*), Hein et al. (2005), Hester et al. (2007).

Evening Bat
Nycticeius humeralis

IDENTIFICATION The hair on the back of this bat is dark brown with grayish tips, and that on the underside is paler brown. Forearm length is about 36 mm. The single upper incisor of evening bats distinguishes them from big brown bats; also, the forearm length of big brown bats is about 45 mm. The tragus is short, broad, blunt, and bent slightly forward, with a distinct lobule on the posterior base.

DENTAL FORMULA i 1/3, c 1/1, p 1/2, m 3/3, total = 30.

SIZE AND WEIGHT Average and range in size of 28 specimens from Alabama:

> total length, 85 (54–102) mm / 3.4 (2.2–4.1) inches
> tail length, 30 (24–43) mm / 1.2 (1.0–1.7) inches
> foot length, 9 (6–14) mm / 0.4 (0.2–0.6) inch

Distribution of the evening bat in Alabama and North America.

ear length of 18 specimens, 12 (11–14) mm / 0.5 (0.4–0.6) inch
weight of 11 specimens, 9.7 (6.9–14.5) g / 0.3 (0.2–0.5) ounce
Females are larger than males.

DISTRIBUTION Statewide in Alabama. Geographic range of the species is from southern Michigan into northeastern Mexico and from western Kansas to the Atlantic Ocean.

ECOLOGY Evening bats occur in urban, suburban, and rural habitats, and they forage most often over agricultural fields, deciduous forests, marshes, and swamps. They almost never enter caves. Roosts are in buildings, under bark, or in cavities of trees, including pines (*Pinus*), maples (*Acer*), hickories (*Carya*), ashes (*Fraxinus*), oaks (*Quercus*), and cypresses (*Taxodium distichum*). Average height of roosts in pine-oak forests is about 8 m, and roosts are usually above an open or grassy understory. In Alabama, colonies are frequently present in attics of houses with openings that allow access by these bats; these colonies may also be occupied by big brown bats or Brazilian free-tailed bats. Diet includes beetles (Carabidae, Chrysomelidae, Scarabaeidae), flies (Drosophilidae), ants (Formicidae), froghoppers (Hemiptera), plant hoppers (Delphacidae), squash bugs (Coreidae), moths, stinkbugs (Pentatomidae), chinch bugs (Lygaeidae), and leafhoppers (Cicadellidae). Predators include black rat snakes (*Elaphe obsoleta*), raccoons, and domestic cats.

BEHAVIOR There is considerable evidence that these bats are migratory: sexes occur together in winter, few males are known in the northern parts

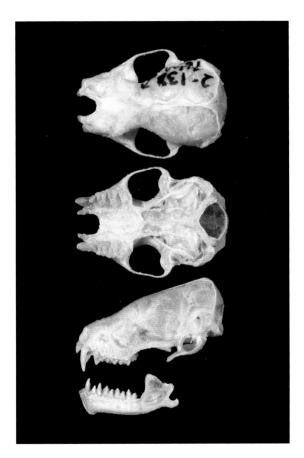

Dorsal, ventral, and lateral views of the cranium, and lateral view of the mandible of a male evening bat. Greatest length of cranium is 14.8 mm

of the geographic range in summer, there are no evening bats in northern areas in winter, northern roosting sites are abandoned by mid-October, and these bats have been recovered 200–550 km south of where they were captured, banded, and released. Colonies in attics of buildings are exposed to a wide range of temperatures; when temperatures are high, spacing is rather even over the ceiling and walls, but when temperatures are low, evening bats form tight clusters. Males and females do not mingle during the maternity season. Young are born in maternity colonies of 2–950 individuals; the average is about 50. Just before giving birth, females move away from clusters of other bats. The mother engages in considerable grooming and cleaning of the newborn and encloses the newborn within her wing and interfemoral membranes. Shortly after birth, the young move to nipples, often aided by the mother, and they become firmly attached. Maternity colonies may be in Spanish moss (*Tillandsia usneoides*), buildings, or cavities. When young are in roosts, mothers may return from foraging periodically to care for their babies. Average foraging time away from the roost is about 90 minutes. There are no stops during this time for night roosting, and foraging ranges are about 2.5–4.5 km for females, regardless of reproductive condition.

LIFE HISTORY Breeding season is unknown but is probably in autumn, with overwinter storage of sperm and fertilization in spring. Young are born from mid-May to early July by breech presentation (tail end exits the birth canal first). Average litters contain 2 young (range is 1–4). All deciduous dentition is present at birth. Newborns are pink and naked, with only a few scattered hairs. Their eyes and ears are closed, but by 12–30 hours after birth, eyes and ears open. Newborns are extremely vocal when separated from their mother. By day 5, short gray fur has appeared on the back, and by about day 20, young can fly. Life span is probably 6–8 years, but this has not been documented.

PARASITES AND DISEASES Ectoparasites include mites (*Acanthophthirius, Androlaelaps, Macronyssus, Steatonyssus*) and bat bugs (*Cimex*). Endoparasites include cestodes (*Vampirolepis*), nematodes (*Allintoshius, Capillaria*), and trematodes (*Acanthatrium, Allassogonoporus, Dicrocoelium, Ochoterenatrema, Paralecithodendrium, Urotrema*). In Alabama, incidence of rabies (*Lyssavirus*) is low; 2 of 175 (1.1%) individuals tested positive for this viral disease.

CONSERVATION STATUS Lowest conservation concern in Alabama.

COMMENTS *Nycticeius* is from the Greek *nykteus*, meaning "nocturnal"; *humeralis* is from the Latin *humerale*, meaning "a cape for the shoulders."

REFERENCES Watkins (1972), Lotz and Font (1991), Whitaker and Clem (1992), M. A. Menzel et al. (1999), Best (2004*a*), Duchamp et al. (2004), Hester et al. (2007), Whitaker et al. (2007).

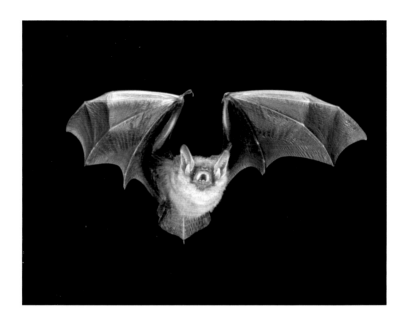

Tri-colored Bat
Perimyotis subflavus

IDENTIFICATION This small bat has distinctly tri-colored hairs that are dark at the base, paler and yellowish-brown in the middle, and dark at the tip. The back varies from pale yellowish orange to dark reddish brown. The underside is pale brownish. The tragus is rounded and the forearm is pinkish. There is no keel on the calcar.

DENTAL FORMULA i 2/3, c 1/1, p 2/2, m 3/3, total = 34.

SIZE AND WEIGHT Average and range in size of 35 specimens from Alabama:

    total length, 79 (72–89) mm / 3.2 (2.9–3.6) inches
    tail length, 37 (30–41) mm / 1.5 (1.2–1.6) inches
    foot length, 8 (6–10) mm / 0.3 (0.2–0.4) inch
    weight, 4.2 (3.2–5.5) g / 0.2 (0.1–0.2) ounce
    wingspan, 236 (220–250) mm / 9.4 (8.8–10.0) inches

Females are larger than males; perhaps their larger size is related to the high energetic costs of lactation and the production of 2 large embryos.

Distribution of the tri-colored bat in Alabama and North America.

DISTRIBUTION Statewide in Alabama. Geographic range of the species is from southeastern Canada into Central America and from eastern Wyoming to the Atlantic Ocean.

ECOLOGY In Canada and the United States, habitats occupied by tri-colored bats include deciduous and mixed pine-oak-hickory-maple (*Pinus-Quercus-Carya-Acer*) forests, marshes, swamps, and scrublands, and in Mexico and Central America, they inhabit tropical lowlands and mountainous regions. In Georgia, roosts were in the understory about 4 m above the ground in Spanish moss (*Tillandsia usneoides*), eastern red cedars (*Juniperus virginianus*), live oaks (*Quercus virginiana*), yaupons (*Ilex vomitoria*), sparkleberries (*Vaccinium arboreum*), red bays (*Persea borbonia*), and slash pines (*Pinus elliottii*). In Alabama, these bats occur in all habitats. Tri-colored bats forage over waterways and edges of forests and may travel more than 4 km from roost to foraging site. Diet includes primarily Coleoptera (Carabidae—beetles), Hemiptera (Cicadellidae—leafhoppers), Diptera (Culicidae—mosquitoes, Chironomidae—midges), Hymenoptera (Formicidae—ants), and Lepidoptera (moths). Documented predators include a prairie vole, a hoary bat, and a leopard frog (*Rana pipiens*), but other predators probably include raccoons and a variety of snakes and owls.

BEHAVIOR Tri-colored bats are among the first to appear in the evening and they have a slow, erratic, fluttery flight when foraging. On average, they catch a 4–10-mm-long food item about every 2 seconds. Even in

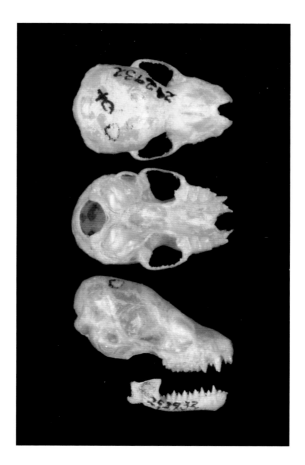

Dorsal, ventral, and lateral views of the cranium, and lateral view of the mandible of a female tri-colored bat. Greatest length of cranium is 12.9 mm.

warmer parts of their range, tri-colored bats are obligate hibernators. They are among the first bats to enter hibernation in autumn and the last to emerge in spring. Both sexes hibernate together in caves, mines, and human-made structures. Tri-colored bats usually roost singly during hibernation. Following hibernation, sexes separate. Males become solitary and often roost in trees. Females may migrate up to 50 km from hibernacula to form small maternity colonies of 2–3 adults. Pregnant and lactating females often occupy clusters of dead leaves in terminal branches of deciduous trees (especially oaks and maples), dense live foliage, or occasionally, nests of eastern gray squirrels. These roosting sites are well below the forest canopy and protect tri-colored bats from predators and provide shelter from wind and rain. Most females move about 60 m (range is 19–139 m) to new roost trees every 3–5 days, but some in late pregnancy or lactation use the same roost tree for more than 2 weeks.

LIFE HISTORY Mating occurs in autumn and possibly in spring. Females store sperm and fertilization occurs in spring. Gestation period from implantation to parturition is about 44 days. Two young are born from late May to early July depending on latitude. At birth, young may weigh 33–52% as much as their mother. Newborns are pink and hairless, their eyes and ears are closed, and they can make loud clicks that may aid their mother in retrieving them. By 3 weeks old, young begin to fly, and by 4 weeks, they achieve flying and foraging abilities comparable to adults. In southern latitudes, the epiphyses of long bones fuse before hibernation, but in northern latitudes, young may enter hibernation before epiphyses close, making them identifiable as young-of-the-year during and immediately following hibernation. Females may reach sexual maturity during the year of birth, but first reproduction usually occurs the following spring at an age of 10–12 months. Average maximum life span may be 8–10 years, but some have lived nearly 15 years in the wild.

PARASITES AND DISEASES Ectoparasites include mites (*Acanthophthirius*, *Euschoengastia*, *Macronyssus*, *Neomyobia*, *Perissopalla*, *Pteracarus*, *Steatonyssus*). Endoparasites include protozoans (*Eimeria*), cestodes, nematodes (*Capillaria*), and trematodes (*Acanthatrium*, *Allassogonoporus*, *Limatulum*, *Ochoterenatrema*, *Paralecithodendrium*, *Prosthodendrium*, *Plagiorchis*, *Urotrema*). Generally, the incidence of rabies (*Lyssavirus*) is low in this species, but in Alabama, 12 of 55 (22%) tri-colored bats tested positive for rabies. Because this species hibernates in the most humid parts of caves, it is susceptible to infection with white-nose syndrome (*Geomyces destructans*).

CONSERVATION STATUS Lowest conservation concern in Alabama.

COMMENTS *Perimyotis* is from the Greek *peri*, meaning "near," *mys*, meaning "mouse," and *ōtos*, meaning "ear"; *subflavus* is from the Latin *sub*, meaning "below," and *flavus*, meaning "yellow," referring to the yellowish ventrum.

REFERENCES Fujita and Kunz (1984), Lotz and Font (1991), M. A. Menzel et al. (1999), Hilton and Best (2000), Veilleux et al. (2003), Best (2004*a*), White et al. (2006), Hester et al. (2007), Whitaker et al. (2007), Cryan et al. (2010), Harvey et al. (2011).

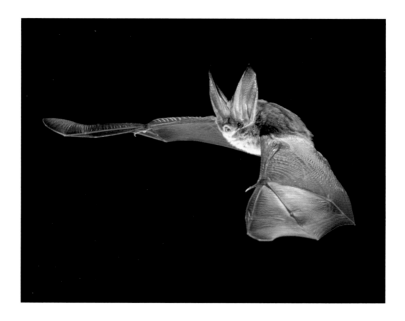

Rafinesque's Big-eared Bat

Corynorhinus rafinesquii

IDENTIFICATION This bat has ears that are more than 2.5 cm long. There are 2 large lumps on the face between the tip of the nose and the ears. Pelage on the back is grayish brown, while hairs on the underside are blackish to black at the base and whitish to white at the tips. Hairs on the feet extend beyond the toenails.

DENTAL FORMULA i 2/3, c 1/1, p 2/3, m 3/3, total = 36.

SIZE AND WEIGHT Range in size:
  total length, 80–110 mm / 3.2–4.4 inches
  tail length, 42–54 mm / 1.7–2.2 inches
  foot length, 8–13 mm / 0.3–0.5 inch
  ear length, 27–37 mm / 1.1–1.5 inches
  forearm length, 39–44 mm / 1.6–1.8 inches
  wingspan, 265–301 mm / 10.6–12.0 inches
  weight of males, 7.9–9.5 g (0.3–0.3 ounce)
  weight of females, 7.9–13.6 g (0.3–0.5 ounce)
There is no significant sexual dimorphism.

This abandoned house in southern Alabama was the summer home of a Rafinesque's big-eared bat.

DISTRIBUTION Statewide, with one subspecies (*C. r. rafinesquii*) northward from about the Tennessee River and the other (*C. r. macrotis*) throughout the remainder of Alabama. Geographic range of the species is from southern Indiana to the Gulf of Mexico and from southeastern Oklahoma to the Atlantic Ocean.

ECOLOGY Rafinesque's big-eared bats occur in nearly every type of forested habitat within their range. Roosts used most often include partially lit caves, abandoned buildings, hand-dug wells, cisterns, grain silos, attics of occupied houses, highway culverts, bridges, and other human-made structures. These bats also roost in caves, trees, and other natural places. Foraging areas are usually 0.1–1.1 km from the roost and are within 1 km of permanent water; they may be in hardwood hammocks, pine (*Pinus*) flatwoods, and ridgelines in oak-hickory (*Quercus-Carya*) forests. Diet is primarily night-flying moths (Lepidoptera—Arctiidae, Geometridae, Megalopygidae, Noctuidae, Notodontidae, and Sphingidae), but other orders of insects also consumed are Coleoptera (beetles), Hemiptera (true bugs), Diptera (flies and midges), Hymenoptera (ants), and Trichoptera (caddisflies). Predators include red corn snakes (*Elaphe guttata*), eastern diamondback rattlesnakes (*Crotalus adamanteus*), cottonmouths (*Agkistrodon piscivorus*), raccoons, Virginia opossums, and

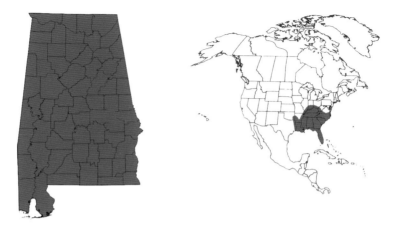

Distribution of the Rafinesque's big-eared bat in Alabama and North America.

domestic cats. Because Rafinesque's big-eared bats inhabit buildings and other human-made structures, they are also susceptible to disturbance by humans. A maternity colony in a deep rock shelter next to a hiking trail in Kentucky was disturbed by flash photography, loud vocalizations, flashlights, and discarded debris, and by people eating inside the shelter, building fires, and urinating at the entrance. Rafinesque's big-eared bats abandoned the roost in response to the increased use and disturbance of the shelter by humans. They have also abandoned roosts in response to logging in adjacent forests.

BEHAVIOR Rafinesque's big-eared bats may be solitary or occur in colonies of up to 100 individuals. During summer, they are active and can take immediate flight. If disturbed, Rafinesque's big-eared bats turn their head and begin to move their ears as if they are surveying the threat. In winter, and sometimes in summer, they coil their ears alongside their head; coiled ears usually lie between the head and folded wing. When disturbed in this position, during rest or torpor, Rafinesque's big-eared bats may take several minutes to arouse and erect their ears. They emerge to forage after twilight and return to the roost before daylight. Flight is agile and varies from fast to nearly hovering. There is no evidence that mothers take their young with them to forage, but females will carry their young away from roosts when they perceive danger. Rafinesque's big-eared bats frequently move among sites within roosts during summer and winter. However, they tend to occupy specific places within roosts. In Baldwin and Clarke counties, Rafinesque's big-eared bats were observed to return to exactly the same sites within roosts in 2 abandoned houses.

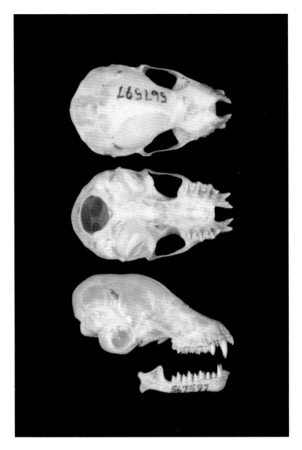

Dorsal, ventral, and lateral views of the cranium, and lateral view of the mandible of a male Rafinesque's big-eared bat. Greatest length of cranium is 15.7 mm.

LIFE HISTORY Mating takes place in autumn and winter. Length of gestation is unknown. From late May to early June, 1 young is born. Newborns are hairless and weigh 2.3–2.6 g. At 15–18 days of age, young are capable of straight-line, nonagile flight, usually landing with their head up. At about 21 days old, young have permanent dentition and they are able to fly very well. By 30 days of age, young are the same size as adults. Life span may be 8–10 years in the wild.

PARASITES AND DISEASES Ectoparasites include mites (*Chiroptoglyphus, Macronyssus, Teinocoptes*). Endoparasites include cestodes (*Vampirolepis*) and nematodes (*Capillaria, Physaloptera*). Neither of the 2 Rafinesque's big-eared bats tested for rabies in Alabama was rabid.

CONSERVATION STATUS Highest conservation concern in Alabama.

COMMENTS *Corynorhinus* is from the Greek *korynē*, meaning "a club-shaped bud," and *rhinos*, meaning "nose"; *rafinesquii* is a patronym honoring Constantine S. Rafinesque (1783–1840), an American naturalist. The subspecific name *macrotis* is from the Greek *makros*, meaning "long," and *ōtos*, meaning "ear."

REFERENCES George and Strandtmann (1960), Barbour and Davis (1969), Jones (1977), Best et al. (1993), Hurst and Lacki (1997, 1999), Lacki (2000), Lacki and Ladeur (2001), Best (2004*a*), Trousdale and Beckett (2004), McAllister et al. (2005), Hester et al. (2007), Whitaker et al. (2007), Kilgore (2008), Beolens et al. (2009).

Silver-haired Bat

Lasionycteris noctivagans

IDENTIFICATION This bat is easily identifiable by its dark brown to black pelage with white-tipped hairs, which give it a frosted appearance. Wings are black. Ears are rounded and black, and the tragus is broad and blunt. Some old individuals may not have white-tipped hairs in the pelage.

DENTAL FORMULA i 2/3, c 1/1, p 2/3, m 3/3, total = 36.

SIZE AND WEIGHT Average and range in size of 12 males and 9 females, respectively, from Alabama:

> total length, 101 (96–106), 103 (97–108) mm / 4.0 (3.8–4.2), 4.1 (3.9–4.3) inches
>
> tail length, 39 (34–45), 39 (35–42) mm / 1.6 (1.4–1.8), 1.6 (1.4–1.7) inches

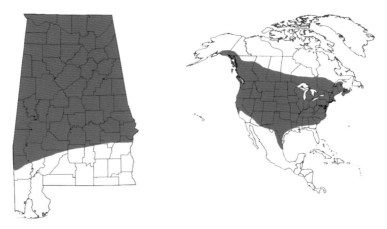

Distribution of the silver-haired bat in Alabama and North America.

foot length, 9 (8–10), 9 (9–11) mm / 0.4 (0.3–0.4), 0.4 (0.4–0.4) inch

ear length, 13 (12–19), 14 (13–15) mm / 0.5 (0.5–0.8), 0.6 (0.5–0.6) inch

forearm length, 42 (39–44), 42 (40–44) mm / 1.7 (1.6–1.8), 1.7 (1.6–1.8) inches

weight, 12.6 (12.0–13.5), 13.2 (12.0–14.5) g / 0.4 (0.4–0.5), 0.5 (0.4–0.5) ounce

There is no sexual dimorphism.

DISTRIBUTION Northern Alabama during cooler months. Geographic range of the species is from southeastern Alaska to northeastern Mexico and from the Pacific Ocean to the Atlantic Ocean across the United States and southern Canada.

ECOLOGY In Alabama, silver-haired bats are present only in cooler months (autumn–spring). During October (in Jackson County), they fly along streams and over ponds in forested areas, where up to 39 individuals have been captured in mist nets, examined, and released in a single night. In winter, silver-haired bats roost in caves, mines, hollow trees, under loose bark, in rocky crevices, and in buildings. In spring (April–June), silver-haired bats migrate to summer habitats in Canada and in the northern and western United States. In summer, they usually roost in folds of bark, in narrow crevices in trees, and in narrow spaces formed between 2 touching trunks. Silver-haired bats are well concealed when they wedge themselves into these crevices so tightly that their back and

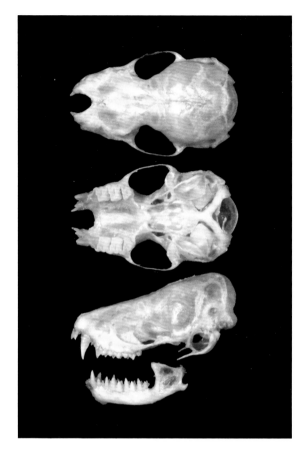

Dorsal, ventral, and lateral views of the cranium, and lateral view of the mandible of a female silver-haired bat. Greatest length of cranium is 16.3 mm.

underside are in contact with the substrate. Roosts in crevices provide shelter from rain and wind, and probably protection from predators. Typically, silver-haired bats forage in or near mixed coniferous or deciduous forests that are adjacent to ponds, streams, and other bodies of water. Silver-haired bats are opportunistic foragers. They consume spiders and insects including Lepidoptera (moths), Hemiptera (true bugs; Corixidae, Cicadellidae), Coleoptera (beetles; Carabidae, Scarabaeidae), Diptera (flies and midges), Trichoptera (caddisflies), and Isoptera (termites). Predators include great horned owls (*Bubo virginianus*) and striped skunks; a silver-haired bat was attacked in flight by a rabid hoary bat.

BEHAVIOR Silver-haired bats are generally solitary, but some roost in pairs and in groups of 3–6 individuals. Even when roosting in contact with other bats, they are rarely packed close together. Maternity colonies of 3–11 adult females and their young are formed in summer; 1 colony was in an abandoned nesting cavity used by common flickers (*Colaptes*) and was about 5 m above the ground in the dead section of a tree. Silver-haired bats may not leave the roost for 3–4 days during cold weather, and they reduce the duration of their foraging activity when it is raining and during times of bright moonlight. When these bats roost, their temperature varies with that of the environment; thus, they become torpid when the temperature is low. When torpid, they are sluggish, cold to the touch, and unable to fly. When silver-haired bats emerge from their roost, they often fly directly to water. They are rather slow fliers, foraging along erratic courses and taking many twists and short glides. Silver-haired bats may migrate long distances in spring and autumn. A study of silver-haired bats in Alabama revealed that this species arrives during autumn after a long migration, possibly from the western United States or from southeastern Canada.

LIFE HISTORY Breeding is probably in autumn, sperm are stored over winter, and fertilization occurs in spring. Gestation is 50–60 days. Usually, 2 young are born from May to July. Young are born tail end first, they are pink and hairless, their eyes and ears are closed, and they weigh about 1.8–1.9 g. When exposed to a threat, mothers may carry both of their young as they fly away. At about 36 days old, young are weaned. Both males and females reach sexual maturity during their first summer, and females give birth to their first litter the following spring. Life span may be 4–6 years, but silver-haired bats have lived to 12 years in the wild.

PARASITES AND DISEASES Ectoparasites include mites (*Acanthophthirius, Cryptonyssus, Leptotrombidium, Macronyssus*), fleas, bat flies (*Trichobius*), and bat bugs (*Cimex*). Endoparasites include cestodes (*Hymenolepis*), nematodes (*Capillaria*), and trematodes (*Acanthatrium, Plagiorchis, Prosthodendrium, Urotrema, Urotrematulum*). Diseases include rabies (*Lyssavirus*), but none of the 6 silver-haired bats that were tested in Alabama had evidence of the virus.

CONSERVATION STATUS Moderate conservation concern in Alabama.

COMMENTS *Lasionycteris* is from the Greek *lasios*, meaning "hairy," and *nykteris*, meaning "bat"; *noctivagans* is from the Latin *noctis*, meaning "night," and *vagans*, meaning "wandering."

REFERENCES Macy (1933), Sperry (1933), Blankespoor and Ulmer (1970), Webster and Casey (1973), Turner (1974), Rausch (1975), Kunz (1982), Parsons et al. (1986), Barclay et al. (1988), Best (2004*a*), McAllister et al. (2005), Hester et al. (2007), Whitaker et al. (2007), Hirt (2008), Whitaker and Mumford (2009).

Southeastern Myotis

Myotis austroriparius

IDENTIFICATION This bat has a long, pointed tragus, a pelage with a woolly appearance, hairs that are nearly uniform in color from base to tip, toes with long hairs that extend beyond the toenails, and a calcar with no keel. Pelage on the back varies from bright cinnamon brown to pale brown. Underside of the body is grayish.

DENTAL FORMULA i 2/3, c 1/1, p 3/3, m 3/3, total = 38.

SIZE AND WEIGHT Average and range in size of 18 males and 29 females, respectively:

> total length, 84 (77–89), 87 (80–97) mm / 3.4 (3.1–3.6), 3.5 (3.2–3.9) inches
>
> tail length, 37 (26–44), 38 (29–42) mm / 1.5 (1.0–1.8), 1.5 (1.2–1.7) inches
>
> foot length, 10 (7–11), 10 (8–12) mm / 0.4 (0.3–0.4), 0.4 (0.3–0.5) inch
>
> ear length, 12 (11–16), 13 (9–15) mm / 0.5 (0.4–0.6), 0.5 (0.4–0.6) inch

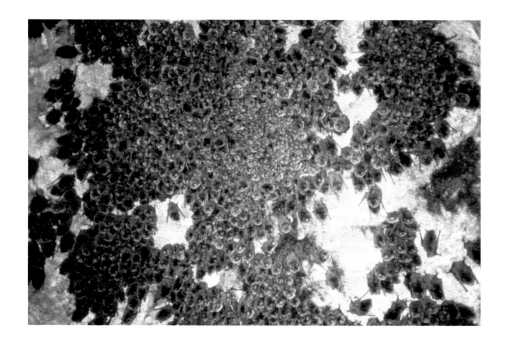

forearm length, 36 (33–39), 39 (34–40) mm / 1.4 (1.3–1.6), 1.6
   (1.4–1.6) inches
weight of 9 males, 5.9 (5.1–6.8) g / 0.2 (0.2–0.2) ounce
weight of 19 females, 6.9 (5.2–8.1) g / 0.2 (0.2–0.3) ounce
Females are larger than males.

Only one maternity colony of the southeastern myotis is known in Alabama.

DISTRIBUTION Statewide in Alabama. Geographic range of the species is from southern Illinois southward to the Gulf of Mexico and from southeastern Oklahoma to the Atlantic Ocean.

ECOLOGY In summer, roosts are usually in caves or mines, but they may be in buildings and hollow trees. In winter, southeastern myotis may occupy caves, hollow dead trees, crevices between wooden bridge timbers, storm sewers, culverts beneath roadways, vertical drainpipes of concrete railroad bridges, and boat houses. Diet is night-flying insects. Predators include rat snakes (*Elaphe obsoleta*), owls, raccoons, and Virginia opossums; cockroaches (*Periplaneta*) prey on newborns. By destroying roost sites and killing bats, humans are the major threat to the southeastern myotis. Humans have vandalized the only maternity roost in Alabama by shooting and by throwing mud-balls at the bats.

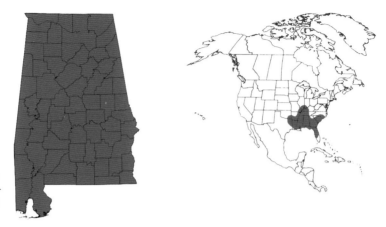

Distribution of the southeastern myotis in Alabama and North America.

BEHAVIOR In spring, the southeastern myotis forms large maternity colonies and small bachelor colonies. During autumn, the only maternity colony in Alabama (Conecuh County) abandons its cave and disperses in small groups to small caves or elsewhere to spend the winter. In Alabama, the southeastern myotis actively forages during winter and does not hibernate. These bats become torpid when temperatures are near freezing and become active again when the weather warms. When torpid, they are easily aroused. In northern parts of the range, the southeastern myotis hibernates in caves or under buildings for up to 7 months (September–March). While hibernating, it may hang from the ceiling and walls of caves in clusters of up to 50 individuals.

LIFE HISTORY Breeding season is unknown but is probably in autumn or early spring, depending on latitude. Usually, 2 young (range is 1–3) are born during late April and early May. During parturition, females hang from the ceiling by their widely spaced feet and thumbs. Young are born tail end first. When birth is about 25% complete, the mother may tear the membrane that covers the young, allowing it to pop out. Her curved tail and slightly spread wings form a pocket for the young as it is born. The newborn crawls about in the pocket as far as its umbilical cord allows. Young are born 10–15 minutes apart. By the time the second is born, the first is attached to a nipple. The mother licks and cleans both young and pushes the second toward the unoccupied nipple. About 1–2 hours after parturition, the mother resumes the usual head-down position and hangs only by her feet, a baby tucked under the base of each wing. Newborns

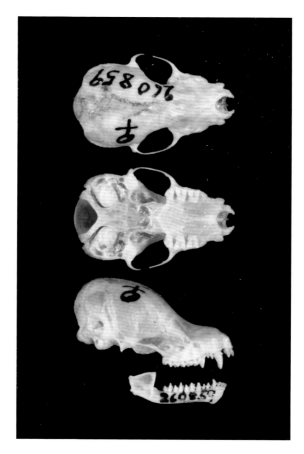

Dorsal, ventral, and lateral views of the cranium, and lateral view of the mandible of a female southeastern myotis. Greatest length of cranium is 14.6 mm.

are nearly hairless and their eyes and ears are closed. Weight is 1.1–1.2 g, length of head and body is 25–26 mm, tail length is 11.5–13.0 mm, foot length is 8.1–8.9 mm, forearm length is 9.8–11.2 mm, and all deciduous teeth are erupted. Young remain attached to the mother for the first day but only occasionally afterward. By 2–7 days, eyes and ears are open. When females leave the roost to forage, they do not take their young with them. Older babies form clusters by themselves when their mothers are away. At 5–6 weeks, young can fly. They attain sexual maturity in the spring following their birth. Life span is unknown but is probably 10–12 years.

PARASITES AND DISEASES Ectoparasites include mites (*Euschoengastia, Ichoronyssus, Macronyssus, Olabidocarpus, Spinturnix*), ticks (*Ornithodoros*), fleas (*Sternopsylla*), and flies (*Basilia, Trichobius*). Endoparasites include protozoans (*Polychromophilus*), cestodes, nematodes, and trematodes. Diseases include histoplasmosis (*Histoplasma*) and rabies (*Lyssavirus*). As in most other bats, the incidence of rabies is low. Of the 6 southeastern myotis tested for rabies in Alabama, none was rabid.

CONSERVATION STATUS High conservation concern in Alabama.

COMMENTS *Myotis* is from the Greek *mys*, meaning "mouse," and *ōtos*, meaning "ear"; *austroriparius* is from the Latin *austri*, meaning "of the south wind," and *riparius*, which means "frequenting or belonging to a stream bank."

REFERENCES Sherman (1930), Rice (1957), Jones and Manning (1989), Forrester (1992), Best et al. (1993), Hilton and Best (2000), Best (2004*a*), Hester et al. (2007), Whitaker et al. (2007).

Gray Myotis
Myotis grisescens

IDENTIFICATION This bat has a long, pointed tragus, hair that is uniformly gray from base to tip, a wing membrane that is attached to the ankle rather than to the base of the toe, a notch on the claws of the feet, and a calcar with no keel. Pelage on the back is brownish to grayish brown. Underside of the body is pale gray.

DENTAL FORMULA i 2/3, c 1/1, p 3/3, m 3/3, total = 38.

SIZE AND WEIGHT Average size of males and females, respectively:
head and body length, 49, 50 mm / 2.0, 2.0 inches
tail length, 38, 40 mm / 1.5, 1.6 inches
foot length, 10, 10 mm / 0.4, 0.4 inch
ear length, 14, 14 mm / 0.6, 0.6 inch
forearm length, 43, 44 mm / 1.7, 1.8 inches
weight of males in Alabama, 7.9–9.1 g (0.3–0.3 ounce)
weight of females in Alabama, 8.0–13.5 g (0.3–0.5 ounce).
Females are slightly larger than males.

Steel bars and a chute-style flyway were installed at the opening of a cave that contains a large colony of gray myotis in northern Alabama. These structures protect endangered bats from disturbance and allow bats to easily enter and exit the cave.

DISTRIBUTION Nearly statewide in Alabama, with largest populations near the Tennessee River. Geographic range of the species is from northern Illinois to the panhandle of Florida and from northeastern Oklahoma to the eastern side of the Appalachian Mountains.

ECOLOGY Gray myotis usually live in caves during summer and winter. Habitat around the caves is usually a mosaic of pasturelands, agricultural croplands, and forested and riparian areas closely associated with open water. Caves occupied in summer are usually not occupied in winter. The 2 largest summer colonies of gray myotis occur in Marshall and Jackson counties, Alabama. The largest winter hibernaculum of gray myotis is in Jackson County. Rarely, gray myotis may live in other structures, such as

A sign informs visitors that endangered gray myotis roost in this cave and that there are significant penalties for molesting these bats.

storm sewers, barns, or dams. Gray myotis forage in open, riparian, and forested areas and over open water. In northern Alabama, there are several colonies of gray myotis near impoundments of the Tennessee River and its tributaries. Gray myotis that live in caves distant from open water may travel over land but often follow streams as they fly back and forth between roosts and foraging sites. Diet in northern Alabama includes a variety of night-flying insects such as caddisflies (Trichoptera), bugs (Hemiptera), leafhoppers (Homoptera), wasps (Hymenoptera), mayflies (Ephemeroptera), and lacewings (Neuroptera), but moths (Lepidoptera), flies and midges (Diptera), and beetles (Coleoptera) are the most common items consumed. Predators include screech owls (*Otus asio*), black rat snakes (*Elaphe obsoleta*), raccoons, and Virginia opossums.

BEHAVIOR Gray myotis hibernate in fewer than 10 caves, thus making them extremely vulnerable to harm from vandals or natural disasters. Migration averages about 200 km, but some gray myotis move 775 km from summer roosts in Florida to hibernacula in Tennessee. Within roosts, they form dense clusters (average is 1,828 bats/m², range is 999–2,557/m²), depending on temperature and the roughness of the walls and ceiling in the cave. In summer, gray myotis may move among 5–6 roosts,

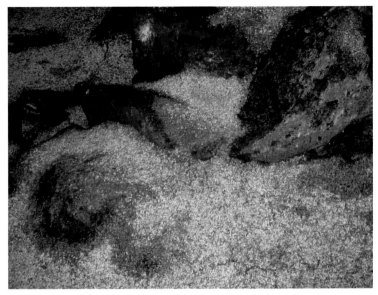

Beneath this roost of gray myotis is a large deposit of guano. The whitish color is from fungus growing on the guano in winter. Guano was used in manufacturing gunpowder during the Civil War and it is a valuable fertilizer in many parts of the world today. In the early 1900s, it was mined in New Mexico and transported by rail to California for application on agricultural fields.

forming maternity colonies of females and their young and bachelor colonies of males and nonreproductive females. Females with young usually return to the same roost until young can fly, but they move among roosts thereafter. After young begin to fly, there is significant intermingling of sexes and ages among roosts. Gray myotis are especially sensitive to disturbance during parturition and lactation. If disturbed by humans or other intruders, young may be dropped or dislodged from the ceiling by other young or fleeing mothers. Young are not retrieved, and they die on the floor of the cave. Gray myotis usually forage over water and in adjacent riparian areas up to 70 km from their roost. Minimum home range is about 100 km². In summer, they emerge from caves shortly after sunset, often forming a stream of bats from the opening of the cave to, or over, open-water and riparian foraging areas. Gray myotis frequently forage in groups and the same individuals tend to return to the same sites each night, even at about the same time each night. Gray myotis are especially sensitive to artificial lights; beams from flashlights or spotlights should not be directed toward caves as bats are emerging or entering.

LIFE HISTORY Breeding season is primarily in late summer and autumn, but copulation also occurs in winter and spring. Sperm are stored until spring and fertilization occurs at about the time females leave hibernac-

Distribution of
the gray myotis
in Alabama and
North America.

ula. Implantation of the embryo is usually in the right uterine horn. A single young is born each year during May or June. Newborns are hairless, their eyes are closed, and their average weight is 2.9 g (range is 2.4–3.4 g). Depending on the size of the colony, young begin to fly at 24–33 days old; bats from larger colonies begin to fly sooner than those from smaller colonies. By autumn, subadults are indistinguishable from adults. They attain sexual maturity in their first year of life. Average life span is probably 9–10 years, but gray bats have lived more than 16 years in the wild.

PARASITES AND DISEASES Ectoparasites include mites (*Macronyssus, Olabidocarpus, Spinturnix, Trombicula*), fleas (*Myodopsylla*), and bat flies (*Trichobius*). Endoparasites include cestodes (*Vampirolepis*), nematodes (*Allintoshius, Capillaria, Trichuroides*), and trematodes (*Allassogonoporus, Limatulum, Plagiorchis, Urotrema*). There is a low incidence of rabies (*Lyssavirus*); 4 of 40 (10%) of these bats tested positive for rabies in Alabama.

CONSERVATION STATUS Highest conservation concern in Alabama, listed as endangered by the United States Fish and Wildlife Service, and listed on the International Union for Conservation of Nature and Natural Resources Red List of Threatened Species as near threatened.

COMMENTS *Myotis* is from the Greek *mys*, meaning "mouse," and *ōtos*, meaning "ear"; *grisescens* is from the Latin *griseus*, meaning "gray."

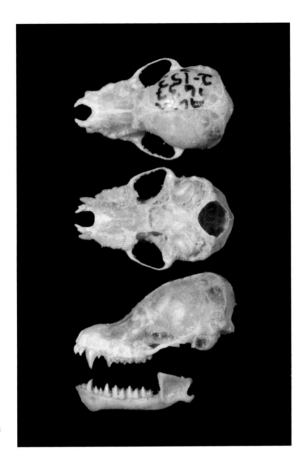

Dorsal, ventral, and lateral views of the cranium, and lateral view of the mandible of a male gray myotis. Greatest length of cranium is 16.2 mm.

REFERENCES Decher and Choate (1995), Best and Hudson (1996), Best et al. (1997), Thomas and Best (2000), Best (2004*a*), McAllister et al. (2005), Hester et al. (2007), Whitaker et al. (2007).

Eastern Small-footed Myotis
Myotis leibii

IDENTIFICATION This small bat has a long, pointed tragus, a keeled calcar, feet that are less than 8 mm long, ears that are less than 15 mm long, forearms that are about 32 mm long, a black facial mask, and black ears. When laid forward, the ears extend slightly beyond the nose. Pelage is dark yellowish brown on the back and pale yellowish brown on the underside, and the wings are nearly black.

DENTAL FORMULA i 2/3, c 1/1, p 3/3, m 3/3, total = 38.

SIZE AND WEIGHT Average size:
> total length, 83 mm / 3.3 inches
> tail length, 36 mm / 1.4 inches
> foot length, 7 mm / 0.3 inch
> forearm length, 32 mm / 1.3 inches
> wingspan, 210–250 mm / 8.4–10.0 inches
> weight, 3.8–5.0 g / 0.1–0.2 ounce (range is 3.2–5.5 g / 0.1–0.2 ounce)

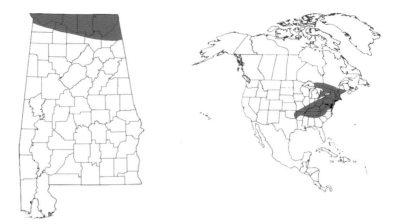

Distribution of the eastern small-footed myotis in Alabama and North America.

DISTRIBUTION Maps depicting the geographic range of this species have frequently included northern Alabama. However, the eastern small-footed myotis has not been verified to occur in the state. Geographic range of the species is from southeastern Canada to eastern Oklahoma and from southern Illinois to northern Georgia.

ECOLOGY The eastern small-footed myotis usually occurs in mountainous areas, but not always. It may occur in caves, buildings, turnpike tunnels, on rocky cliffs, and under rocks. It hibernates in caves and abandoned mines, where it hangs on the wall or from the ceiling or occupies narrow crevices on the floor or under rocks. In Arkansas, eastern small-footed myotis also hibernate in large piles of sandstone rocks that are 8–16 hectares in area and 15 m deep from the top down to the bedrock. The insulating effect of these massive structures and the numerous openings that lead to chambers with suitable temperature and protection afford these bats a place to hibernate in locations without fractures and solution caves. Diet is probably night-flying insects, and predators probably include snakes, owls, and raccoons.

BEHAVIOR The eastern small-footed myotis is usually a solitary species, but groups of up to 50 have been found packed closely together in crevices. It emerges from its day roost at dusk. The eastern small-footed myotis flies slowly, often erratically, up to 3 m above the ground. Apparently, the eastern small-footed myotis uses ceilings about 3 m above the floor as night roosts in caves. It has been observed flying in and out of caves

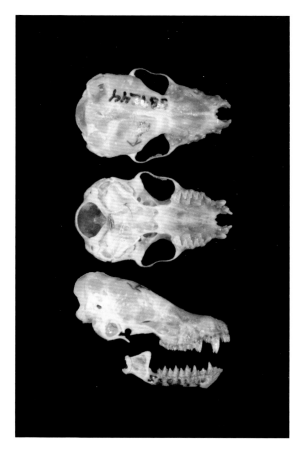

Dorsal, ventral, and lateral views of the cranium, and lateral view of the mandible of a male eastern small-footed myotis. Greatest length of cranium is 13.8 mm.

and in open fields, and it has been captured in butterfly nets. Compared to other species of bats, the eastern small-footed myotis enters hibernation later (mid-November) and leaves hibernacula earlier (March–April). During winter, it roosts near openings of caves or mines where temperatures may be below freezing, but during especially cold weather, the eastern small-footed myotis may move farther back into caves. Although not verified, individuals may move among hibernacula as evidenced by their movements in and out of caves in winter and the variation in the size of hibernating colonies through winter. Although torpid during hibernation, eastern small-footed myotis can become active quickly. They are active nearly continually throughout the winter. Instead of hanging by their feet during hibernation, eastern small-footed myotis occupy depressions or crevices where they lie flat on their undersides.

LIFE HISTORY Breeding probably occurs in late summer to autumn. Fertilization occurs the following spring. One young is born in May or June. Life span may be 6–12 years in the wild.

PARASITES AND DISEASES Ectoparasites include mites (*Euschoengastia*, *Trombicula*). Nothing is known about diseases or internal parasites of the eastern small-footed myotis. Because this species hibernates in caves, it is susceptible to infection with white-nose syndrome (*Geomyces destructans*). However, the incidence of infection of the eastern small-footed myotis is low, possibly because it roosts in drier parts of the hibernaculum where it is seldom covered with condensation during hibernation.

CONSERVATION STATUS Moderate conservation concern in Alabama.

COMMENTS Although not verified from Alabama, the eastern small-footed myotis is included here because it probably occurs in counties north of the Tennessee River. *Myotis* is from the Greek *mys*, meaning "mouse," and *ōtos*, meaning "ear"; *leibii* is a patronym honoring George C. Leib (1809–1888), a physician and the collector of the type specimen.

REFERENCES Best and Jennings (1997), Best (2004*a*), Beolens et al. (2009), Cryan et al. (2010), Harvey et al. (2011).

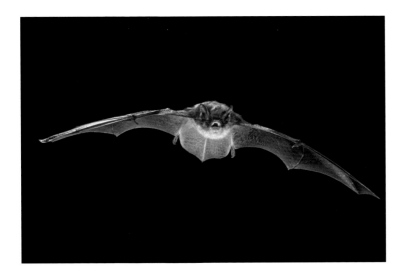

Little Brown Myotis
Myotis lucifugus

IDENTIFICATION This bat has a long, pointed tragus, a calcar with no keel, and hairs on the feet that extend beyond the toenails. Pelage on the back is pale yellowish brown to dark brown. Underside of the body is pale brown and wings are pale to dark brown.

DENTAL FORMULA i 2/3, c 1/1, p 3/3, m 3/3, total = 38.

SIZE AND WEIGHT Range in size:
head and body length, 54–57 mm / 2.2–2.3 inches
foot length, 8–10 mm / 0.3–0.4 inch
ear length, 11–16 mm / 0.4–0.6 inch
tragus length, 7–9 mm / 0.3–0.4 inch
forearm length, 33–41 mm / 1.3–1.6 inches
weight, 5–14 g / 0.2–0.5 ounce
Females are slightly larger than males.

DISTRIBUTION Possibly statewide in Alabama, but the only 2 specimens are from Conecuh and Morgan counties. Geographic range of the species is from northern Alaska to Florida and from the Pacific to Atlantic coasts in Canada and the United States.

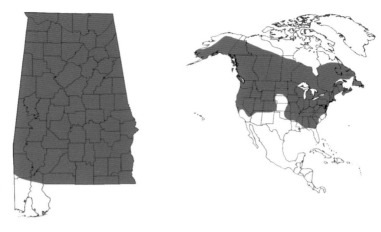

Distribution of the little brown myotis in Alabama and North America.

ECOLOGY Day roosts may be in buildings, in trees, between narrow joints in bridges, under rocks, in piles of wood, and occasionally in caves. Night roosts are usually in the same structures as day roosts, but at different places within the roosts. Nursery roosts are often in buildings, but they may be in hollow trees or other natural crevices. Roosts occupied by males and nonreproductive females are away from nursery roosts and are usually occupied by only 1 or a few bats, and they may be in stacks of lumber, behind sheets of tar paper, under rocks, and occasionally in caves. When young are weaned and become independent, they often roost on the outsides of buildings and in other open spaces. Hibernation sites are usually in caves or abandoned mines; some of which are several hundred kilometers from summer roosts. During late summer to autumn, hibernacula are swarming centers for bats. Diet is a variety of 3–10-mm-long aquatic flying insects, especially midges (Chironomidae). Little brown myotis chew rapidly and food passes through their digestive system within 35–54 minutes. Little brown myotis may die from vandalism or pesticides, from becoming impaled on barbed wire or burdocks (*Arctium*), or from drowning in floods during hibernation. Their numerous predators include snakes, owls, deermice, and carnivores.

BEHAVIOR There is no obvious social organization within a colony; most aggregations seem to be associated with limited resources, such as roosting sites or proximity to food. Foraging behavior varies during the night and may begin by zigzagging through vegetation 2–5 m above the ground along the margins of streams and lakes and end with groups of bats flying 1–2 m above open water. They do not exhibit territoriality, but individuals

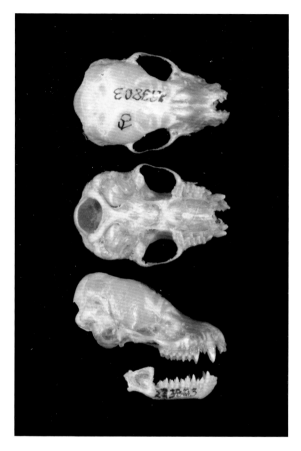

Dorsal, ventral, and lateral views of the cranium, and lateral view of the mandible of a male little brown myotis. Greatest length of cranium is 15.2 mm.

return time after time to the same foraging sites. When 2 little brown myotis are on a collision course during foraging, they will vocalize (honk) toward each other. This honk is the only nonecholocation call they make during foraging. Like other species of bats in Alabama, the little brown myotis makes echolocation calls that are so loud that it temporarily deafens itself when it calls. These calls may exceed 100 decibels (nearly the loudness of the warning whistle of a locomotive). Humans cannot hear these calls because they are beyond the range of our hearing. Vocalizations between mother and young are complex and variable, but the vocal repertoire of little brown myotis is not large. Swarming begins in and around the hibernaculum in late summer and continues well into autumn. It is associated with mating and provides an opportunity for various populations to intermingle, but swarming may also provide a rendezvous site for migrating bats or familiarize subadults with the locations of hibernacula.

Life History Most mating occurs from mid-August to November. The mating system appears random, unstructured, and promiscuous. When females leave hibernation in spring, ovulation and fertilization occur. Gestation is 50–60 days. Young are born in maternity colonies consisting almost entirely of adult females, but occasionally there may be a few adult males. During birthing, females reverse their usual head-down position and the emerging newborn is born into the interfemoral membrane. One young is born each year. Newborns are pinkish and covered with fine hairs, their eyes and ears are closed, and their deciduous teeth are almost fully erupted. Within a few hours, their eyes and ears open. Young use their incisors, thumbs, and feet to cling to their mother. By 3 weeks old, young can fly, their permanent teeth are almost fully erupted, they begin to eat insects, and they are weaned. By late autumn, subadults and adults are nearly indistinguishable. Females mate during their first year, but males do not. In the wild, a life span of more than 30 years has been recorded, but average life span is probably 10–15 years.

Parasites and Diseases Ectoparasites include mites (*Acanthophthirius, Alabidocarpus, Chiroptoglyphus, Cryptonyssus, Euschoengastia, Macronyssus, Pygmephorus, Spinturnix, Steatonyssus*), ticks (*Ornithodorus*), fleas (*Myodopsylla, Nycteridopsylla*), and bat bugs (*Cimex*). Endoparasites include cestodes (*Hymenolepis*), nematodes (*Capillaria, Rictularia*), and trematodes (*Allassogonoporus, Limatulum, Prosthodendrium, Urotrema*). As in most bats, the incidence of rabies (*Lyssavirus*) is less than 1%.

Conservation Status Because this species hibernates in caves, it is susceptible to infection with white-nose syndrome (*Geomyces destructans*). Populations of the little brown myotis have been so severely impacted that the species has been proposed for listing as endangered. High conservation concern in Alabama.

Comments *Myotis* is from the Greek *mys*, meaning "mouse," and *ōtos*, meaning "ear"; *lucifugus* is from the Latin *lucis*, meaning "light," and *fugio*, meaning "to flee."

References La Val (1967), Blankespoor and Ulmer (1970), Williams and Findley (1979), Fenton and Barclay (1980), Best (2004a), Whitaker et al. (2007), Whitaker and Mumford (2009), Cryan et al. (2010), Harvey et al. (2011).

Northern Myotis
Myotis septentrionalis

IDENTIFICATION This small, brown to pale brown bat has long ears that, when pushed forward, usually extend beyond the nose. The tragus is long and pointed.

DENTAL FORMULA i 2/3, c 1/1, p 3/3, m 3/3, total = 38.

SIZE AND WEIGHT Average and range in size:
    total length, 86 (77–95) mm / 3.4 (3.1–3.8) inches
    tail length, 38 (35–42) mm / 1.5 (1.4–1.7) inches
    foot length, 9 (8–10) mm / 0.4 (0.3–0.4) inch
    ear length, 16 (14–19) mm / 0.6 (0.6–0.8) inch
    forearm length, 36 (34–38) mm / 1.4 (1.4–1.5) inches
    weight, 5–8 g / 0.2–0.3 ounce
Females average larger than males.

DISTRIBUTION The northern myotis occurs in northern and eastern Alabama. Geographic range of the species is from northwestern Canada to the Atlantic Ocean and southward into northern Florida.

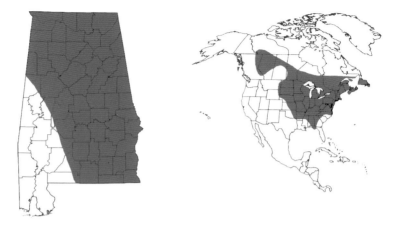

Distribution of the northern myotis in Alabama and North America.

ECOLOGY Habitat is primarily coniferous and deciduous forests. In summer, northern myotis occupy a variety of roosts that may be in dead, standing, tall, large-diameter hardwood trees (e.g., maples, *Acer*, and ashes, *Fraxinus*), or in living trees with less canopy cover than nearby trees. They have also occupied human-made structures. They roost in crevices and cavities and under the exfoliating bark of standing trees that are dead or alive. Northern myotis forage under the forest canopy, over small streams and ponds, along paths or roads in forests, and along edges of forests. Foraging may last throughout the night but peaks shortly after sunset and before sunrise. Diet includes Lepidoptera (moths), Coleoptera (beetles), Trichoptera (caddisflies), and Diptera (flies and mosquitoes), which are captured in the air, and spiders and larval lepidopterans, which are gleaned from vegetation. Predators probably include snakes, owls, and raccoons.

BEHAVIOR In winter, the northern myotis hibernates in caves and abandoned mines that may be 50–60 km from habitats occupied in summer. Hibernation is preceded by swarming, which is characterized by bats flying within and around the opening of the hibernaculum. Mating occurs during swarming from midsummer to autumn. Depending on latitude, hibernation begins from September to early November and lasts until March to May. Northern myotis hibernate in deep crevices in the walls or ceilings of the hibernaculum. During winter, northern myotis may move among hibernation sites. In summer, sexes have separate roosts. Males and nonreproductive females usually roost singly or in groups of

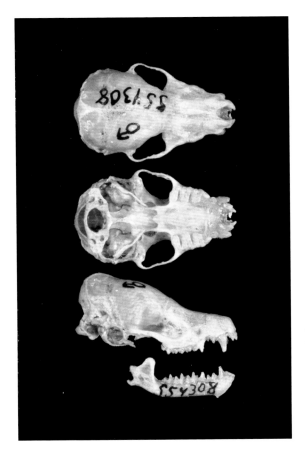

Dorsal, ventral, and lateral views of the cranium, and lateral view of the mandible of a male northern myotis. Greatest length of cranium is 15.2 mm.

fewer than 10 individuals in caves, trees, or buildings. Females may form groups of up to 60 individuals. Maternity colonies may be in trees, under shingles, or in buildings. Individual northern myotis occupy a roost for about 2 days and then move an average of 200 m to another roost. All roosts occupied by an individual are usually within the same cluster of suitable trees. Foraging sites are often within 600 m of roosting sites. Northern myotis may use caves as temporary roosts between foraging bouts at night. When foraging, northern myotis often fly in cluttered environments where they capture flying insects, or insects and spiders that are sitting or moving among foliage. These bats use passive listening and echolocation to find prey that is sitting on leaves, branches, trees, or buildings.

LIFE HISTORY From July to early October, mating occurs at swarming sites outside hibernacula. Females probably store sperm over winter and fertilize a single ovum in spring. Depending on latitude, young are born from mid-May to mid-July. Young grow rapidly and begin to fly 3–5 weeks after birth. By late summer, young-of-the-year cannot be distinguished from adults. They probably reach sexual maturity by the second autumn following birth. Life span is unknown but may average 12–15 years in the wild. Some northern myotis have lived more than 18 years in the wild.

PARASITES AND DISEASES Ectoparasites include mites (*Acanthophthirius, Euschoengastia, Leptotrombidium, Macronyssus, Olabidocarpus, Spinturnix, Steatonyssus*) and bat bugs (*Cimex*). Endoparasites include cestodes (*Hymenolepis, Vampirolepis*) and trematodes (*Plagiorchis, Prosthodendrium*). Diseases include rabies (*Lyssavirus*) and equine encephalomyelitis (*Alphavirus*). Rabies has not been detected in this species in Alabama. Because the northern myotis hibernates in the most humid parts of caves, it is especially susceptible to infection with white-nose syndrome (*Geomyces destructans*).

CONSERVATION STATUS High conservation concern in Alabama.

COMMENTS *Myotis* is from the Greek *mys*, meaning "mouse," and *ōtos*, meaning "ear"; *septentrionalis* is Latin for "belonging to the north."

REFERENCES Caceres and Barclay (2000), Brack and Whitaker (2001), Best (2004*a*), Hester et al. (2007), Whitaker et al. (2007), Cryan et al. (2010).

Indiana Myotis
Myotis sodalis

IDENTIFICATION This bat has a long, pointed tragus, a keeled calcar, and hairs on the feet that do not extend beyond the toenails. Pelage on the back is grayish to dark brown, fine, and fluffy. Underside of the body is pale brown and wings are pale to dark brown.

DENTAL FORMULA i 2/3, c 1/1, p 3/3, m 3/3, total = 38.

SIZE AND WEIGHT Range in size:
    head and body length, 41–49 mm / 1.6–2.0 inches
    ear length, 10–15 mm / 0.4–0.6 inch
    forearm length, 36–41 mm / 1.4–1.6 inches
    wingspan, 240–267 mm / 9.6–10.7 inches
    weight of males, 5.8–7.3 g (0.2–0.3 ounce)
    weight of females, 6.4–9.5 g (0.2–0.3 ounce)

calcar length, about 16.5 mm / 0.7 inch
tail is 80–81% as long as head and body
Females average larger than males.

DISTRIBUTION Northern and eastern Alabama. Geographic distribution of the species is from New Hampshire southwestward into eastern Oklahoma and from central Iowa southeastward into northern Florida.

ECOLOGY Optimal foraging habitat of the Indiana myotis is the foliage of trees in riparian and floodplain areas. Populations have declined more than 50% in recent years due to natural hazards, disturbance by humans, and changes in microhabitat in caves where they hibernate. In winter, caves and mines with cool, stable temperatures are used for hibernation. In summer, females and young roost in hollow trees and under the exfoliating bark of trees. These roost trees are usually standing dead oaks (*Quercus*) or hickories (*Carya*), but roosts may also be in standing dead maples (*Acer*), ashes (*Fraxinus*), or cottonwoods (*Populus*). Males stay near winter hibernacula in summer. Diet of females varies with stage of reproduction. During pregnancy, 90% of the diet is composed of small, soft-bodied insects such as flies, moths, and caddisflies. During lactation, 70% of the diet is moths. After lactation, 54% of the diet of females and young is moths, but they also consume hard-bodied flies and beetles.

Distribution of the
Indiana myotis
in Alabama and
North America.

Predators include black rat snakes (*Elaphe obsoleta*), screech owls (*Otus asio*), deermice, American minks, raccoons, and domestic cats.

BEHAVIOR Indiana myotis tend to take direct routes to and from foraging areas. Foraging is at night, but some Indiana myotis may be active within roosts during daylight hours. Females segregate from males in summer and form small maternity colonies of up to about 30 adults beneath the bark of trees. Individual Indiana myotis tend to return to the same caves to hibernate each year. Swarming occurs at hibernation sites from mid-August to late October, with peaks in early September and mid-October. Most mating occurs at this time. Indiana myotis form large, dense clusters of individuals in hibernation sites. Clusters, which are 1 layer deep and may contain thousands of bats, are usually formed on relatively flat surfaces on the ceilings and walls of caves. Positions of bats within and among clusters change as winter progresses. Smaller and more transient clusters are usually in warmer areas of the cave. Because Indiana myotis awake periodically during winter (possibly every 8–10 days), there are usually a few active individuals within a hibernaculum. During winter, there is also movement among hibernation sites, but this may be related to disturbance by researchers. Females enter and leave hibernation before males. Males may be active longer in autumn to increase their chances of mating with females. Hibernation ends during April or May.

LIFE HISTORY Mating occurs at caves in autumn, but occasionally in winter and spring. Ovulation, fertilization, and implantation probably do

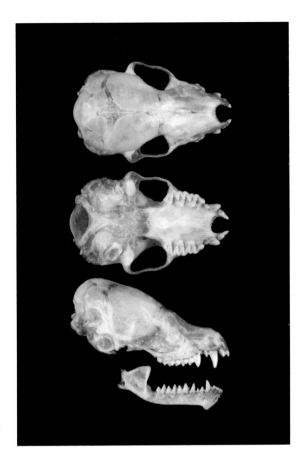

Dorsal, ventral, and lateral views of the cranium, and lateral view of the mandible of a male Indiana myotis. Greatest length of cranium is 14.1 mm.

not occur until females leave hibernation. During late June or July, a single young is born. At 25–37 days old, young can fly and they are weaned shortly thereafter. Parental behaviors include mothers moving the young to more suitable parts of the roost tree and making foraging flights in tandem with their young as the young become able to fly. Life span is about 13–14 years, with life spans of more than 20 years recorded in the wild.

PARASITES AND DISEASES Ectoparasites include mites (*Acanthophthirius, Cryptonyssus, Euschoengastia, Macronyssus, Spinturnix, Steatonyssus*), fleas (*Myodopsylla*), and bat bugs (*Cimex*). Endoparasites include protozoans (*Klossiella*), cestodes, nematodes, and trematodes (*Allassogonoporus, Limatulum, Plagiorchis, Prosthodendrium*). Rabies (*Lyssavirus*) has not been reported in this species. Because the Indiana myotis hibernates in caves, it is susceptible to infection with white-nose syndrome (*Geomyces destructans*). However, the incidence of infection is low, possibly because the Indiana myotis roosts in drier parts of the hibernaculum where it is seldom covered with condensation during hibernation.

CONSERVATION STATUS Highest conservation concern in Alabama, listed as endangered by the United States Fish and Wildlife Service, and listed on the International Union for Conservation of Nature and Natural Resources Red List of Threatened Species as endangered.

COMMENTS *Myotis* is from the Greek *mys*, meaning "mouse," and *ōtos*, meaning "ear"; *sodalis* is Latin for "a comrade," a reference to the habit of roosting in large colonies.

REFERENCES Williams (1962), Thomson (1982), Callahan et al. (1997), Best (2004*a*), Pruitt and TeWinkel (2007), Whitaker et al. (2007), Whitaker and Mumford (2009).

Carnivores

Order Carnivora

The order Carnivora contains 15 families, 126 genera, and 286 species. Although the term *carnivore* implies an animal-eating diet, there is a great diversity of diets in this order. Also, numerous other orders of mammals contain animal-eating members, such as Didelphimorphia, Cingulata, Rodentia, Soricomorpha, and Chiroptera. Members of Carnivora occur in all major habitats worldwide. Of the 19 species of carnivores documented from Alabama, the fisher and jaguar are known only from prehistoric remains. It is likely that several other species of carnivores once occurred in Alabama. The red wolf and cougar have been extirpated in recent times, the jaguarundi and California sealion are of questionable or accidental occurrence, and the ringtail and American badger do not have verified breeding populations.

Cats

Family Felidae

There are 14 genera and 40 species in the family Felidae. Cats range in total length from about 75 to 370 cm and weigh from 1 to 275 kg. Likewise, other characters vary widely: most cats are carnivorous on vertebrates but some eat fruits, fish, and mollusks; some are solitary and others occur in family groups; some have long tails and others have short tails; some have long legs and others have short legs; and some may have a uniform color and others may have spots or stripes. Most stalk their prey, and their senses of vision, smell, and hearing are well developed. Cats occur in most of North America and throughout most of South America, Europe, Asia, and Africa, but they do not occur in polar regions and they are not native to the Australian region. Beginning in prehistoric times, there are records of 4 species of cats occurring in Alabama (bobcat, cougar, jaguarundi, jaguar).

Bobcat
Lynx rufus

IDENTIFICATION This large cat is at least twice the size of a domestic cat. Bobcats are spotted and their tail has black rings and a tip that is black only on the upper side. The pelage is dense, short, and soft, and is yellowish brown to reddish brown with black spots and black-tipped hairs. Underside of the body is white with black spots.

DENTAL FORMULA i 3/3, c 1/1, p 2/2, m 1/1, total = 28.

SIZE AND WEIGHT Average and range in size of 6 specimens from Alabama:

 total length, 710 (560–838) mm / 28.4 (22.4–33.5) inches
 tail length, 127 (91–178) mm / 5.1 (3.6–7.1) inches
 hind foot length of 5 specimens, 146 (130–172) mm / 5.8 (5.2–6.9) inches
 ear length of 4 specimens, 61 (53–64) mm / 2.4 (2.1–2.6) inches
 weight of 5 specimens, 5.7 (2.6–8.2) kg / 12.5 (5.7–18.0) pounds
Males are significantly larger than females.

Facial markings of a bobcat are distinct.

DISTRIBUTION Statewide in Alabama. Geographic range of the species is southern Canada southward into southern Mexico.

ECOLOGY Bobcats occupy habitats that range from arid deserts, grasslands, coniferous forests, and deciduous forests to swampy lowlands. Habitats used vary with sex, age, and season. Dens where young are born may be used for many years. These dens are often in rocky areas or caves but may be in abandoned lodges constructed by American beavers, in storage sheds, or even in cooling towers of nuclear reactors. Bobcats are carnivores and can kill healthy adult white-tailed deer. Diet may also include insects, fish, a wide variety of birds and reptiles, bats, rabbits, squirrels, deermice, voles, hispid cotton rats, woodrats, common muskrats, American beavers, wild boars, foxes, domestic animals, and carrion. Some deer that are consumed may be roadkill or animals wounded by hunters. In Alabama, hispid cotton rats and rabbits are major components of the diet, but white-tailed deer become more common prey in winter. Predators include humans, cougars, coyotes, and domestic dogs.

BEHAVIOR Bobcats may be active at any time during the day or night, but most activity is at night. Daily resting sites are usually on steep, rocky

slopes with dense vertical cover and sparse undergrowth. Other resting sites may be piles of rocks or brush, abandoned burrows, hollow trees, or rocky cliffs. Bobcats are usually solitary hunters, although family groups may travel together. The most common hunting tactic consists of a stealthy approach followed by a pounce and strike. During approach, bobcats take advantage of all cover and topography. Bobcats may also wait, crouched on a log or any object that provides them good visibility, until the prey approaches close enough to be reached in a few short leaps. Deer are often attacked while they are bedded, and most kills are in bedding areas. Deer are killed by choking following a bite to the throat. Often, bobcats cover the remains of their prey, but they do not always finish consuming it. Bobcats rarely go up into trees to hunt, but they may climb trees to escape from predators. Bobcats scent-mark using urine, feces, scrapes, and glandular secretions. Vocalizations include caterwauls (especially during the breeding season) and spits, growls, puffs, and hisses when they are threatened.

Bobcats are good swimmers and may enter water to hunt, defecate, or play.

LIFE HISTORY Breeding season may be throughout the year but depends on latitude, longitude, elevation, and climate. Most mating occurs from December to July. Occasionally, females produce 2 litters within a year,

Distribution of the bobcat in Alabama and North America.

but usually only 1 litter is born each year. Although litters may be born during any month, most are born from late April to June. Gestation averages 63 days (range is 50–70 days). Young are born in a dry, well-hidden, and relatively inaccessible den. Litters average 2–4 young (range is 1–6). At birth, newborns weigh 280–340 g and their eyes are closed. At 3–11 days, eyes open; at 11–14 days, teeth begin to erupt; and at about 9 weeks, young are eating solid food. At 3 months old, young begin accompanying their mother away from the den; by 4 months, weaning is complete; by 6 months, the young travel alone but stay close to the den; and by about 10–12 months, young disperse before their mother has her next litter. Most females mate during their second spring. Life span is 10–12 years in the wild, but some live more than 15 years in the wild. In captivity, life span has reached 32 years.

PARASITES AND DISEASES Ectoparasites include mites (*Lynxacarus, Notoedres*), ticks (*Amblyomma, Dermacentor, Haemaphysalis, Ixodes*), lice (*Felicola*), and fleas (*Echidnophaga, Euhoplopsyllus, Pulex*). Endoparasites include protozoans (*Sarcocystis, Toxoplasma*), acanthocephalans (*Centrorhynchus*), cestodes (*Mesocestoides, Spirometra, Taenia*), nematodes (*Anafilaroides, Ancylostoma, Angiostrongylus, Capillaria, Citellinema, Cyathospirura, Cylicospirura, Cyrnea, Dirofilaria, Gongylonema, Metathelazia, Molineus, Oesophagostomum, Oslerus, Physaloptera, Pterygodermatites, Rictularia, Trichostrongylus, Toxascaris, Toxocara, Troglostrongylus, Uncinaria, Vigisospirura, Vogeloides*), and trematodes (*Alaria,*

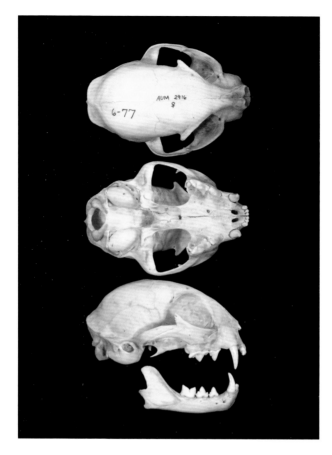

Dorsal, ventral, and lateral views of the cranium, and lateral view of the mandible of a female bobcat. Greatest length of cranium is 102.6 mm.

Paragonimus). Diseases and disorders include cat-scratch fever (*Bartonella*), cytauxzoonosis (*Cytauxzoon*), gastric enteritis, pneumonia, rabies (*Lyssavirus*), and respiratory infections.

CONSERVATION STATUS Lowest conservation concern in Alabama.

COMMENTS *Lynx* is Greek for "lynx"; *rufus* is Latin for "reddish."

REFERENCES Davis (1955), Young (1958), Miller and Harkema (1968), Stone and Pence (1977), Miller and Speake (1980), Watson et al. (1981), Tiekotter (1985), Anderson et al. (1992), Smith et al. (1995), Wehinger et al. (1995), Larivière and Walton (1997), Best (2004*a*), Whitaker et al. (2007).

Cougar
Puma concolor

IDENTIFICATION This large cat has a slender body and a long cylindrical tail that is about one-third the total length of the animal. The legs are muscular and relatively short, and ears are short and rounded. The back is pale grayish brown to dark reddish brown; the sides of the muzzle, backs of the ears, and tip of the tail are dark brown or black; and the chin, middle of the muzzle, and underside of the body are creamy white.

DENTAL FORMULA i 3/3, c 1/1, p 3/2, m 1/1, total = 30.

SIZE AND WEIGHT Average and range in size of adult males and females, respectively:

> total length, 2,215 (1,850–2,743), 1,908 (1,626–2,134) mm / 88.6
> (74.0–109.7), 76.3 (65.0–85.4) inches
> tail length, 791 (686–965), 667 (610–737) mm / 31.6 (27.4–38.6),
> 26.7 (24.4–29.5) inches
> shoulder height, 705 (686–787), 640 (584–762) mm / 28.2 (27.4–
> 31.5), 25.6 (23.4–30.5) inches

Distribution of the cougar in Alabama and North America.

weight, 64 (46–102), 44 (32–60 g) kg / 140.8 (101.2–224.4), 96.8
    (70.4–132.0) pounds
Males are larger than females.

DISTRIBUTION Listed as extirpated in Alabama, but reports of sightings continue from throughout the state. Although extirpated over much of eastern North America, cougars once ranged from central Alaska across almost all of North America to the southern tip of South America.

ECOLOGY The common name "mountain lion" is widely used to refer to this animal because most places where it occurs in North America are mountainous. Because humans exterminated populations of cougars to protect themselves, their families, and their livestock in much of North America in the 1800s and 1900s, most surviving populations remain in mountainous areas; a notable exception was a population in the swamplands of the southeastern United States. Cougars once occurred in the contiguous United States at elevations ranging from sea level to 4,000 m and in habitats ranging from arid deserts in the West and Southwest into the rainforests of the Pacific Northwest, throughout the mountain and Intermountain West, and across the Great Plains and eastern deciduous forests to the Atlantic Ocean and Gulf of Mexico. Diet is extremely varied and may include larger species such as deer, wild boars, elk, North American porcupines, American beavers, armadillos, raccoons, bobcats, and coyotes; smaller species such as rabbits, skunks, squirrels, pocket gophers, woodrats, and deermice; domestic species such as sheep, cattle, dogs, cats, turkeys, chickens, and horses; and other items such as fruits,

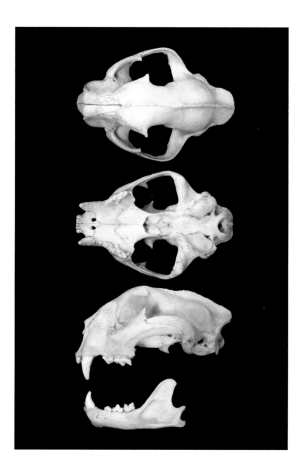

Dorsal, ventral, and lateral views of the cranium, and lateral view of the mandible of a male cougar. Greatest length of cranium is 221.0 mm.

berries, fish, and birds. Other than humans with their guns, poisons, traps, and motor vehicles, adult cougars have no significant predators.

BEHAVIOR Cougars are usually solitary, except during the breeding season or when mothers are with their young. Communication between cougars is usually visual or olfactory but also includes auditory, visual, and tactile cues during the mating season and during interactions of mothers with young. Vocalizations include yowls, purrs, low-pitched squeals, hisses, growls, and loud, chirping whistles. Cougars can swim and readily climb trees, especially when being pursued by hunters. To kill larger prey, a cougar will maneuver to within about 15 m, leap onto the back of the animal within a few strides, and break the neck with a powerful bite below the base of the skull. It then drags its prey to a secluded spot before beginning to eat. The unconsumed portion is covered with whatever is available, usually sticks, pine needles, and leaves. The cougar may return later to uncover and consume more of the prey. As do other cats, cougars lick themselves clean with their rough tongue. They express annoyance or anger by hissing or growling while flattening their ears against their head. Cougars remain playful throughout life, especially adult females during the mating season and juveniles.

LIFE HISTORY Breeding may be year-round in some populations but usually peaks from February to June. Because of the stability of their home ranges, the same animals may mate year after year. Gestation is 82–96 days, average litters have 2–3 young (range is 1–6), and newborns are born with closed eyes and ears, a densely spotted pelage, and a weight of about 400 g. By 7–14 days, the eyes and ears open. At 10–20 days, deciduous incisors erupt and young weigh about 1,000 g; at 20–30 days, canines erupt; and at 30–50 days, premolars erupt. Until they are 3–4 months old, young have irregular lines of distinct black spots along their body. Also at 3–4 months of age, young begin accompanying their mother on hunts. At about 5–6 months, permanent incisors appear, and by 8 months, permanent canines are present. Young stay with their mother for 1–2 years. By 2–4 years, they are fully grown. Females reach sexual maturity 2–3 years following their birth but usually do not mate until they have established a home territory. Males can remain reproductively active until at least 20 years old, and females until at least 12 years old. Life span in the wild is about 8–12 years, but cougars have lived 20 years in captivity.

PARASITES AND DISEASES Ectoparasites include mites (*Eutrombicula, Lynxacarus*), ticks (*Amblyomma, Dermacentor, Ixodes*), lice (*Felicola*), fleas (*Ctenocephalides, Pulex*), and louseflies (*Lipoptena*). Endoparasites include protozoans (*Babesia, Cytauxzoon, Eimeria, Isospora, Sarcocystis*), cestodes (*Mesocestoides, Spirometra, Taenia*), nematodes (*Ancylostoma, Capillaria, Dirofilaria, Dracunculus, Filaroides, Gnathostoma, Molineus, Strongyloides, Toxocara, Trichinella*), and trematodes (*Alaria, Heterobilharzia, Heterophyes*). Diseases include anthrax (*Bacillus*), feline calicivirus (*Vesivirus*), feline enteric coronavirus (Coronaviridae), feline immunodeficiency virus (*Lentivirus*), feline infectious peritonitis virus (Coronaviridae), feline panleukopenia virus (*Parvovirus*), feline rhinotracheitis virus (*Varicellovirus*), pseudorabies virus (*Varicellovirus*), rabies (*Lyssavirus*), and toxoplasmosis (*Toxoplasma*).

CONSERVATION STATUS Extirpated in Alabama. The subspecies *P. c. coryi* is the nearest extant population of cougars to Alabama, but *P. c. couguar* may also have occurred in Alabama. Both subspecies are listed as endangered by the United States Fish and Wildlife Service.

COMMENTS There are many reports of tracks and sightings from throughout Alabama, including documentation in 1984 of a breeding pair of cougars in northern Baldwin County, but whether these individuals are from a natural population or are released captives is unknown. The only preserved specimen from the state is from an archaeological site near the Tennessee River in northern Alabama. *Puma* is from a Peruvian Indian word; *concolor* is Latin for "of the same color." The subspecific name *coryi* is from the Greek *korys*, meaning "a helmet," and *couguar* is probably derived from a South American Indian language.

REFERENCES Young and Goldman (1946), Barkalow (1972), Currier (1983), Mount (1984), Forrester (1992), Harveson et al. (2000), Best (2004*a*), Durden et al. (2006), Whitaker et al. (2007).

Jaguarundi
Puma yagouaroundi

IDENTIFICATION A medium-sized cat with an elongate body, small, rounded ears, relatively short legs, and a uniformly colored, unspotted pelage. The tail is about two–thirds the length of the head and body. Color may be a uniform brownish black, gray, or reddish yellow.

DENTAL FORMULA i 3/3, c 1/1, p 3/2, m 1/1, total = 30.

SIZE AND WEIGHT Average and range in size of 28–33 males and 26–31 females, respectively, from throughout the geographic range:

    total length, 1,120 (865–1,404), 1,003 (763–1,250) mm / 44.8
        (34.6–56.2), 40.1 (30.5–50.0) inches
    head and body length, 661 (555–832), 596 (430–735) mm / 26.4
        (22.2–33.3), 23.8 (17.2–29.4) inches
    tail length, 455 (310–572), 419 (275–515) mm / 18.2 (12.4–22.9),
        16.8 (11.0–20.6) inches
    weight of 13 males, 5.6 (3.0–7.6) kg / 12.3 (6.6–16.7) pounds
    weight of 8 females, 4.5 (3.5–5.4) kg / 9.9 (7.7–11.9) pounds
Males are larger than females.

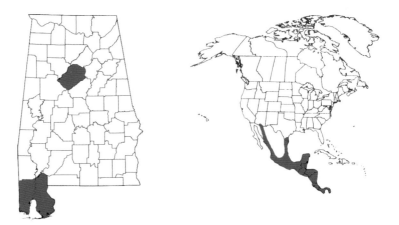

Distribution of the jaguarundi in Alabama and North America.

DISTRIBUTION Accidental in Alabama. Geographic range of the species is along the Pacific and Atlantic coasts of Mexico, south-central Arizona, the southern tip of Texas, and southward from central Mexico through most of South America east of the Andes into southern Argentina.

ECOLOGY Jaguarundis occur in a wide range of habitats including tropical rainforests, deciduous, semideciduous, and thorny forests, dry and humid subtropical forests, subalpine Andean scrub, palmetto (*Sabal*) thickets, scrub and wet-swampy savannas, swampy grasslands, semiarid thorn scrub, dense chaparral, and scrub desert. They rarely occur in dense vegetation but are often in more open habitats such as edges of forests, thickets, hedge-like strips of scrub intermingled with spiny bromeliads, fallow fields, and near streams. Dens may be in fallen logs, dense thickets, or hollow living trees. Diet consists primarily of birds, reptiles, and small mammals but may occasionally include carrion, poultry, fish, arthropods, small deer (*Mazama*), short-tailed opossums (*Monodelphis*), marmosets (*Callithrix*), and several kinds of cavies (*Cavia, Galea, Kerodon*). Other than humans, there are no significant predators. Their pelts have no commercial value, but the species is vulnerable to destruction of habitat.

BEHAVIOR Most activity is during daylight hours, especially in early morning and late afternoon, but nighttime activity is common. Although mostly active on the ground, jaguarundis take refuge and move about in trees with great agility, especially when pursued. Although they may occur in pairs, they are mostly solitary. Vocalizations are similar to those of

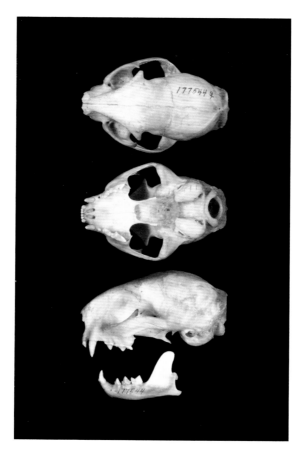

Dorsal, ventral, and lateral views of the cranium, and lateral view of the mandible of a female jaguarundi. Greatest length of cranium is 89.6 mm.

most small cats and include a gurgle, which is a short, noisy, low-intensity, rhythmic sound used for friendly close contact, courtship, mating, and communication between mother and young. Males may spray urine. Other scent-marking behaviors include scraping, rubbing with the head, and raking with the claws. Jaguarundis are easily tamed and can become good pets.

LIFE HISTORY Breeding season is year-round. Gestation is 63–75 days. Average litters contain 2 young (range is 1–4). At 28 days old, young start leaving the den; at 21–30 days, mothers provide small amounts of solid food to the young; and by 42 days, young forage on their own. Females reach sexual maturity at 18–20 months old. In the wild, life span is probably 8–10 years. In captivity, life span is up to 15 years.

PARASITES AND DISEASES Ectoparasites include fleas (*Ctenocephalides*). Endoparasites include protozoans (*Toxoplasma*), acanthocephalans (*Oncicola*), cestodes (*Echinococcus, Spirometra*), nematodes (*Ancylostoma, Molineus, Toxocara*), and trematodes (*Paragominus*). Diseases and disorders in captivity include pneumonia, disorders of the urogenital and digestive systems, and cancer.

CONSERVATION STATUS Accidental in Alabama. Listed as endangered by the United States Fish and Wildlife Service and listed on the International Union for Conservation of Nature and Natural Resources Red List of Threatened Species as least concern.

COMMENTS Although there have been sightings in Jefferson, Mobile, and Baldwin counties, there is no evidence that a breeding population exists in Alabama. Fossil remains of the jaguarundi have been reported from Florida and they have been observed there. Jaguarundis may have been intentionally released in Florida during the early 1900s. It is likely that sightings in Alabama and Florida are of escaped or released captive animals. *Puma* is from a Peruvian Indian word; *yagouaroundi* is derived from the Guaraní word *yaguarundi*, or it may be a compilation of *yagua* (or *jagua, yaguara*, or *jaguara*), referring to all meat-eating animals, and *hung'i*, meaning "brownish" or "tannish."

REFERENCES Neill (1961), Ray (1964), Mount (1984, 1986), de Oliveira (1998), Best (2004*a*), de S. Pinto et al. (2009).

Jaguar
Panthera onca

IDENTIFICATION A large, spotted cat with relatively short, muscular legs, a deep-chested body, and a large, rounded head. This is the largest species of cat in the Western Hemisphere.

DENTAL FORMULA i 3/3, c 1/1, p 3/2, m 1/1, total = 30.

SIZE AND WEIGHT Range in size:
    total length of males, 1,720–2,410 mm / 68.8–96.4 inches
    total length of females, 1,570–2,190 mm / 62.8–87.6 inches
    head and body length, 1,120–1,850 mm / 44.8–74.0 inches
    tail length, 450–750 mm / 18.0–30.0 inches
    shoulder height, 680–750 mm / 27.2–30.0 inches
    weight of males, 57–95 kg / 125.4–209.0 pounds
    weight of females, 56–78 kg / 123.2–171.6 pounds
Males average larger than females.

DISTRIBUTION Jaguars may have been present statewide in what is now Alabama 11,000 or more years ago. Present geographic range of the spe-

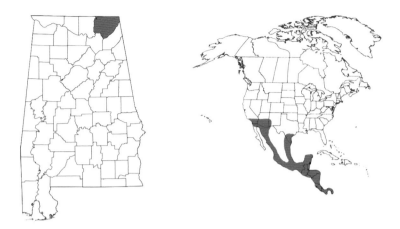

Distribution of the jaguar in Alabama and North America.

cies is from the southwestern United States across most of Mexico, all of Central America, most of northern South America, and east of the Andes southward into central Argentina.

ECOLOGY Jaguars can tolerate a wide variety of environmental conditions from arid deserts to dense tropical rainforests. They occur most frequently in areas with considerable vegetative cover, water, and sufficient prey. Jaguars are opportunistic in foraging habits, and although the diet is dominated by diurnal terrestrial mammals that weigh more than 1 kg, other sizes and kinds of organisms are also consumed as encountered. Jaguars may kill domestic livestock for food. In addition to large mammals, many other foods are consumed including fruits, frogs, fish, lizards, turtles, tortoises, snakes, numerous kinds of birds, armadillos, skunks, monkeys, opossums, and coypus. Humans are the only significant predator of adults, but young may be killed by other jaguars, crocodilians, and large snakes. Some deaths may be caused by injuries inflicted by prey or by bites from venomous snakes.

BEHAVIOR Jaguars are active primarily at night, but they may hunt during any time of the day or night. They are excellent climbers and often swim, hunt, and play in water. Although several jaguars may be present in the same area, they are solitary except during the breeding season, when traveling with siblings, or while young are still dependent on their mother. They may have dens in caves, canyons, and the remains of ancient civilizations. Vocalizations vary by age, but a grunting sound consisting of 5–6 guttural notes is common in adult males and females. When attempting to lure a jaguar close enough to shoot it, hunters often mimic

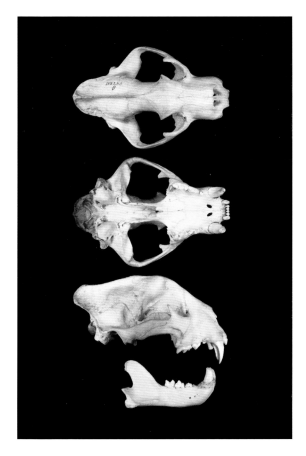

Dorsal, ventral, and lateral views of the cranium, and lateral view of the mandible of a male jaguar. Greatest length of cranium is 278.0 mm.

the guttural sound by pulling a long strip of rawhide through a hole cut into a large gourd; as the rawhide is slowly pulled back and forth, it produces the deep, grunting call. Jaguars may also use scraping, urination, scent-marking, defecation, and tree raking for communication; however, tree raking may be used simply to remove loose sheaths from claws.

LIFE HISTORY Breeding occurs throughout the year, but peaks occur during different months depending on latitude. Peaks in births seem to be related to times when prey is abundant, such as during the rainy season in tropical areas. Gestation is 91–111 days (average is 101 days). Usually, 2 young are born (range is 1–4) in a sheltered place, such as a cave, under uprooted trees, among rocks, in thickets, or under riverbanks. At birth, length is about 400 mm, weight is 700–900 g, and eyes are closed. The pelage is long, coarse, and woolly and has clearly marked round, black spots, some with pale centers. There are also stripes on the face. At 3–13 days (average is 8 days), the eyes open. At 9–19 days, lower incisors erupt; at 11–23 days, upper incisors erupt; at about 30 days, upper canines erupt; and lower canines erupt at 36–37 days. By about 18 days, young can walk, and at about 10–11 weeks, they begin eating meat. At 5–6 months old, young are weaned. Young remain with the mother for 1.5–2 years. Males become sexually mature at 3–4 years and females reach sexual maturity at 2–3 years. In the wild, average life span may be 7–8 years, but some have lived more than 11 years. In captivity, life span has reached 22 years.

PARASITES AND DISEASES Ectoparasites include ticks (*Amblyomma*, *Dermacentor*) and fleas (*Pulex*). Jaguars may also be parasitized by screwworms (*Cochliomyia*), warble fly larvae (*Dermatobia*), and fungi (*Trychophyton*). Endoparasites include protozoans (*Hammondia*, *Isospora*, *Trypanosoma*, *Toxoplasma*), acanthocephalans (*Oncicola*), cestodes (*Diphyllobothrium*, *Echinococcus*, *Spirometra*, *Taenia*, *Toxascaris*), nematodes (*Aelurostrongylus*, *Ancylostoma*, *Capillaria*, *Physaloptera*), and trematodes (*Paragonimus*). In captivity, jaguars may contract anthrax (*Bacillus*), poxviruses (Poxviridae), diabetes, and various types of tumors.

CONSERVATION STATUS There is no evidence that jaguars currently occur in Alabama. Listed as endangered by the United States Fish and Wildlife Service and listed on the International Union for Conservation of Nature and Natural Resources Red List of Threatened Species as near threatened.

COMMENTS In Alabama, the jaguar is known from a single specimen discovered in a cave in Jackson County. Although that jaguar probably died during the Pleistocene (3,000,000–11,000 years ago), it is included here because it is an extant species, it may have occurred in what is now Alabama more recently, and it once inhabited much of the southern contiguous United States. Remains of a red wolf and an extinct short-faced bear (*Arctodus simus*) were with the fragmentary remains of this jaguar. The short-faced bear became extinct about 12,500 years ago. *Panthera* is from the Greek *panthēr*, meaning "a panther"; *onca* is from the New Latin *onca*, which may have been derived from the Portuguese *onça*. Jaguars are referred to in Portuguese as "onça verdadeira" or "onça pintada" and in Spanish as "el tigre."

REFERENCES Paradiso and Nowak (1973), Daggett and Henning (1974), Seymour (1989), Durden et al. (2006), Schubert et al. (2010).

Wolves, Dogs, Foxes, and Jackals

Family Canidae

There are 13 genera and 35 species of Canidae. Total length ranges from 500 to 1,900 mm and weight is 1.5 to 80 kg. Canine teeth are long and powerful, premolars are sharp, carnassials are well developed, and the remaining molars possess surfaces used for crushing. Although canids are mostly carnivorous, they eat many other types of foods including insects, mollusks, crustaceans, fruits, and carrion. They have well-developed senses of vision, smell, and hearing, and they are intelligent, alert, and cunning. Canids are distributed nearly worldwide, except for Antarctica and most oceanic islands. In Alabama, the coyote, gray fox, and red fox are widely distributed, but the red wolf has been extirpated.

Coyote
Canis latrans

IDENTIFICATION In Alabama, this large canid is usually distinguished by its smaller size compared to the extirpated red wolf. Color is usually yellowish brown interspersed with reddish, white, and black hairs, but dark brown to nearly black individuals occur occasionally, especially in northern Alabama.

DENTAL FORMULA i 3/3, c 1/1, p 4/4, m 2/3, total = 42.

SIZE AND WEIGHT Average and range in size of 29 males and 26 females, respectively, from throughout the range:

    total length, 1,207 (1,070–1,320), 1,142 (1,030–1,281) mm / 48.3 (42.8–52.8), 45.7 (41.2–51.2) inches

    tail length, 347 (275–394), 321 (255–368) mm / 13.9 (11.0–15.8), 12.8 (10.2–14.7) inches

    hind foot length of 28 males, 198 (180–250) mm / 7.9 (7.2–10.0) inches

hind foot length of 25 females, 188 (169–220) mm / 7.5 (6.8–8.8) inches

weight of 8 males, 14.7 (12.2–16.2) kg / 32.3 (26.8–25.6) pounds

weight of 2 females, 14.2 (12.6–15.8) kg / 31.2 (27.7–34.8) pounds

track length from back of heel pad to end of longest claw, 66 (57–72) mm / 2.6 (2.3–2.9) inches

stride length, 414 (324–483) mm / 16.6 (13.0–19.3) inches

Males average larger than females.

Coyotes are common in Alabama, including many urban areas.

DISTRIBUTION Statewide in Alabama. Geographic range of the species includes almost all of North America from northern Alaska southward into Costa Rica.

ECOLOGY Coyotes occur in nearly every habitat in Alabama from pine-hardwood forests, agricultural fields, pastures, and swamps to most suburban and urban areas. In Alabama, home range is about 20 km² for males and 41 km² for females. Elsewhere, average home range is 42 km² for adult males and 10 km² for adult females. Reasons for these geographic

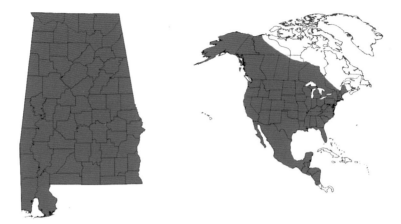

Distribution of the coyote in Alabama and North America.

differences in home range are unknown. Dens may be on brush-covered slopes, on steep banks, in thickets, in hollow logs, under rock ledges, or in abandoned dens of other animals. Dens may be used year after year and may be shared. They are usually about 0.3 m in diameter and 1.5–7.5 m long, and they may have more than 1 opening. Diet of this opportunistic omnivore includes various berries, fruits such as watermelons (*Citrullus lanatus*), cantaloupes (*Cucumis melo*), apples (*Malus domestica*), and pears (*Pyrus*), snails, crustaceans, fish, lizards, ground-nesting birds such as northern bobwhite quail (*Colinus virginianus*) and Canada geese (*Branta canadensis*), deer, domestic livestock, rabbits, domestic dogs and cats, numerous kinds of rodents, and carrion. They also often ingest non-food items such as canvas, cloth, and paper. Because the diet frequently contains seeds, coyotes aid the dispersal of several kinds of plants by depositing the seeds in their feces. In some parts of the geographic range of coyotes, gray wolves and cougars are significant predators. In Alabama, humans kill coyotes by trapping, shooting, and poisoning them and by hitting them with motor vehicles.

BEHAVIOR Coyotes may be active at any time during the day but are most active at dusk, at night, and near sunrise. They spend much time trotting and exploring. They can run up to 48 km/hour but usually trot at less than 32 km/hour. Like other canids, they use olfactory cues and scent-mark with urine, feces, and glandular secretions. Scent stations are conspicuous objects that may mark territories. Scratching the ground is commonly associated with urination and defecation, and it is possible that

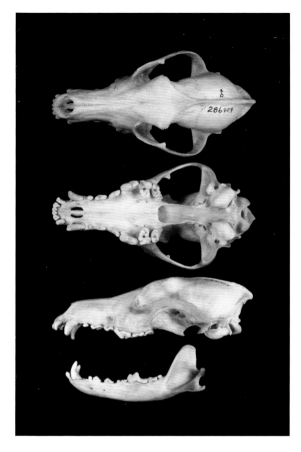

Dorsal, ventral, and lateral views of the cranium, and lateral view of the mandible of a male coyote. Greatest length of cranium is 208.5 mm.

glands between the toes are used in scent-marking. Coyotes use visual, olfactory, auditory, and probably tactile signals in communication. Their numerous vocalizations include squeaks, yelps, howls, and distress calls. Coyotes will hunt alone, in pairs, or in larger groups. They have been known to hunt with other species such as eagles, ravens, and American badgers. When threatened by another coyote, coyotes will approach each other with a stiff-legged gait, ears forward and erect, hairs on the neck and back erect, and tail at about a 45° angle from vertical, usually snarling and exposing the teeth by vertically retracting the lips. A coyote may communicate submission by running away, or it may roll onto its back, flatten its ears against its head, retract its lips into a submissive grin, and possibly urinate and whine, or it may approach in a low crouch-walk with the tail tucked or held low and may perform face licking and face pawing.

LIFE HISTORY Breeding season is from January to March and is usually preceded by a 2–3-month courtship. The same individuals may mate from year to year, but not necessarily for life. One litter averaging 4–7 young is born each year in an excavated den. Gestation averages 63 days (range is 58–65 days). Neonates are helpless, their eyes are closed, their weight is 250–275 g, and their head and body length is about 160 mm. At about 2 weeks, eyes open. At 2–3 weeks, young begin to emerge from the den. At about 3 weeks old, young begin eating solid food and the female (and possibly the male) begins regurgitating semisolid food at this time. Males play a role in rearing the young by bringing food to the lactating female and to the young. Weaning occurs 5–7 weeks after birth. At 6–9 months, young leave the den and disperse. By 9 months old, young are fully grown. Both sexes reach sexual maturity at about 1 year old. In the wild, life span is probably 6–8 years, but some live longer than 14 years. In captivity, they have lived for 18 years. Coyotes can mate successfully with domestic dogs, red wolves, and gray wolves (*Canis lupus*).

PARASITES AND DISEASES Ectoparasites include mites (*Eutrombicula*), ticks (*Dermacentor*, *Ixodes*), lice (*Linognathus*, *Trichodectes*), and fleas (*Cediopsylla*, *Chaetopsylla*, *Hoplopsyllus*, *Pulex*). Endoparasites include protozoans (*Isospora*), acanthocephalans (*Oncicola*), cestodes (*Mesocestoides*, *Multiceps*, *Taenia*), nematodes (*Ancylostoma*, *Capillaria*, *Dioctophyma*, *Dirofilaria*, *Filaroides*, *Mastophorus*, *Molineus*, *Physaloptera*, *Protospirura*, *Rictularia*, *Spirocerca*, *Toxascaris*, *Toxocara*, *Trichuris*), and trematodes (*Alaria*, *Troglotrema*). Diseases and disorders include aortic aneurysms, bubonic plague (*Yersinia*), cancer, canine distemper (*Morbillivirus*), cardiovascular disorders, mange, rabies (*Lyssavirus*), and tularemia (*Francisella*).

CONSERVATION STATUS Lowest conservation concern in Alabama.

COMMENTS Extirpation of the red wolf from the eastern United States may have allowed the coyote to expand its range, and releases of coyotes by humans may have helped establish new populations. *Canis* is Latin for "dog"; *latrans* is Latin for "a barker."

REFERENCES Erickson (1944*a*), Young and Jackson (1951), Bekoff (1977), Best et al. (1981), Sumner et al. (1984), Forrester (1992), Whitaker and Hamilton (1998), Best (2004*a*), Whitaker and Mumford (2009).

Red Wolf
Canis rufus

IDENTIFICATION A large canid; this species is usually larger than the coyote and has longer and more slender legs, a broader muzzle, and usually a darker color. The tail extends horizontally or higher when traveling. Red wolves are usually smaller than gray wolves (*Canis lupus*).

DENTAL FORMULA i 3/3, c 1/1, p 4/4, m 2/3, total = 42.

SIZE AND WEIGHT Average and range in size of 26 males and 26 females, respectively, from eastern Texas:

 total length, 1,420 (1,321–1,600), 1,341 (1,219–1,422) mm / 56.8 (52.8–64.0), 53.6 (48.8–56.9) inches

 hind foot length, 231 (211–249), 221 (203–241) mm / 9.2 (8.4–10.0), 8.8 (8.1–9.6) inches

 ear length, 127 (114–140) mm / 5.1 (4.6–5.6) inches

 weight, 23 (17–35), 20 (16–25) kg / 50.6 (37.4–77.0), 44.0 (35.2–55.0) pounds

 track length from back of heel pad to end of longest claw, 102 (89–127) mm / 4.1 (3.6–5.1) inches

A red wolf with a radiotransmitter in North Carolina. Data on habitats occupied, diet, and movement patterns are essential to reintroduction and recovery efforts for this critically endangered species.

stride length, 658 (553–762) mm / 26.3 (12.1–30.5) inches

Males average larger than females.

DISTRIBUTION Extirpated in Alabama but once occurred statewide. The subspecies *C. r. floridanus* was probably present throughout most of Alabama, except for the southwestern section where *C. r. gregoryi* occurred. The last stronghold of red wolves in Alabama was the rough, hilly region from Walker County northwestward to Colbert County. Geographic range of red wolves once included the southeastern United States from central Texas to the Atlantic Ocean and from central Illinois to the Gulf of Mexico. Red wolves have been reintroduced into Tennessee and North Carolina.

ECOLOGY Red wolves occupy warm, moist, densely vegetated habitats. They occur in pine forests, bottomland hardwood forests, and coastal prairies and marshes. Average home range is about 90 km². Dens are in hollow logs, stumps, road culverts, sandy knolls, and banks of canals, ditches, ponds, and lakes. Dens average 2.4 m long and 0.6–0.8 m in diameter and are usually hidden by vines, piles of brush, trees, or other veg-

etation. Diet includes white-tailed deer, wild boars, domestic livestock, rabbits, rodents, and other small prey. Humans killed red wolves by poisoning, trapping, and shooting them and by hitting them with motor vehicles.

Red wolves show little fear of humans, a behavioral trait that probably accelerated their extirpation.

BEHAVIOR Although red wolves may be active at any time of the day, most activity is at night. During the day, they rest in weedy fields or in grassy and brushy pastures. Red wolves investigate noises by standing up on their hind legs, especially in tall vegetation. Threat behavior is characterized by growling, raising the tail, elevating the hairs on the neck and shoulders, baring the canine teeth, and lowering the head. Red wolves are good swimmers. They have been known to bed down in herds of cattle at night. Rarely, red wolves may gather in groups of 5–11 animals, greet each other, and disperse, but mated pairs are often seen together, sometimes with a second male. Pairs travel around their range using established runways marked by scent posts and scratch marks. Red wolves have a long, smooth howl that ends on a slightly higher note. They also have

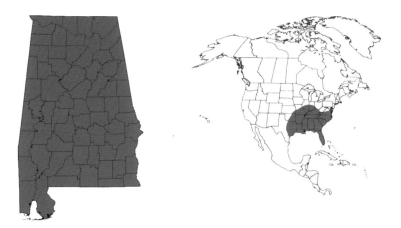

Distribution of the red wolf in Alabama and North America.

many yodeling calls that sound exactly like those of coyotes. They are not wary of humans and were easily poisoned, shot, and trapped. Red wolves have been observed traveling near farm buildings and residential areas, following farm tractors in search of rodents that are plowed up from the ground, and showing little fear of humans on horseback. Both parents care for the young. Young from the previous year frequently live in a den near where they were born, but they do not participate in guarding, feeding, or training the new litter.

LIFE HISTORY Breeding season is probably from December to early March. Gestation is 60–62 days. From April to early June, 6–7 young are born (range is 1–10). After about 6 weeks, young spend more time bedding in areas with good cover rather than in the den. First matings occur at about 3 years old. Life span may be 4–5 years in the wild and about 7–10 years in captivity.

PARASITES AND DISEASES Ectoparasites include severe infestations of sarcoptic mange mites (*Sarcoptes scabei*) and ticks (*Dermacentor, Ixodes*). Endoparasites include acanthocephalans, cestodes (*Taenia*), and nematodes (*Ancylostoma, Dirofilaria*). Diseases include canine distemper (*Morbillivirus*).

CONSERVATION STATUS Extirpated in Alabama. The last time a red wolf was positively identified in Alabama was in 1917. Whether humans caused the demise of red wolves in Alabama directly or indirectly by destroying their habitat and poisoning, trapping, and shooting them, or

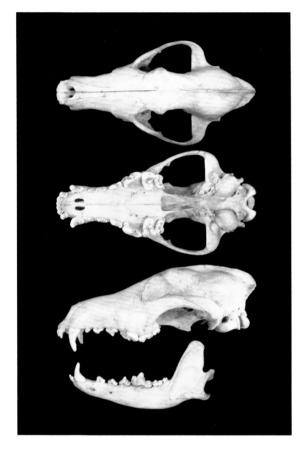

Dorsal, ventral, and lateral views of the cranium, and lateral view of the mandible of a male red wolf. Greatest length of cranium is 234.6 mm.

whether they hybridized themselves nearly to extinction by interbreeding with coyotes, or both, may remain a mystery. Listed as endangered by the United States Fish and Wildlife Service and listed on the International Union for Conservation of Nature and Natural Resources Red List of Threatened Species as critically endangered.

COMMENTS *Canis* is Latin for "dog"; *rufus* is Latin for "reddish." The subspecific name *floridanus* is Latin for "of Florida," and *gregoryi* is a patronym honoring Arthur T. Gregory (1886–1961), a mammalogist and photographer.

REFERENCES Howell (1921), Young and Goldman (1944), Nowak (1972), Paradiso and Nowak (1972), Riley and McBride (1975), Best (2004*a*, 2004*b*), Beolens et al. (2009), Dellinger (2011).

Gray Fox
Urocyon cinereoargenteus

IDENTIFICATION This medium-sized carnivore has a grizzled, dark gray back and sides, reddish-orange pelage along the neck, shoulders, and underside, and a black-tipped tail with a black stripe on top that extends from the base to the tip. The grizzled appearance of the back and sides is due to individual guard hairs that have white, gray, and black bands. The throat and chest are white and the underside is grayish.

DENTAL FORMULA i 3/3, c 1/1, p 4/4, m 2/3, total = 42.

SIZE AND WEIGHT Average and range in size of 44 males and 43 females, respectively, from Alabama:

    total length, 921 (820–983), 894 (821–940) mm / 36.8 (32.8–39.3), 35.8 (32.8–37.6) inches

    tail length, 336 (280–365), 332 (287–360) mm / 13.4 (11.2–14.6), 13.3 (11.5–14.4) inches

    hind foot length, 138 (118–155), 132 (126–142) mm / 5.5 (4.7–6.2), 5.3 (5.0–5.7) inches

    ear length, 69 (63–75), 68 (62–74) mm / 2.8 (2.5–3.0), 2.7 (2.5–3.0) inches

    weight of 37 males, 3.7 (3.1–5.0) kg / 8.1 (6.8–11.0) pounds

    weight of 34 females, 3.5 (2.8–4.1) kg / 7.7 (6.2–9.0) pounds

Males average larger than females.

Distribution of the gray fox in Alabama and North America.

DISTRIBUTION Statewide in Alabama. Geographic range of the species is from southern Canada to northern Venezuela and Colombia, except parts of the mountainous northwestern United States, parts of the Great Plains, and eastern Central America.

ECOLOGY Gray foxes inhabit wooded, brushy, and rocky habitats in both deciduous and pine forests, and they rarely occur on cleared cultivated lands. However, they often occupy fragmented habitats with rocky outcrops, brush, forest, and agricultural fields. Dens are usually in brushy and wooded habitats and may be in piles of brush, under abandoned buildings, in burrows excavated by other mammals, in burrows of gopher tortoises (*Gopherus polyphemus*), in hollow logs, or in holes on rocky ledges. Gray foxes may also dig their own dens. One den was 7.6 m above the ground in a hollow tree. In Alabama, dens are usually within 0.4 km of a permanent water source, and they are often in piles of sawdust at abandoned sawmills or in hollow logs, cavities under rocks, rotten stumps, or abandoned dens of red foxes. Diet includes small mammals such as rabbits, deermice, voles, tree squirrels, woodrats, and hispid cotton rats. Other foods include carrion, insects (especially grasshoppers), persimmons (*Diospyros virginiana*), grapes (*Vitis*), apples (*Malus*), corn (*Zea mays*), birds and their eggs, and poultry. Predators include coyotes and bobcats. Humans kill gray foxes for their pelts or for sport, sometimes regarding them as pests.

BEHAVIOR Although gray foxes may be active at any time during the day, most activity is near sunset, at night, and near dawn. These foxes appear

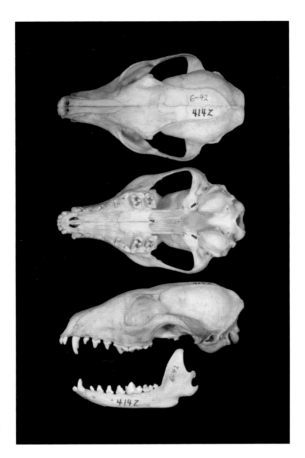

Dorsal, ventral, and lateral views of the cranium, and lateral view of the mandible of a male gray fox. Greatest length of cranium is 118.4 mm.

to be monogamous. Urine and feces are apparently important in communication as evidenced by deposition in conspicuous locations such as along trails and on logs, bare ground, rocks, and other elevated places. Gray foxes are well known for their tree-climbing abilities, and they climb trees to forage, rest, and escape. They have been observed climbing up to 18 m on vertical, branchless trunks of trees. Gray foxes can also climb by jumping from branch to branch. They may descend by backing down vertical trees or by running headfirst down slanting trees.

LIFE HISTORY In Alabama, breeding season peaks in February, gestation is about 53–63 days, and an average of 4 young (range is 1–5) are born in March or April. At birth, gray foxes weigh about 86 g; at 11 days, 129 g; at 23 days, 234 g; at 51 days, 588 g; and at 210 days, 3,700 g (fully grown). By 3 months old, young accompany their mother on trips away from the den, and by 4 months old, young forage on their own. Juveniles may disperse up to 84 km, but they usually establish territories near where they were born. Males are sexually mature at about 4 months old but probably do not reproduce until at least 10–12 months of age. Most females are sexually mature at 10 months old. Life span is probably 5–6 years in the wild, but some have lived 14–15 years in the wild.

PARASITES AND DISEASES Ectoparasites include mites (*Androlaelaps, Atricholaelaps, Eulaelaps, Glycyphagus, Ornithonyssus*), ticks (*Amblyomma, Dermacentor, Ixodes*), lice (*Trichodectes*), and fleas (*Cediopsylla, Ctenocephalides, Echidnophaga, Hoplopsyllus, Orchopeas, Pulex*). Endoparasites include acanthocephalans (*Macracanthorhynchus, Pachysentis*), cestodes (*Mesocestoides, Multiceps, Spirometra, Taenia*), nematodes (*Ancylostoma, Capillaria, Dirofilaria, Eucoleus, Haemonchus, Molineus, Physaloptera, Spirocerca, Strongyloides, Trichuris*), and trematodes (*Alaria, Eurytrema, Procyotrema, Sellacotyle*). Diseases include canine distemper (*Morbillivirus*), canine parvovirus (*Parvovirus*), Chagas disease (*Trypanosoma*), eastern equine encephalitis (*Alphavirus*), histoplasmosis (*Histoplasma*), infectious canine hepatitis (*Mastadenovirus*), leptospirosis (*Leptospira*), listeriosis (*Listeria*), Lyme disease (*Borrelia*), Q fever (*Coxiella*), rabies (*Lyssavirus*), Rocky Mountain spotted fever (*Rickettsia*), salmonellosis (*Salmonella*), toxoplasmosis (*Toxoplasma*), tularemia (*Francisella*), and Tyzzer's disease (*Clostridium*). Gray foxes are highly resistant to saprophytic mange (*Sarcoptes*).

CONSERVATION STATUS Lowest conservation concern in Alabama.

COMMENTS *Urocyon* is from the Greek *oura*, meaning "tail," and *kyōn*, meaning "a dog"; *cinereoargenteus* is from the Latin *cinereus*, meaning "ash-colored," and *argenteus*, meaning "silvery."

REFERENCES Buechner (1944), Erickson (1944a), Eads and Menzies (1950), Chandler and Melvin (1951), Sullivan (1956), Miller and Harkema (1968), Coultrip et al. (1973), Fritzell and Haroldson (1982), Forrester (1992), Whitaker and Hamilton (1998), Oliver et al. (1999), Best (2004a), Whitaker et al. (2007).

Red Fox
Vulpes vulpes

IDENTIFICATION A medium-sized, slender carnivore that varies from pale reddish to dark reddish in Alabama. Underside of the body and tip of the tail are white. Lower legs and feet are black, and ears are tipped with black. Legs and ears are long, and tail is round and bushy.

DENTAL FORMULA i 3/3, c 1/1, p 4/4, m 2/3, total = 42.

SIZE AND WEIGHT Measurements of 2 males from Alabama:
    total length, 935 and 959 mm / 37.4 and 38.4 inches
    tail length, 332 and 350 mm / 13.3 and 14.0 inches
    hind foot length, 146 and 160 mm / 5.8 and 6.4 inches
    ear length of 1 specimen, 80 mm / 3.2 inches
    weight, 3.1 and 3.9 kg / 6.8 and 8.6 pounds
Males average larger than females.

DISTRIBUTION Statewide in Alabama. Red foxes have the largest geographic range of any carnivore in the world. They occur throughout most of North America north of Mexico, all of Europe, nearly all of Asia, and in northern Africa. They have also been widely introduced in other parts of the world.

Distribution of the red fox in Alabama and North America.

ECOLOGY Red foxes live in a wide variety of habitats ranging from semi-arid deserts to tundra, farmlands, forests, and urban areas. Heterogeneous and fragmented habitats with abundant prey support the largest populations of these foxes. Releases for hunting and fragmentation of forested habitats have allowed populations to increase as well as to become more widely distributed. Diet is varied and often includes many kinds of ground-dwelling mammals, rabbits, and squirrels. In Alabama, hispid cotton rats are a common food item. Red foxes also eat food discarded by humans, as well as many kinds of birds and their eggs, and they may consume earthworms, insects, fish, reptiles, fruits, seeds, carrion, raccoons, Virginia opossums, mustelids, and common muskrats. Predation by this species may be significant in reducing the reproductive success of waterfowl on nesting areas in the Great Plains of Canada and the United States. Predators may include bobcats, cougars, coyotes, and domestic dogs. Most human-induced mortality is from trapping, shooting, and roadkill.

BEHAVIOR Red foxes may be active at any time during the day, but they are usually nocturnal. These foxes have keen senses of vision, smell, and hearing. Often, the foxes locate small mammals by sound and pounce upon them with an aerial jump that may be more than 4 m long. They pin the prey with their forefeet and kill it with a bite. Both young and adult foxes may play with live prey. When abundant, prey is stored for later consumption. Red foxes are sleek and agile, with the usual gait being a spritely walk that intermittently breaks into a running trot. Red foxes are fast runners and can travel many kilometers when pursued. They are also good swimmers, and they occasionally climb trees. Daily hunting

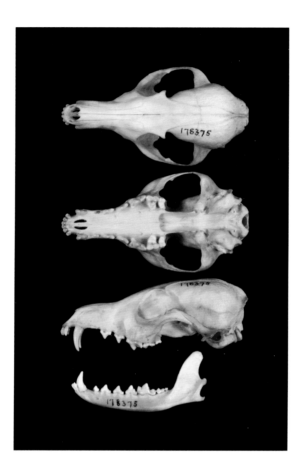

Dorsal, ventral, and lateral views of the cranium, and lateral view of the mandible of a red fox of unknown sex. Greatest length of cranium is 136.6 mm.

forays often cover distances of more than 10 km. They communicate by facial expression, vocalization, and scent-marking. Most dens have several openings and are in sandy soil in pastures and agricultural lands. Dens may be up to 23 m long, with openings usually about 40 cm tall. The basic social unit is the monogamous pair. The male provides parental care and the male-female association lasts until the young are reared. Occasionally, a female without young may be present in a group and she may assist in rearing the young of another female.

LIFE HISTORY Red foxes are seasonally monogamous. Breeding season is from December to April, with a peak in January and February. Following a gestation of 52 days, 3–6 young (range is 1–12) are born. Newborns are dark gray on the back (hairs are 7–10 mm long) and paler on the underside (hairs are 3–5 mm long). Feet are tan with creamy-white foot pads and toenails. For 1 litter with 7 newborns, total length averaged 211 mm, tail length was 67 mm, hind foot length was 32 mm, and ear length was 13 mm. Weight of newborns in the same litter averaged 118 and 105 g for 3 males and 4 females, respectively. Eyes are closed at birth but open at about 3 weeks. By 1–2 weeks, color has changed to pale tan; by 3 weeks, young can walk; and by 9–14 weeks, color is reddish, as in adults. Weaning occurs gradually and lactation ceases at about 5 weeks. Females may mate during their first autumn. Life span is probably 4–6 years, but some may live 8–9 years in the wild.

PARASITES AND DISEASES Ectoparasites include mites *(Echinonyssus, Sarcoptes, Scalopacarus)*, ticks *(Amblyomma, Dermacentor, Ixodes)*, lice *(Neotrichodectes)*, and fleas *(Cediopsylla, Pulex)*. Endoparasites include fungi *(Microsporum)*, protozoans *(Isospora, Sarcocystis, Toxoplasma)*, cestodes *(Amoebotaenia, Cittotaenia, Diphyllobothrium, Dipylidium, Echinococcus, Mesocestoides, Multiceps, Taenia)*, nematodes *(Anafilaroides, Ancylostoma, Angiostrongylus, Capillaria, Crenosoma, Dirofilaria, Molineus, Physaloptera, Pterygodermatites, Spirocerca, Toxascaris, Toxocara, Trichinella, Trichostrongylus, Trichuris, Uncinaria)*, and trematodes *(Alaria, Apophallus, Cryptocotyle, Istmiophora, Metorchis, Opisthorchis, Paragonimus, Pseudamphistomum)*. Diseases and disorders include adenovirus (Adenoviridae), canine distemper *(Morbillivirus)*, canine parvovirus *(Parvovirus)*, chronic interstitial nephritis, herpes virus (Herpesviridae), Lyme disease *(Borrelia)*, parainfluenza virus (Paramyxoviridae), rabies *(Lyssavirus)*, ringworm (fungal dermatophytes), and rotavirus *(Rotavirus)*.

CONSERVATION STATUS Lowest conservation concern in Alabama.

COMMENTS *Vulpes* is Latin for "fox."

REFERENCES Erickson (1944*a*), Miller and Harkema (1968), Forrester (1992), Larivière and Pasitschniak-Arts (1996), Oliver et al. (1999), Best (2004*a*), Whitaker et al. (2007).

Bears and Giant Pandas

Family Ursidae

Worldwide, there are 5 genera and 8 species of bears and giant pandas. Total length ranges from 1.2 to 3.0 m and weight ranges from 27 to 760 kg. Smell is the dominant sense, with sight and hearing usually poorly developed. Bears are relatively quiet and sluggish, and they make few sounds. They may hibernate in caves or burrows. Most are omnivorous and eat insects, eggs, fruits, berries, carrion, or whatever happens to be available. Ursids occur throughout most of North America, Europe, and Asia, including the Malay Peninsula. They also occur in the Andes of South America and the Atlas Mountains of northwestern Africa. One species (American black bear) is present in Alabama.

American Black Bear
Ursus americanus

IDENTIFICATION A large carnivore with long, black to cinnamon-colored hairs (usually black), small eyes, rounded and erect ears, short tail, elongated and nonretractile claws, and no prominent shoulder hump. There may be white markings on the chest.

DENTAL FORMULA i 3/3, c 1/1, p 4/4, m 2/3, total = 42.

SIZE AND WEIGHT Measurements of a 17-year-old male from Alabama:
 total length, 1,830 mm / 73.2 inches
 tail length, 30 mm / 1.2 inches
 hind foot length, 210 mm / 8.4 inches
 weight, 201.9 kg / 444.2 pounds
 Average and range in size of 4 females from Alabama:
 total length, 1,445 (1,220–1,670) mm / 57.8 (48.8–66.8) inches
 tail length, 58 (35–70) mm / 2.3 (1.4–2.8) inches
 hind foot length, 170 (160–180) mm / 6.8 (6.4–7.2) inches
 ear length of 1 specimen, 105 mm / 4.2 inches
 weight of 3 females, 48.4 (45.4–54.4) kg / 106.5 (99.9–119.7)
  pounds
Males are larger than females.

Climbing trees may allow young American black bears to escape from predators.

DISTRIBUTION Prior to the mid-1900s, American black bears probably occurred statewide in Alabama, as evidenced by reports of sightings, tracks, and other evidence. There are occasional reports of bears wandering into eastern and northern Alabama from populations in Tennessee and Georgia. The only breeding populations currently in Alabama are probably limited to southwestern and northeastern counties. Geographic range of the species is from northern Alaska across most of North America to southern Mexico, except for the most arid sections of the southwestern United States and western and eastern regions of Mexico.

ECOLOGY American black bears require a variety of habitats that produce seasonal foods, as well as large and secluded areas for denning. Diet is omnivorous but centers on vegetation. Food discarded by humans may be important in some areas. In spring, American black bears consume new vegetation and winter-killed carrion. In summer, they consume herbaceous vegetation and fruits, and in autumn, they feed primarily on berries, acorns, and other mast. Because American black bears feed opportunistically, insects, reptiles, birds, and mammals are consistently a part of the diet. American black bears are capable of catching and killing domestic livestock and young white-tailed deer. Adults have few preda-

Scats (fecal deposits) are good evidence that American black bears are in the area. Most confirmed sightings in Alabama are in northern, eastern, and southwestern counties.

tors, but young may be preyed upon by bobcats, coyotes, and other American black bears. Most mortality is caused by humans through shooting, trapping, and collisions with vehicles. American black bears may attack humans as prey or in defense of their young or food.

BEHAVIOR American black bears may be active at any time during the day or night, but they are usually active in daylight hours. Daily activity may be reduced by rain, snow, or extreme temperatures. American black bears may become secretive and nocturnal in human-altered habitats such as orchards, campgrounds, garbage dumps, or urban areas. Although often solitary, they may occur in mother-young groups or in aggregations in places where food is superabundant, such as garbage dumps or streams. American black bears do not mark territories using scent glands, but they mark trees by clawing and biting. Defensive behaviors include huffing, jaw popping, charging, slapping one or both forefeet on the ground, walking with stiff legs, standing, and running away. Vocalizations include bellows, bawls, and woofing noises. They are true hibernators, but some may not hibernate in southern areas of the geographic range. Winter dens may be in cavities, crevices, caves, brush piles, underground burrows, or aboveground beds. Denning chambers are lined with vegetation, and openings to chambers are plugged completely or partially. In southern areas, aboveground beds are used more often than in north-

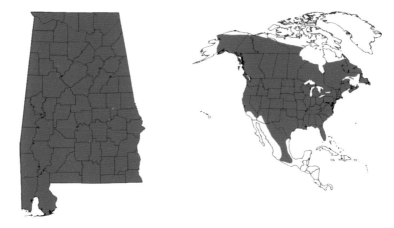

Distribution of the American black bear in Alabama and North America.

ern areas, and some American black bears may change denning sites 3–4 times during the winter. Hibernation usually begins in January or February, depending on food and weather, and emergence is in March or April. During hibernation, American black bears do not eat, drink, urinate, or defecate, but they do maintain a near-normal temperature and do not accumulate toxic metabolic wastes.

LIFE HISTORY Breeding season is from June to September, but most mating takes place in June or July. Both sexes may mate with multiple partners during a breeding season. Implantation is delayed and occurs from mid-November to early December. Actual gestation is 60–70 days and 1–4 young are born in January or early February. Young are born fully furred, toothless, and with eyes closed. Occasionally, females mate while raising young. Young remain with the mother for about 16 months. Females reach sexual maturity at 2–8 years old. Pregnancies and births occur at 1–4-year intervals. Life span is probably 15–18 years, but some have lived 23 years in the wild and 24 years in captivity.

PARASITES AND DISEASES Ectoparasites include mites (*Demodex, Ursicoptes*), ticks (*Amblyomma, Dermacentor, Ixodes*), lice (*Trichodectes*), fleas (*Chaetopsylla, Orchopeas, Pulex*), and larval flies (*Cochliomyia*). Endoparasites include protozoans (*Eimeria, Toxoplasma*), acanthocephalans (*Macracanthorhynchus*), cestodes (*Anacanthotaenia, Diphyllobothrium, Mesocestoides, Multiceps, Taenia*), nematodes (*Ancylostoma, Baylisascaris, Capillaria, Crenosoma, Dirofilaria, Gnathostoma, Gongylonema, Lagochilascaris, Molineus, Physaloptera, Placoconus, Strongyloides, Thela-*

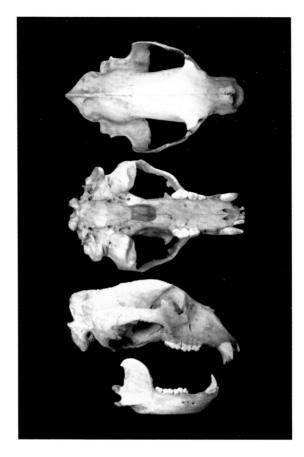

Dorsal, ventral, and lateral views of the cranium, and lateral view of the mandible of a male American black bear. Greatest length of cranium is 316.3 mm.

zia, Toxascaris, Trichinella, Uncinaria), and trematodes (*Heterobilharzia, Nanophyetus, Pharyngostomoides*). Diseases and disorders include California encephalitis (*Orthobunyavirus*), dental caries, eastern equine encephalomyelitis (*Alphavirus*), Elokomin fluke fever (*Neorickettsia*), heart disease, leptospirosis (*Leptospira*), periodontal disease, pseudorabies virus (*Varicellovirus*), Saint Louis encephalitis (*Flavivirus*), salmonellosis (*Salmonella*), and vesicular stomatitis-Indiana virus (*Vesiculovirus*).

CONSERVATION STATUS Highest conservation concern in Alabama. It is not clear whether the subspecies *U. a. floridanus* or *U. a. luteolus*, or both, occur in Alabama. The United States Fish and Wildlife Service lists *U. a. luteolus* as threatened. *Ursus americanus* is listed on the International Union for Conservation of Nature and Natural Resources Red List of Threatened Species as least concern.

COMMENTS *Ursus* is Latin for "bear," *americanus* is Latin for "of America," *floridanus* is Latin for "of Florida," and *luteolus* is Latin for "yellowish."

REFERENCES Forrester (1992), Larivière (2001), Best (2004*a*), Whitaker et al. (2007), Yabsley et al. (2009).

Eared Seals, Fur Seals, and Sealions

Family Otariidae

Seven genera and 16 species are in the family Otariidae. Males are much larger than females and weigh 270 to 1,000 kg, while females weigh 60 to 270 kg. Range in total length is 1.5–3.5 m. Some species congregate into large herds during the breeding season where males guard harems; they are pelagic the rest of the year. Some may migrate more than 9,000 km. Otariids inhabit marine waters and coastal areas in subpolar, temperate, and subtropical regions, but they do not occur in Antarctica or the Arctic Ocean. One species (California sealion) has been documented as an accidental occurrence in Alabama; the Gulf of Mexico is not included in the natural range of this family.

California Sealion
Zalophus californianus

IDENTIFICATION This is a large sealion with blackish limbs modified as flippers. Small external ears are present. Hind limbs can be turned forward under the body and used for movement on land. There is pronounced sexual dimorphism; males are much larger than females.

DENTAL FORMULA i 3/2, c 1/1, molariform teeth 5/5, total = 34.

SIZE AND WEIGHT Average length and weight of adult males is about 2,400 mm (96.0 inches) and 350 kg (770.0 pounds), respectively, compared to 1,800 mm (72.0 inches) and 100 kg (220.0 pounds) for adult females.

DISTRIBUTION Accidental occurrence in Alabama. California sealions occur in coastal regions of the Pacific Ocean from Vancouver Island in Canada to southern Mexico.

ECOLOGY Habitat includes oceanic islands and remote mainland beaches, but California sealions may occur in freshwater rivers upstream from the

Distribution of the California sealion in Alabama and North America.

ocean. Common in coastal California, where they may be so numerous as to be considered pests. California sealions bask on buoys, piers, and other human-made structures in coastal areas. Diet varies greatly among years, seasons, locations, and individuals but frequently includes fish such as northern anchovies (*Engraulis mordax*), Pacific and jack mackerels (*Scomber, Trachunus*), salmon (*Salmo*), and rock fish (*Sebastes*), as well as cephalopods, including squids (*Loligo*). Predators include great white sharks (*Carcharodon carcharias*), bull sharks (*Carcharhinus leucas*), and killer whales (*Orcinus orca*). Demand for hides, blubber, meat, predator control, whiskers, and bacula led to significant reduction in populations in the 1800s and early 1900s. In addition to marine predators and shooting by humans, deaths may be caused by starvation, infection, disease, blooms of toxic phytoplankton, and entanglement in marine debris.

BEHAVIOR Highly intelligent, California sealions are able to comprehend symbols and sign language in a way similar to chimpanzees (*Pan*) and dolphins (Delphinidae). California sealions are colonial and usually live in moderate-sized to large groups, with females living in dense aggregations on sandy or rocky beaches. Most males make long-distance seasonal migrations northward following the breeding season; females remain in southern parts of the range year-round. All ages produce barking noises. Young are playful and spend much time chasing each other, bodysurfing, mock fighting, and playing with objects such as kelp and feathers.

LIFE HISTORY In May, males begin to establish mating territories by fighting with other males. Females give birth to 1 young during May or

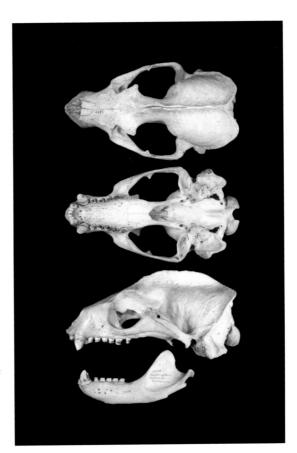

Dorsal, ventral, and lateral views of the cranium, and lateral view of the mandible of a male California sealion. Greatest length of cranium is 270.1 mm.

June and begin mating again about 27 days after parturition. Some females do not breed every year. Implantation is delayed about 3.5 months and usually occurs during September or October when there are fewer than 11 hours of daylight. Gestation is 11 months; lactation continues for several months to more than 2 years. At birth, males weigh about 9 kg and females about 6 kg. Males reach sexual maturity at 4–5 years, although they are usually not yet large enough to defend a territory. Average life span is 15–24 years in the wild, and California sealions have lived 34 years in captivity.

PARASITES AND DISEASES Ectoparasites include mites (*Demodex, Orthohalarachne*) and lice (*Antarctophthirus*). Endoparasites include acanthocephalans (*Corynosoma*), cestodes (*Diphyllobothrium*), nematodes (*Anisakis, Contracaecum, Dipetalonema, Dirofilaria, Dujardinia, Para-*

filaroides, Porrocaecum, Uncinaria), and trematodes (*Pricetrema, Stephanoprora, Stictodora*). Diseases include dermatitis (*Malassezia*) and meningoencephalitis (*Zalophotrema*).

CONSERVATION STATUS Accidental in Alabama.

COMMENTS Known in Alabama from a sighting and photographs of an adult female at Sand Point Light, the lighthouse for Mobile Bay, in July 1966. The same animal was found dead from a gunshot wound on a beach in the Chandeleur Islands, Louisiana, in August 1966. This female was probably a released captive animal; California sealions are often used as performing seals because they are intelligent and easily trained. The only seal native to the Gulf of Mexico was the now-extinct West Indian monk seal (Phocidae; *Monachus tropicalis*), which was last observed in 1952 in the southern Caribbean Sea. *Zalophus* is from the Greek *za*, meaning "intense," and *lophos*, meaning "crest"; *californianus* is Latin for "of California."

REFERENCES Gunter (1968), Margolis and Dailey (1972), Riedman (1990), Guillot et al. (1998), Boyd et al. (1999), Nowak (1999), Wilson and Ruff (1999), Würsig et al. (2000), Heath (2002), Best (2004a), Fauquier et al. (2004).

Weasels, Badgers, and Otters

Family Mustelidae

Mustelids include 22 genera and 59 species. The facial region of the skull is shortened and the legs are usually short relative to the length of the body. Total length varies from 16 to 220 cm, with the least weasel (*Mustela nivalis*) being the smallest, weighing 35–70 g, and the sea otter (*Enhydra lutris*) being the largest at about 45 kg. Although primarily flesh eaters, they also eat fruits, nuts, insects, and honey. Dens are in trees, crevices, or holes in the ground. Mustelids generally occur worldwide, except Madagascar, Australia, Antarctica, and most oceanic islands. Five genera (*Lontra*, *Martes*, *Mustela*, *Neovison*, *Taxidea*) representing five species (North American river otter, fisher, long-tailed weasel, American mink, American badger) have been recorded in Alabama.

North American River Otter
Lontra canadensis

IDENTIFICATION North American river otters can be distinguished from all others in the family Mustelidae in the United States by their fully webbed feet, long, tapered tail, and short, dark, glossy fur. Other characteristics include small eyes, short legs, a neck no smaller than the head, an elongate body with broad hips, a flattened head with a broad muzzle, and ears that are round and inconspicuous. Pelage is short, dense, and usually pale brown to nearly black.

DENTAL FORMULA i 3/3, c 1/1, p 4/3, m 1/2, total = 36, but extra premolars may be present.

SIZE AND WEIGHT Average and range in size of 3 males and 3 females, respectively, from Louisiana:
> total length, 1,129 (1,118–1,150), 978 (900–1,113) mm / 45.2 (44.7–46.0), 39.1 (36.0–44.5) inches
> tail length, 444 (420–470), 358 (317–400) mm / 17.8 (16.8–18.8), 14.3 (12.7–16.0) inches
> hind foot length, 129 (115–140), 112 (101–126) mm / 5.2 (4.6–5.6), 4.5 (4.0–5.0) inches
> ear length, 24 (23–25), 23 (22–23) mm / 1.0 (0.9–1.0), 0.9 (0.9–0.9) inch

North American river otters often live in groups, and they are highly mobile in water and on land.

weight for males and females, respectively, from various localities: 8.1 (7.7–9.4), 7.6 (7.3–8.4) kg / 17.8 (16.9–20.7), 16.7 (16.1–18.5) pounds

Males are larger than females.

DISTRIBUTION Statewide in Alabama. Geographic range of the species includes most of Alaska and Canada, as well as the northwestern and eastern United States.

ECOLOGY North American river otters are highly aquatic and closely associated with permanent water. They favor habitats with banked shores, burrows of semiaquatic mammals, and lodges of American beavers. North American river otters avoid bodies of water with gradually sloping shorelines of sand or gravel. In Florida, abundance is lowest in freshwater marshes, intermediate in salt marshes, and greatest in swampy forests. Latrines may be on conifers, points of land, dens and lodges of American beavers, isthmuses, mouths of permanent streams, or any object that protrudes above water. Dens are often burrows excavated by American beavers, woodchucks, coypus, or common muskrats, but they may also be in hollow trees or logs, undercut banks, rock formations, backwater sloughs, or flood debris. Diet is primarily fish, but North American river otters

also consume aquatic insects, fruits, crustaceans, amphibians, mollusks, snakes, birds, American beavers, common muskrats, American minks, and small mammals. Predators in the water include alligators (*Alligator mississippiensis*), American crocodiles (*Crocodylus acutus*), and killer whales (*Orcinus orca*). On land, North American river otters are more vulnerable and predators may include cougars, bobcats, coyotes, and domestic dogs. Most mortality is human related and includes trapping, illegal shooting, collisions with vehicles, and accidental killing in fishing nets or set lines.

North American river otters are superb swimmers.

BEHAVIOR Active at any time of the day or night throughout the year, but most active at night, early in the morning, and near sundown. American river otters are skilled swimmers. They can remain underwater for nearly 4 minutes, dive to depths of nearly 20 m, and swim underwater for up to 400 m. On land, they can walk, run, bound, and slide. Sliding occurs mostly on level surfaces of snow and ice, but also on grassy slopes and muddy banks. Sliding across snow and ice is an efficient means of travel, and North American river otters traveling over mountain passes or between drainages or descending from mountain lakes often slide contin-

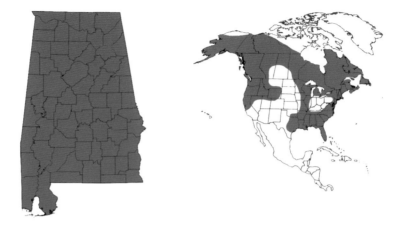

Distribution of the North American river otter in Alabama and North America.

uously for hundreds of meters. Play behavior is well developed and may consist of sliding, chasing the tail, swimming, juggling sticks or pebbles, wrestling, or playing with captured prey or other North American river otters. Communication is primarily by scent from feces, urine, and glandular secretions, and by vocalizations, which may include low-frequency chuckling, explosive snorts, snarling growls, hissing barks, shrill whistles, low purring grunts, and birdlike chirps. North American river otters are not territorial. They often form family groups consisting of a mother and her offspring, but some family groups may contain unrelated helpers, and groups of up to 17 adult males have been documented. They are highly mobile and frequently move 2–5 km each day. North American river otters have been trained to catch and retrieve fish and to hunt and retrieve ducks (Anatidae) and pheasants (*Phasianus colchicus*) from land or water.

LIFE HISTORY Breeding season is from December to April. Delayed implantation of 8 or more months occurs, gestation is 61–63 days, time from mating to birth of litter may be 10–12 months, and litters are usually born from February to April. Although litters may contain 5 young, usually litters of 1–3 young are born. At birth, North American river otters are fully furred, claws are well formed, facial vibrissae are present (about 5 mm long), eyes are closed, no teeth have erupted, total length is about 275 mm, and weight is about 132 g. At about 5 weeks, eyes open; at 9–10 weeks, young begin eating solid food; and at 12 weeks, weaning occurs. Females provide solid food for their young until they are 30–38 weeks old. Juveniles disperse when 12–13 months old; they may disperse up to

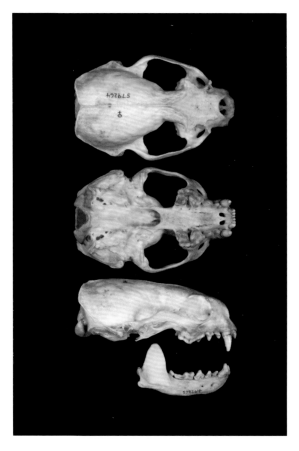

Dorsal, ventral, and lateral views of the cranium, and lateral view of the mandible of a female North American river otter. Greatest length of cranium is 101.5 mm.

200 km and travel at a rate of 3–4 km/day. Although females may have a litter when 1 year old, they usually produce their first litter when 2 years old. Males become sexually mature at 2 years old. Both sexes reach their maximum length and weight at 3–4 years old. Average life span in the wild is probably 10–13 years. In captivity, North American river otters have lived 25 years.

PARASITES AND DISEASES Ectoparasites include mites (*Lutracarus, Lynxacarus*) and ticks (*Amblyomma, Dermacentor, Ixodes*). Endoparasites include protozoans (*Isospora*), crustaceans (*Sebekia*), acanthocephalans (*Corynosoma, Oncicola*), cestodes (*Schistocephalus, Spirometra*), nematodes (*Anisakis, Capillaria, Contracaecum, Crenosoma, Cnathostoma, Dirofilaria, Eustrongylides, Gnathostoma, Hedruris, Metabronema, Spinitectus, Strongyloides*), and trematodes (*Baschkirovitrema,*

Enhydridiplostomum, Euparyphium). Diseases and disorders include canine distemper (*Morbillivirus*), feline panleukopenia (*Parvovirus*), hepatitis, jaundice, parvovirus (*Parvovirus*), pneumonia, rabies (*Lyssavirus*), respiratory tract disease, salmonellosis (*Salmonella*), and urinary infection. North American river otters are highly sensitive to pollution and readily accumulate high levels of mercury, organochloride compounds, and other chemicals.

CONSERVATION STATUS Low conservation concern in Alabama.

COMMENTS *Lontra* is from the Latin *lutra*, meaning "otter"; *canadensis* is Latin for "of Canada."

REFERENCES Greer (1955), Miller and Harkema (1968), Forrester (1992), Serfass et al. (1992), Hoberg et al. (1997), Larivière and Walton (1998), Best (2004a), Whitaker et al. (2007).

Fisher
Martes pennanti

IDENTIFICATION A medium-sized carnivore with a tubular body, a generally weasel-like shape, and short legs. Back, rump, and legs are black, and the head, anterior part of the back, and shoulders are a grizzled golden to silver color. Underside is generally brown, but with varying amounts of creamy white pelage.

DENTAL FORMULA i 3/3, c 1/1, p 4/4, m 1/2, total = 38.

SIZE AND WEIGHT Average and range in size of males and females, respectively:

> total length, 940 (847–1,073), 808 (703–946) mm / 37.6 (33.9–42.9), 32.3 (28.1–37.8) inches
> body length, 593 (519–700), 502 (450–582) mm / 23.7 (20.8–28.0), 20.1 (18.0–23.3) inches
> tail length, 350 (321–378), 306 (240–364) mm / 14.0 (12.8–15.1), 12.2 (9.6–14.6) inches
> hind foot length, 118 (97–125), 100 (87–109) mm / 4.7 (3.9–5.0), 4.0 (3.5–4.4) inches
> weight, 3.7 (2.6–5.5), 2.1 (1.3–3.1) kg / 8.1 (5.7–12.1), 4.6 (2.9–6.8) pounds

Males are larger than females.

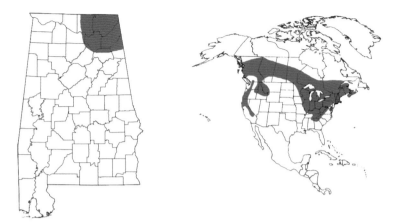

Distribution of
the fisher in
Alabama and
North America.

DISTRIBUTION Possibly occurred in what is now northern Alabama until about 300 years ago. Although more widely distributed in the past, current geographic range of the fisher is from coast to coast across southern Canada, and in the United States, it occurs southward into California and Wyoming in the West, in Michigan, Minnesota, and Wisconsin in the Midwest, and from northern New England into West Virginia in the East.

ECOLOGY Fishers occur in habitats with extensive and continuous canopy cover. Dense lowland forests and spruce-fir (*Picea-Abies*) forests with high canopy cover are favored. Forests with little overhead cover and with open areas are avoided. Fishers can turn their hind feet almost 180°, allowing them to descend trees headfirst. Prey is usually small to medium-sized birds and mammals, but carrion is also consumed. The most common species of prey are the snowshoe hare (*Lepus americanus*) and North American porcupine, but other species consumed include tree squirrels, flying squirrels, chipmunks, deermice, voles, and shrews. North American porcupines must be attacked and killed on the ground; the fisher accomplishes this by repeatedly attempting to bite the head and face of the porcupine, which is the only area not covered with quills. If the porcupine attempts to flee up a tree, the fisher will go up the tree above the porcupine and force it to retreat back to the ground. Once the fisher is able to seize the head, it bites until the porcupine is so weakened that it dies from the wounds or can be killed with a bite to the head. The fisher begins eating on the underside of the porcupine, neatly removing the skin to avoid the quills and consuming nearly all of it during the next 2–3 days

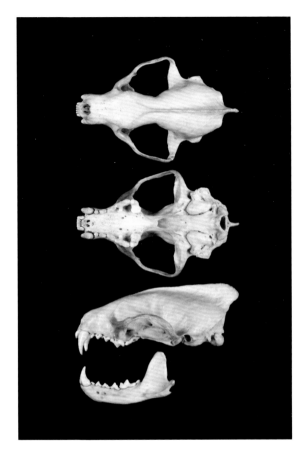

Dorsal, ventral, and lateral views of the cranium, and lateral view of the mandible of a male fisher. Greatest length of cranium is 128.2 mm.

(except bones, skin with quills, and feet). Fishers are not usually subject to predation, but humans trap and kill them for their fur.

BEHAVIOR Fishers may be active at any time during the day or night, but the amount of activity varies within days and among seasons. Fishers are primarily terrestrial, but in coniferous forests, they may travel from tree to tree without descending to the ground. Except for brief periods during the mating season, fishers are solitary. They are territorial, as evidenced by the generally nonoverlapping home ranges of animals of the same sex, but home ranges of males and females frequently overlap. Fishers may use foot, abdominal, and anal glands to communicate information regarding territories and reproductive status. Vocalizations include chuckles, hisses, and growls. Fishers use a variety of sites for sleeping, including hollow trees, logs and stumps, piles of brush and rocks, abandoned lodges of American beavers, and snow dens. Most sleeping sites are temporary, but some may be used more than once. Maternity dens may be in hollow trees.

LIFE HISTORY Breeding season is from March to May. Fishers have a delayed implantation of about 10–11 months. Implantation can occur anytime from January to April. Postimplantation lasts about 30 days and births usually occur from February to May. Average litters contain 3 young (range is 1–6). Newborns are long and slender and sparsely covered with fine, pale gray hair, and their eyes and ears are closed. By 3 days old, they are covered with fine gray fur and weigh about 40 g. By 4–12 weeks, they become chocolate brown and long guard hairs begin to appear. By 8–10 weeks, weaning begins but may not be complete until young are 16 weeks old. By 16 or more weeks old, young are able to kill prey on their own. Young disperse at about 5 months old. Females are fully grown in 5–6 months, become sexually mature at 1 year old, and have their first litter when 2 years old. Males are fully grown by about 12 months and reach sexual maturity at 1–2 years old. Life span may be 7–10 years in the wild. Fishers have lived more than 10 years in captivity.

PARASITES AND DISEASES Ectoparasites include ticks (*Ixodes*) and fleas (*Oropsylla*). Endoparasites include cestodes (*Mesocestoides, Taenia*), nematodes (*Arthrocephalus, Ascaris, Baylisascaris, Capillaria, Crenosoma, Dioctophyma, Dracunculus, Molineus, Physaloptera, Soboliphyme, Trichinella, Uncinaria*), and trematodes (*Alaria, Metorchis*). Diseases include canine distemper (*Morbillivirus*) and canine parvovirus (*Parvovirus*).

CONSERVATION STATUS There is no evidence that fishers currently occur in Alabama.

COMMENTS In Alabama, the fisher is known from a bone fragment that was discovered in Marshall County during excavation of an archaeological site, which was dated to about the year 1700. Reports of fishers in Tennessee and another specimen from an archaeological site in Bartow County, northwestern Georgia, indicate that fishers may have occurred in what is now northern Alabama and Georgia until about 300 years ago. Of course, these remains could have been brought to Alabama and Georgia from elsewhere. *Martes* is Latin for "marten." The specific epithet *pennanti* honors Thomas Pennant (1726–1798), an early European naturalist.

REFERENCES Meyer and Chitwood (1951), Hamilton and Cook (1955), Barkalow (1961), Dick and Leonard (1979), Powell (1981), Brown et al. (2006), Lubelczyk et al. (2007), Beolens et al. (2009).

Long-tailed Weasel
Mustela frenata

IDENTIFICATION A small carnivore with a narrow, elongated body, a long neck, short legs, a brown to yellowish-brown back, and a yellowish to white underside. Tail is about one-half as long as the head and body, and tip of tail is black.

DENTAL FORMULA i 3/3, c 1/1, p 3/3, m 1/2, total = 34.

SIZE AND WEIGHT Average and range in size of 11 specimens from Alabama:

    total length, 394 (300–440) mm / 15.8 (12.0–17.6) inches
    tail length, 138 (107–166) mm / 5.5 (4.3–6.6) inches
    hind foot length, 46 (36–49) mm / 1.8 (1.4–2.0) inches
    ear length of 8 specimens, 23 (17–26) mm / 0.9 (0.7–1.0) inch
    weight of 7 specimens, 322.1 (260.0–393.5) g / 11.3 (9.1–13.8)
      ounces
Males are larger than females.

DISTRIBUTION Statewide in Alabama. Geographic range of the species is the largest of any mustelid in the Western Hemisphere and extends from

western Canada southward through most of North America and into northern South America.

Long-tailed weasels forage in burrows, on the ground, and in trees and shrubs.

ECOLOGY Long-tailed weasels occur in most habitats from arid grasslands and tropical forests to rocky alpine slopes and high-elevation meadows. They commonly occupy habitats that are close to cover and have abundant signs of prey, such as birds and rodents. Dens are usually in burrows excavated by other animals, under piles of rocks, or in brush piles. Within the den, nests are usually constructed inside enlarged chambers 3.5–5 cm in diameter, and they are made of dried grasses and sometimes lined with the fur of prey. Diet includes a wide variety of birds and their eggs, as well as small and medium-sized mammals including voles, deermice, harvest mice, cotton rats, woodrats, common muskrats, meadow jumping mice, pocket gophers, chipmunks, tree squirrels, rabbits, moles, and shrews. Long-tailed weasels die from human-related trapping, shooting, and collisions with vehicles, and their predators include great horned owls (*Bubo virginianus*), barred owls (*Strix varia*), gray and red foxes, coyotes, bobcats, and domestic dogs and cats. Although farmers occasionally consider them to be pests, the removal of large numbers of rodents by long-tailed weasels outweighs the harm that they may inflict on most agricultural endeavors.

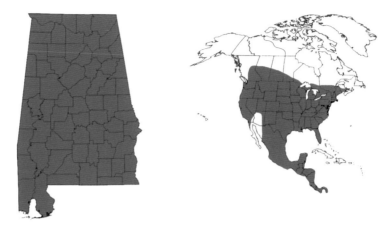

Distribution of the long-tailed weasel in Alabama and North America.

BEHAVIOR Long-tailed weasels are highly active and may forage and feed at any time during the day or night. Predatory behavior consists of searching for, pursuing, attacking, subduing, killing, and eating prey. In searching for prey, long-tailed weasels actively hunt under the ground, on the ground, and arboreally. They have keen senses of smell and hearing, but their vision appears to be relatively poor. Long-tailed weasels run in a series of bounds with their back arched. They are good swimmers and often splash and run through water. To escape predators by climbing a tree, long-tailed weasels will jump up about 1 m from the ground onto the trunk, then spiral around the tree about 1 m at a time as they ascend. They do not hibernate and will aggressively dig into snow in search of prey in winter. Males and females live separately most of the year, but their home ranges may overlap, and they sometimes live together in the nonbreeding season. Nonpredatory aggressive behaviors may consist of loud, high-pitched screeching or stamping the hind feet. Other vocalizations include a trill, which is used during periods of calm investigation, hunting, or play; and squeals, which are distress calls.

LIFE HISTORY Mating season is July and August. Fertilization usually occurs 53–80 hours after mating. There is a lengthy period of delayed implantation beginning about 68 days after mating and lasting about 180 days. Including delayed implantation, average gestation is 279 days (range is 205–337). From mid-April to early May, 1 litter of 4–5 young (range is 1–9) is born each year. Lactation begins soon after young are born and lasts up to 5 weeks. Newborns have closed eyes and ears and only a few hairs, and they weigh about 3 g. By 2 days old, young are covered with

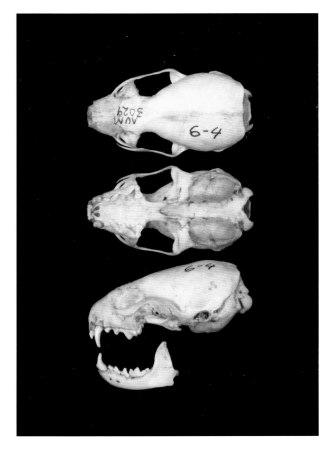

Dorsal, ventral, and lateral views of the cranium, and lateral view of the mandible of a long-tailed weasel of unknown sex. Greatest length of cranium is 46.7 mm.

a fine, long, white pelage that does not completely cover the skin. By 3 weeks old, young are well furred on the back, hair is beginning to turn grayish, and the tip of the tail is black. Eyes open at 5 weeks. Weaning occurs about 5 weeks after birth. By 6 weeks, the characteristic weasel odor has developed. Females usually mate during their first summer. Males reach sexual maturity by about 15 months old. Life span is unknown but may be 5–6 years or more.

Parasites and Diseases Ectoparasites include mites (*Androlaelaps, Aplodontopus, Echinonyssus, Euschoengastia, Eutrombicula, Glycyphagus, Haemogamasus, Hypoaspsis, Laelaps, Lepidoglyphus, Lynxacarus, Pygmephorus, Scalopacarus, Xenoryctes, Zibethacarus*), ticks (*Amblyomma, Dermacentor, Ixodes*), lice (*Neotrichodectes*), and fleas (*Chaetopsylla,*

Ctenophthalmus, Echidnophaga, Epitedia, Hoplopsyllus, Hystrichopsylla, Megabothris, Monopsyllus, Nosopsyllus, Orchopeas, Peromyscopsylla, Polygenis). Endoparasites include nematodes (*Capillaria, Filaroides, Molineus, Skrjabingylus*) and trematodes (*Alaria*). Diseases include Powassan virus (*Flavivirus*) and plague (*Yersinia*).

CONSERVATION STATUS Long-tailed weasels have become increasingly rare in Alabama and the current status of populations in the state is in need of assessment. High conservation concern in Alabama.

COMMENTS *Mustela* is Latin for "weasel"; *frenata* is from the Latin *frenum*, meaning "a bridle," which refers to the facial markings of the long-tailed weasel.

REFERENCES Forrester (1992), Sheffield and Thomas (1997), Best (2004*a*), Whitaker et al. (2007).

American Mink

Neovison vison

IDENTIFICATION A small carnivore with uniformly dark brown fur, a tail that becomes nearly black toward the tip, a long, tubular body, and short ears. There is usually a white patch of hair on the chin and often some white on the throat, chest, and belly.

DENTAL FORMULA i 3/3, c 1/1, p 3/3, m 1/2, total = 34.

SIZE AND WEIGHT Average and range in size of 8 specimens from Alabama:

> total length, 560 (495–651) mm / 22.4 (19.8–26.0) inches
> tail length, 213 (180–263) mm / 8.5 (7.2–10.5) inches
> hind foot length, 54 (31–70) mm / 2.2 (1.2–2.8) inches
> ear length of 3 specimens, 20 (18–22) mm / 0.8 (0.7–0.9) inch
> weight of 1 specimen, 621 g / 21.7 ounces

Males are larger than females.

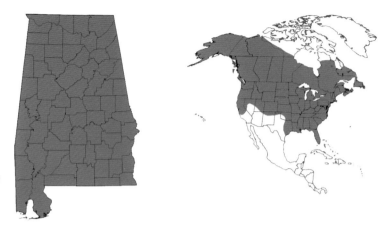

Distribution of the American mink in Alabama and North America.

DISTRIBUTION Statewide in Alabama. Natural geographic range of the species includes Alaska, most of Canada, and most of the United States except for arid grassland and desert regions of the Southwest. The species has been widely introduced into South America, Europe, and Asia, where breeding populations are now established.

ECOLOGY American minks usually occur in freshwater or saltwater habitats in wetlands, where they are semiaquatic. Their streamlined body reduces drag while swimming and enables American minks to enter narrow burrows in search of prey. They forage mostly along waterways, but males may make terrestrial forays in search of rabbits. Diet includes invertebrates, fish, frogs, waterfowl and their eggs, common muskrats, rabbits, squirrels, and mice. American minks do not stalk or ambush prey; they simply rush up to their prey and grab it. They may store food in times of abundance. Predators include American alligators (*Alligator mississippiensis*), great horned owls (*Bubo virginianus*), hawks, coyotes, North American river otters, red foxes, and bobcats. Humans may kill American minks on highways with motor vehicles, in fish cages and gill nets, and in commercial traps. Farm-raised American minks continue to make up a significant portion of the fur-farming industry.

BEHAVIOR Although mostly nocturnal, American minks may be active in the daytime. Nightly movements may be up to 12 km. The farthest-ranging movements are by dispersing juveniles (up to 45 km) and males during the mating season. These excellent swimmers may swim partially or fully

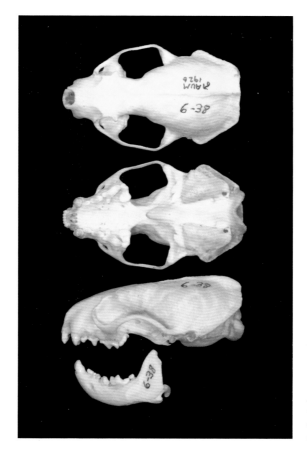

Dorsal, ventral, and lateral views of the cranium, and lateral view of the mandible of a male American mink. Greatest length of cranium is 67.5 mm.

submerged, dive to depths of 5–6 m, and swim underwater for 30–35 m. American minks are excellent climbers and can descend headfirst or leap from tree to tree. They rarely excavate their own burrows. Den sites may be in abandoned burrows of common muskrats, rabbits, or ground squirrels, under piles of rocks or logs, in brush piles, or in culverts. Most dens have 2–5 openings and are less than 2 m from water. Anal scent glands are used to mark territories during defecation or by rubbing the anal region on the ground. American minks do not hibernate and are active all year, but activity is reduced during cold weather. Although usually solitary, they may be in pairs during the mating season; mother and young, or 2 or more siblings, may form temporary family groups. Vocalizations include defensive screams, warning squeaks, hisses, and chuckles. When stressed, American minks may raise their fur, bare their teeth, arch their back, hiss, run rapidly back and forth, empty their scent glands, and move their tail from side to side, possibly to disperse the strong odor released from their scent glands.

LIFE HISTORY Mating occurs from February to April but peaks in March. Because American minks have delayed implantation of the embryo, gestation is 40–75 days (average is 51 days). Actual time for the embryo to develop is 30–32 days. One litter is born each year, usually from April to June. Average litters contain 4 young, but the range is 1–10. Newborns have a coat of fine, silvery hairs, their eyes are closed, and they weigh about 6 g. Eyes open at about 25 days of age. Deciduous teeth erupt 16–49 days after birth and permanent teeth erupt at 44–71 days. Weaning is complete by about 35 days old. Young may remain with the mother until autumn. Sexual maturity is reached at about 10–12 months old. Life span is usually up to 3 years in the wild but about 8 years in captivity.

PARASITES AND DISEASES Ectoparasites include mites (*Androlaelaps, Echinonyssus, Laelaps, Lynxacarus, Marsupialichus, Zibethacarus*), ticks (*Amblyomma, Ixodes*), and fleas (*Ctenophthalmus, Megabothris, Nosopsyllus, Typhloceras*). Endoparasites include protozoans (*Sarcocystis*), acanthocephalans (*Macracanthorhynchus*), cestodes (*Spirometra*), nematodes (*Baylisascaris, Capillaria, Dioctophyma, Filaroides, Gnathostoma, Skrjabingylus*), and trematodes (*Alaria, Brachylaemus, Euparyphium, Fibricola, Rhopalias*). Diseases and disorders include Aleutian disease (*Parvovirus*), amyloidosis, botulism (*Clostridium*), distemper (*Morbillivirus*), hemorrhagic pneumonia, mink virus enteritis (*Parvovirus*), feline panleukopenia (*Parvovirus*), urolithiasis, and canine parvovirus (*Parvovirus*). American minks are sensitive to environmental pollutants such as mercury and, when exposed to high concentrations, may exhibit anorexia, loss of weight, loss of coordination, tremors, and convulsions.

CONSERVATION STATUS Low conservation concern in Alabama.

COMMENTS *Neovison* is from the Greek *neos*, meaning "new," and *vison* is probably Icelandic or Swedish for "a kind of marten or weasel."

REFERENCES Rausch and Tiner (1949), Chandler and Melvin (1951), Lumsden and Zischke (1962), Forrester (1992), Larivière (1999), Best (2004*a*), Whitaker et al. (2007).

American Badger
Taxidea taxus

IDENTIFICATION The body of this medium-sized carnivore is broad and somewhat flattened, and legs and tail are short. Pelage is a coarse, shaggy mixture of yellow, white, brown, and black hairs. Feet are dark brown to black, and there is a white stripe from the top of the muzzle that extends at least to the shoulders. The forefeet have long, curved claws and the hind claws are shovel-like.

DENTAL FORMULA i 3/3, c 1/1, p 3/3, m 1/2, total = 34.

SIZE AND WEIGHT Average and range in size of 11 males and 8 females from Indiana, respectively:

> total length, 754 (605–843), 689 (614–787) mm / 30.2 (24.2–33.7), 27.6 (24.6–31.5) inches
>
> tail length, 148 (115–250), 124 (115–139) mm / 5.9 (4.6–10.0), 5.0 (4.6–5.6) inches
>
> hind foot length, 108 (93–121), 102 (85–115) mm / 4.3 (3.7–4.8), 4.1 (3.4–4.6) inches
>
> weight, 7.4 (3.6–12.1), 7.0 (3.2–9.1) kg / 16.3 (7.9–26.6), 15.4 (7.0–20.0) pounds

Males are larger than females.

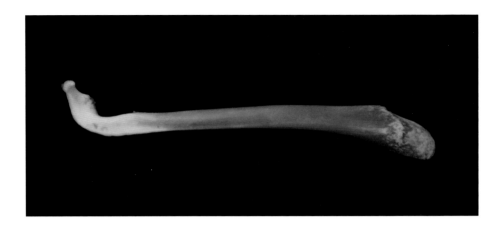

Males of most species of mammals, including the American badger, have a baculum, a bone within the penis. In addition to serving as part of a reproductive organ, bacula have morphological variation that can be used to identify species and to assess taxonomic relationships among groups of mammals.

DISTRIBUTION Although American badgers may occur more widely in northern Alabama, only one record is known for the state; a road-killed American badger was found 7 July 1988 on Highway 72 at Barton, Colbert County. Geographic range of the species is from northern Canada across the western and midwestern United States and southward into central Mexico.

ECOLOGY Grasslands, desert scrub, sagebrush shrublands, and open woodlands are favored, but open areas in a variety of habitats are occupied. Diet includes many kinds of small mammals, snakes, lizards, insects, eggs and broods of birds, and carrion. Food may be stored in old dens for consumption later. Predators include coyotes, cougars, and humans. The coarse fur of American badgers was once used in the manufacture of shaving and paint brushes. Because they have poor vision, are often nocturnal, and tend to rely on olfactory cues, American badgers are especially susceptible to being killed by motor vehicles on roadways.

BEHAVIOR American badgers may be active during the day or night, but activity is usually greatest at night. They are generally solitary except during the mating season and when females are rearing young. When threatened, American badgers growl and bare their teeth, and the hairs on their neck and back stand up. American badgers have well-developed scent glands that they use to mark dens and territories. Their skin is thick and loose, which allows them to turn easily in small spaces. They are good swimmers. They are exceptional diggers and have been reported to dig themselves out of sight in packed soil within minutes. They use the

Distribution of the American badger in Alabama and North America.

burrows they construct for dens, escape, and predation. Females may dig a new den each day in summer, often reuse dens in autumn, and occupy a single den in winter. Dens usually have a mound of soil at the entrance. When American badgers are inactive in cold weather, tunnels within the den are at least partially plugged with loose soil. Dens where females give birth and rear young are structurally more complex than other dens and may include branching main tunnels, side tunnels, and chambers dug into the floor. Some chambers and pockets in dens are used for defecation and are covered with soil. American badgers and coyotes appear to hunt together occasionally; because both species attempt to dig rodents from burrows, each may seek to capture rodents as they escape from burrows that are being excavated by the other species. American badgers are probably not true hibernators but may be torpid for weeks or become less active during cold weather in winter.

LIFE HISTORY Mating occurs in summer and early autumn. Implantation of embryos is delayed until December to February. From March to early April, 1–5 young (average is 2) are born furred and with eyes closed. At 3–4 months old, young usually disperse to live in their own burrow, but some may not disperse until their second year. Males reach sexual maturity at about 12–14 months of age, but females have been known to breed at 4–5 months old. Life span is probably 7–8 years in the wild, but in captivity, some have lived more than 15 years.

PARASITES AND DISEASES Ectoparasites include mites (*Androlaelaps, Echinonyssus, Haemogamasus, Hirstionyssus*), lice (*Neotrichodectes*),

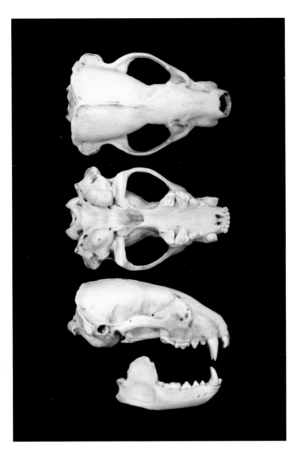

Dorsal, ventral, and lateral views of the cranium, and lateral view of the mandible of a male American badger. Greatest length of cranium is 126.4 mm.

ticks (*Dermacentor, Ixodes*), and fleas (*Echidnophaga, Pulex*). Endoparasites include cestodes (*Atriotaenia, Mesocestoides, Monordotaenia, Taenia*), nematodes (*Ancylostoma, Angiocaulus, Ascaris, Filaria, Molineus, Monopetalonema, Physaloptera, Trichinella*), and trematodes (*Alaria, Euparyphium*). Diseases include canine distemper (*Morbillivirus*), plague (*Yersinia*), rabies (*Lyssavirus*), toxoplasmosis (*Toxoplasma*), and tularemia (*Francisella*).

CONSERVATION STATUS Other than the record from Colbert County, there is no evidence that a breeding population of American badgers has ever existed in Alabama. It is possible that the animal was an escaped or freed pet or that it was intentionally transported to Alabama for release into the wild.

COMMENTS *Taxidea* is from the Greek *taxō*, meaning "to put in order" or "arrange"; *taxus* is New Latin, meaning "badger."

REFERENCES Herman and Goss (1940), Swanson and Erickson (1946), Hubbard (1947), Rausch (1947), Tiner (1953), Ellis (1955), Kalkan and Hansen (1966), Keppner (1969*a*, 1969*b*), Nugent and Choate (1970), Leiby et al. (1971), Long (1973), Whitaker and Goff (1979), Whitaker (1982), Whitaker and Hamilton (1998), Lindzey (2003), Whitaker et al. (2007).

Skunks and Stink Badgers

Family Mephitidae

Mephitidae contains 4 genera and 13 species, all of which possess greatly enlarged anal scent glands that produce noxious odors used to deter threats. Spray from these glands can be projected up to 5–6 m. The body is usually broad and flattened, with a thickly furred tail. Legs are short and claws are well developed for digging. Color is black, occasionally dark brown, with various patterns of white, even within the same species. Diet is omnivorous and includes insects, rodents, lizards, snakes, and eggs. Predators include cougars, coyotes, and foxes. A great diversity of habitats is occupied including forests, riparian areas, grasslands, agricultural areas, and campgrounds. Mephitids in the Americas range from southern Canada into Argentina; stink badgers (*Mydaus*) occur in southeastern Asia in Indonesia, Malaysia, and the Philippines. Two species of skunks occur in Alabama (striped skunk, eastern spotted skunk).

Striped Skunk
Mephitis mephitis

IDENTIFICATION A black animal about the size of a domestic cat, usually with 2 wide white stripes from head to rump, but the pattern and extent of white vary greatly among individuals. Tail is long and bushy and has varying amounts of white hair. Paired anal glands are surrounded by strong muscles that are capable of propelling pungent, yellowish musk for several meters.

DENTAL FORMULA i 3/3, c 1/1, p 3/3, m 1/2, total = 34.

SIZE AND WEIGHT Average and range in size of males and females, respectively, from Illinois:

> total length, 631 (540–765), 603 (520–670) mm / 25.2 (21.6–30.6), 24.1 (20.8–26.8) inches
> body length, 407 (320–478), 378 (302–450) mm / 16.3 (12.8–19.1), 15.1 (12.1–18.0) inches
> tail length, 223 (175–305), 225 (173–307) mm / 8.9 (7.0–12.2), 9.0 (6.9–12.3) inches
> hind foot length, 75 (66–85), 71 (59–78) mm / 3.0 (2.6–3.4), 2.8 (2.4–3.1) inches

Distribution of the striped skunk in Alabama and North America.

ear length, 30 (25–35), 29 (25–33) mm / 1.2 (1.0–1.4), 1.2 (1.0–
1.3) inches

weight, 2.8 (1.4–5.3), 2.0 (1.2–3.9) kg / 6.2 (3.1–11.7), 4.4 (2.6–
25.7) pounds

Males are larger than females.

DISTRIBUTION Statewide in Alabama. Geographic range of the species is from northern Canada throughout most of the United States and into northern Mexico.

ECOLOGY While striped skunks occupy woodlands, woody ravines, and rocky outcrops, they seem most common in intensively cultivated areas and in campgrounds. However, abundance varies greatly on local and regional scales. In intensively cultivated areas, dens tend to be along fence-rows, where there is less likelihood of destruction by livestock or farming activities. Striped skunks use dens extensively during cooler months, and they may excavate these themselves or use abandoned burrows dug by other species. Dens usually have 1 opening, but some may have 2 or more. Diet is primarily insects (mostly grasshoppers, beetles, crickets, and larval moths), but small mammals (deermice and voles), eggs and young of ground-nesting birds, fruits (apples, blueberries, and cherries), corn, and other vegetative materials are also consumed. Predators include great horned owls (*Bubo virginianus*), coyotes, gray and red foxes, cougars, and bobcats. Many are killed by humans on roadways and during farming operations. Striped skunks may become pests around farms because they

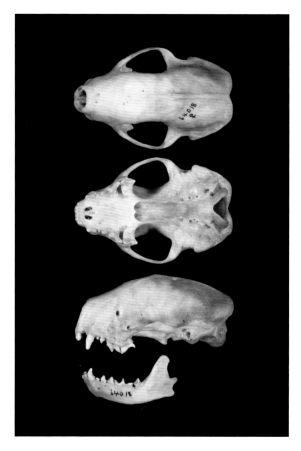

Dorsal, ventral, and lateral views of the cranium, and lateral view of the mandible of a female striped skunk. Greatest length of cranium is 72.1 mm.

raid beehives, inhabit farm buildings, and expel musk when approached by humans or dogs.

BEHAVIOR Striped skunks are usually unobtrusive and docile, except toward perceived threats or other skunks during the breeding season. When threatened, a striped skunk will usually face the intruder, arch its back, elevate its tail, erect the hairs on its tail, stamp the ground with its front feet, and shuffle backward. If the intruder approaches too closely or rapidly, the striped skunk may face the intruder, twist its body into a U shape, and discharge musk toward the intruder. Musk is usually discharged with a slight turning motion, which spreads the musk 30–45° and increases the chance of it hitting the intruder. Human-sized targets can be hit at 6 m, but the greatest accuracy is at about 3 m. Although usually silent, striped skunks can make several vocalizations including squeals, low churrings, growls, dovelike cooings, shrill screeches, birdlike twitters, and hisses. Striped skunks can swim, but they usually avoid water. They commonly occupy the same den for long periods in winter, and communal denning of females or of females with a single male is common. Except for lactating females, striped skunks often use aboveground retreats at least 50% of the day during warmer months. Striped skunks usually leave daytime retreats within an hour of sunset, remain active throughout the night, and retire to a den or aboveground retreat about dawn. Females with young usually make 2–3 excursions at night that increase in duration with time since parturition.

LIFE HISTORY Breeding usually occurs in February or March. Implantation of the embryo occurs by 19 days after mating. Gestation is 59–77 days. A litter of 6–8 young (range is 1–10) is born during late May or June. Newborns weigh 32–35 g and the skin is pinkish, wrinkled, and sparsely covered with hairs 2–3 mm long. Even before birth, the future black and white color pattern is discernible. Eyes and ears open at about 22 and 24–27 days, respectively. Musk is present at birth and emission can occur as early as 8 days of age, but it is not directed at specific objects until after the eyes open. At 34–40 days, teeth begin to erupt. Weaning is complete by 8 weeks. Family units consisting of the female and her offspring may persist until autumn, when juveniles are 3–5 months old. Sexual maturity is reached in about 10 months. Rarely, striped skunks live 5–6 years in the wild, but they have lived 10 years in captivity.

PARASITES AND DISEASES Ectoparasites include mites (*Androlaelaps, Echinonyssus, Hirstionyssus, Ornithonyssus, Pygmephorus*), ticks (*Amblyomma, Dermacentor, Ixodes*), lice (*Neotrichodectes*), and fleas (*Chaetopsylla, Ctenocephalides, Ctenophthalmus, Echidnophaga, Epitedia, Hoplopsyllus, Megabothris, Odontopsyllus, Orchopeas, Polygenis, Pulex*). Striped skunks may be parasitized by larval botflies (*Cuterebra*). Endoparasites include acanthocephalans (*Macracanthorhynchus, Moniliformis, Pachysentis*), cestodes (*Anoplocephala, Mesocestoides, Oochoristica, Taenia*), nematodes (*Arthrocephalus, Ascaris, Capillaria, Crenosoma, Dipetalonema, Dracunculus, Filaria, Filaroides, Gnathostoma, Gongylonema, Molineus, Physaloptera, Skrjabingylus, Strongyloides, Trichinella*), trematodes (*Alaria, Apophallus, Brachylaema, Euryhelmis, Fibricola, Psilostomum, Sellacotyle*), and crustaceans (*Porocephalus*). Diseases and disorders include Chagas disease (*Trypanosoma*), distemper (*Morbillivi-*

rus), histoplasmosis (*Histoplasma*), leptospirosis (*Leptospira*), listeriosis (*Listeria*), nephritis, pancreatitis, pleuritis, pneumonia, Q fever (*Coxiella*), rabies (*Lyssavirus*), and tularemia (*Francisella*). Incidence of rabies in striped skunks is greater than in any other domestic or wild-ranging species in the United States. Rabid skunks may be aggressive and persistent in attacks on humans and other animals.

CONSERVATION STATUS Low conservation concern in Alabama.

COMMENTS *Mephitis* is Latin for "bad odor."

REFERENCES Lumsden and Zischke (1962), Verts (1967), Dyer (1969), Wade-Smith and Verts (1982), Dragoo and Honeycutt (1997), Durden and Richardson (2003), Best (2004*a*), Flynn et al. (2005), Whitaker et al. (2007).

Eastern Spotted Skunk
Spilogale putorius

IDENTIFICATION Pelage is black with a triangular white patch on the nose, a white patch in front of each ear, and 4–6 broken white stripes on the body. The pattern of white on the body is variable and may include spots and vertical stripes.

DENTAL FORMULA i 3/3, c 1/1, p 3/3, m 1/2, total = 34.

SIZE AND WEIGHT Average and range in size of males and females, respectively, in Alabama:

> total length, 519 (463–596), 465 (403–470) mm / 20.8 (18.5–23.8), 18.6 (16.1–18.8) inches
>
> tail length, 202 (193–211), 182 (165–193) mm / 8.1 (7.7–8.4), 7.3 (6.6–7.7) inches

Distribution of the eastern spotted skunk in Alabama and North America.

hind foot length, 47 (43–51), 44 (39–47) mm / 1.9 (1.7–2.0), 1.8 (1.6–1.9) inches

average ear length in Louisiana, males 18 mm (0.7 inch), females 23 mm (0.9 inch)

average weight in Florida, males 399 g (14.0 ounces), females 293 g (10.3 ounces)

Males are larger than females.

DISTRIBUTION Statewide in Alabama. Geographic range of the species is from south-central Pennsylvania southward through the Appalachian region into northern Florida, westward across much of the Great Plains, and southward from near the Canadian border into northeastern Mexico.

ECOLOGY Habitats include brushy, rocky, and wooded areas, as well as farmlands, palmetto (*Sabal*) thickets, and beaches along the ocean, but dense forests and wetlands seem to be avoided. Dens may be in any natural cavity or crevice under a pile of rocks, in hollow logs or stumps, or in cavities of trees 1–7 m above the ground. Eastern spotted skunks can excavate their own burrows or use those dug by pocket gophers, striped skunks, long-tailed weasels, gopher tortoises (*Gopherus polyphemus*), nine-banded armadillos, burrowing owls (*Athene cunicularia*), or other animals. Dens may also be in haystacks, farm buildings, corncribs, grain elevators, or other human-made structures. Eastern spotted skunks are primarily insectivores, but they consume a wide variety of foods. When insects are in short supply, small mammals make up most of the diet,

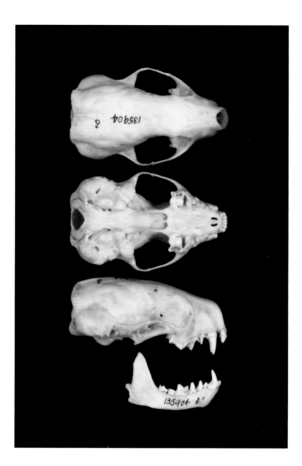

Dorsal, ventral, and lateral views of the cranium, and lateral view of the mandible of a male eastern spotted skunk. Greatest length of cranium is 55.8 mm.

along with birds, eggs, carrion, fruits, and berries. Predators include great horned owls (*Bubo virginianus*), domestic dogs and cats, and bobcats. Collisions with vehicles on roadways, shooting, and trapping are significant causes of mortality.

BEHAVIOR Primarily active at night, eastern spotted skunks are quick, alert, agile, secretive, and rarely seen. Vocalizations include grunts and a high-pitched screech. They are good climbers and diggers, which allows them to forage in both arboreal and terrestrial habitats. Eastern spotted skunks do not hibernate, but activity may be reduced during cold weather. When asleep, they tuck their head and forelegs beneath their abdomen so that the crown of their head, shoulders, hind feet, and tail are in contact with the ground. The striking color of the pelage acts as camouflage when they are foraging at night and as a warning signal for potential predators.

A notable defense is to run at the threatening intruder and then stand on the front feet with the hind legs extended vertically while dancing and foot stamping. This may be accompanied by hissing and twitching of the tail. If the threat approaches closer, the eastern spotted skunk will drop to all fours, assume a U-shaped stance, and point its musk glands and its head toward the intruder before spraying musk. The musk sprayed by eastern spotted skunks is more pungent than that of other types of skunks and can cause temporary blindness, allowing escape from predators. Another ritualistic behavior is associated with breaking eggs. Initially, the eastern spotted skunk straddles the egg with its forelegs and attempts to open it by biting. If unsuccessful, it pushes the egg backward with its forelegs, passes it beneath its hind end, and gives it a quick kick with one hind leg to break it.

LIFE HISTORY Mating and fertilization occur during late March and April. Implantation of embryos occurs 14–16 days after mating. Gestation is 50–65 days. A litter of 5–6 young is born during late May and early June. At birth, young weigh about 10 g and have sparse, fine fur with distinct black and white markings. Their eyes and ears are closed, their claws are well developed, and they are able to vocalize by squealing. By 25 days, the tail can be elevated in warning fashion; by 30 days, teeth have erupted; by 32 days, eyes and ears are open; and by 46 days, eastern spotted skunks are able to discharge musk. Weaning occurs at about 54 days. Life span in the wild is probably 5–6 years, but eastern spotted skunks have lived more than 10 years in captivity.

PARASITES AND DISEASES Ectoparasites include mites (*Androlaelaps, Echinonyssus, Eucheyletia, Eulaelaps, Haemogamasus, Pygmephorus, Xenoryctes*), ticks (*Dermacentor, Ixodes*), lice (*Neotrichodectes*), and fleas (*Ctenocephalides, Echidnophaga, Hoplopsyllus, Polygenis*). Endoparasites include protozoans (*Isospora*), acanthocephalans, cestodes (*Oochoristica*), nematodes (*Capillaria, Skrjabingylus*), and trematodes (*Alaria*). Diseases include coccidiosis (Coccidia), distemper (*Morbillivirus*), histoplasmosis (*Histoplasma*), listeriosis (*Listeria*), microfilaria, pneumonia, Q fever (*Coxiella*), and rabies (*Lyssavirus*).

CONSERVATION STATUS Once locally common, eastern spotted skunks are now considered to be threatened across much of their geographic range. High conservation concern in Alabama.

COMMENTS *Spilogale* is from the Greek *spilos*, meaning "a spot," and *galeē*, meaning "an animal of the weasel kind"; *putorius* is from the Latin *putoris*, meaning "a foul odor."

REFERENCES Swanson and Erickson (1946), Forrester (1992), Kinlaw (1995), Dragoo and Honeycutt (1997), Whitaker and Hamilton (1998), Price et al. (2003), Rosatte and Larivière (2003), Best (2004*a*), Flynn et al. (2005), Whitaker et al. (2007).

Raccoons, Ringtails, and Coatis

Family Procyonidae

Within the family Procyonidae, there are 6 genera and 14 species. Total length varies from 60 to 135 cm and weight ranges from 0.8 to 22 kg. All are excellent climbers and occupy terrestrial and arboreal habitats near water in temperate, arid, and tropical regions. In their omnivorous diet, procyonids consume roots, young shoots, acorns, fruits, insects, crustaceans, reptiles, amphibians, birds, and small mammals. They may travel alone, in family groups, or in larger bands. Geographic range of this family extends from southern Canada to southern South America. Two species have been recorded in Alabama (ringtail, raccoon).

Ringtail

Bassariscus astutus

IDENTIFICATION Tail is about as long as the body, and it has 7–8 alternating black and white rings and is white on the underside. Body is grayish tan to dark brown with a whitish underside, and there are prominent pale facial markings around the eyes.

DENTAL FORMULA i 3/3, c 1/1, p 3/4, m 3/2, total = 40.

SIZE AND WEIGHT Range in size of specimens from throughout the range:
    total length, 616–811 mm / 24.6–32.4 inches
    tail length, 310–438 mm / 12.4–17.5 inches
    hind foot length, 57–78 mm / 2.3–3.1 inches
    ear length, 44–50 mm / 1.8–2.0 inches
    weight, 870–1,000 g / 30.5–35.0 ounces
There is no significant sexual dimorphism.

DISTRIBUTION Accidental in Alabama; known only from Montgomery and Chambers counties. There is no evidence that a viable breeding population has existed in Alabama. The species ranges from southwestern Oregon to northern Louisiana and southward into southern Mexico.

Distribution of the ringtail in Alabama and North America.

ECOLOGY The geographic range of ringtails continues to expand, especially northward. Rocky outcrops, canyons, and talus slopes are occupied in a wide range of habitats including semiarid oak (*Quercus*), pine (*Pinus*), and juniper (*Juniperus*) woodlands, montane coniferous forests, chaparral-covered slopes, deserts, and dry tropical areas. Dens are most often in rocky crevices and piles of boulders but may be in cavities or among roots of trees, in burrows dug by other animals, in brush piles, and in abandoned buildings. Ringtails usually do not construct or modify dens, but they may prepare a nest using dried grasses. They usually use dens for 1–3 days, or for longer periods during inclement weather. Ringtails are omnivorous and diet includes arthropods (grasshoppers, beetles, moths, and spiders), mammals (rodents, rabbits, squirrels, and carrion), birds, snakes, lizards, plants, fruits, and nectar. Predators include great horned owls (*Bubo virginianus*), coyotes, raccoons, and bobcats.

BEHAVIOR Ringtails move in a steady gliding motion with their tail straight out behind and barely clearing the ground. Rather than jumping from rock to rock, they crawl over and around obstacles. Ringtails are excellent climbers and can easily move up and down cacti or other thorny vegetation. Their hind feet can rotate 180° during vertical descents to allow the pads and claws to retain contact with the substrate. Ringtails are rarely active in the daylight, but they may bask in the early morning sun. They may sleep on their side in warm weather, on their back with hind legs spread and forelegs elevated during hot weather, or curled with feet under their body, head against their belly, and tail wrapped around their body when it is cold. Self-grooming consists of a catlike licking of fore-

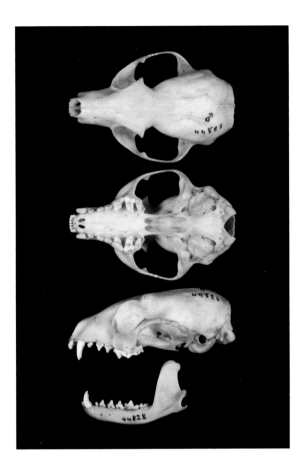

Dorsal, ventral, and lateral views of the cranium, and lateral view of the mandible of a male ringtail. Greatest length of cranium is 80.8 mm.

paws followed by wiping from behind the ears, over the head, and down the muzzle. Vocalizations include metallic chirps, squeaks, and whimpers by infants, chitters (juveniles in distress, females during mating and birthing), chucking, barks (alarm, defensive threats), and aggressive sounds (hisses, grunts, growls, and high-pitched trilling). Scent appears to be as important as vocalization in communication among individuals. Urine is rubbed on the ground and onto raised objects to mark home ranges, and accumulation of feces in latrines is typical.

LIFE HISTORY Breeding season is from February to May but peaks in March and April. Following a gestation of 51–54 days, 1–4 young are born headfirst during May or June. At birth, their eyelids are sealed, their ears are closed, and they weigh 14–40 g and have fuzzy hair on their back. Eyes open at 21–34 days, ears open at 18–30 days, and young are fully furred by 6 weeks old. Deciduous dentition appears at 3–4 weeks, permanent teeth at 17–20 weeks, and solid food is eaten at 4–6 weeks old. Young can walk well at 6 weeks, climb at 8 weeks, begin to forage with their mother at 8–15 weeks, are weaned at about 10 weeks, and are fully grown at about 30 weeks old. Both sexes become sexually mature near the end of their second year. Life span in the wild is probably 5–7 years. In captivity, ringtails have lived more than 16 years.

PARASITES AND DISEASES Ectoparasites include mites (*Androlaelaps, Cheyletus, Dermacarus, Echinonyssus, Eucheyletia, Euschoengastia, Glycyphagus, Hirstionyssus, Pseudoschoengastia*), ticks (*Amblyomma, Dermacentor, Haemaphysalis, Ixodes*), lice (*Neotrichodectes*), and fleas (*Hoplopsyllus, Orchopeas, Polygenis, Pulex*). Endoparasites include acanthocephalans (*Macracanthorhynchus*), cestodes (*Mesocestoides, Taenia*), and nematodes (*Physaloptera, Pneumospirura, Uncinaria*). Diseases include feline and canine panleukopenia (*Parvovirus*) and rabies (*Lyssavirus*).

CONSERVATION STATUS Accidental in Alabama.

COMMENTS Ringtails are similar in size to a small housecat, but they are not closely related to either cats (Felidae) or civets (Viverridae). Because ringtails have a few behaviors and some physical features that are similar to cats and civets, they have been known by a variety of common names, such as ring-tailed cat, cat squirrel, coon cat, miner's cat (historically, ringtails were good mousers in mining camps), and civet cat (possibly because of their musky odor). *Bassariscus* is from the Greek *bassaris*, meaning "a fox"; *astutus* is Latin for "cunning."

REFERENCES Brannon (1923), Poglayen-Neuwall and Toweill (1988), Best (2004a), Whitaker et al. (2007).

Raccoon
Procyon lotor

IDENTIFICATION This medium-sized carnivore is distinguished by a pale band of hairs across the forehead, a black mask across the face and eyes, a nose with pale hairs, and a tail that is shorter than the body and banded with 5–7 brownish-black rings.

DENTAL FORMULA i 3/3, c 1/1, p 4/4, m 2/2, total = 40.

SIZE AND WEIGHT Average and range in size of 14 specimens from Alabama:

    total length, 713 (628–787) mm / 28.5 (25.1–31.5) inches
    tail length, 228 (162–263) mm / 9.1 (6.5–10.5) inches
    hind foot length, 102 (93–114) mm / 4.1 (3.7–4.6) inches
    ear length of 10 specimens, 55 (45–64) mm / 2.2 (1.8–2.6) inches
    weight of 4 specimens, 4.1 (3.6–4.8) kg / 9.0 (7.9–10.6) pounds
Males are usually larger than females.

DISTRIBUTION Statewide in Alabama. The species ranges from northern Canada southward into Panama but is uncommon or absent in some

desert and mountainous regions of the western United States. It has been introduced at several localities across Europe and Asia, as well as onto islands in Alaska.

ECOLOGY Raccoons occupy temperate deciduous and mixed-conifer forests, prairies, mountainous and arid regions, swamps, coastal marshes, subtropical marshes and forests, and a variety of tropical habitats. In urban areas, raccoons are often considered pests. Dens may be in roadway culverts, barns, attics, other human-made structures, cavities, crevices in rock ledges, or abandoned burrows of Virginia opossums, woodchucks, striped skunks, or other mammals. Included in its omnivorous diet is a wide range of plant and animal materials including nuts (such as acorns and walnuts), berries and fruits (such as persimmons and hackberries), seeds, corn, worms, arthropods (such as crabs, crayfish, beetles, grasshoppers), fish, toads, frogs, eggs of turtles and birds, young and adult waterfowl (including those crippled during the hunting season), mice and other young mammals, and foods provided to domestic animals such as horses, cattle, swine, chickens, dogs, and cats. Predators include humans

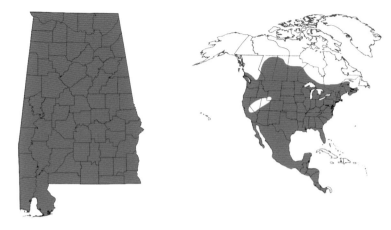

Distribution of the raccoon in Alabama and North America.

(trapping, hunting, and roadkill), bobcats, red foxes, coyotes, American alligators (*Alligator mississippiensis*), eagles, and several species of owls that prey mainly on young raccoons.

BEHAVIOR Raccoons are active primarily at night. They forage on the ground and in trees and shrubs. They are excellent swimmers. Usually, they travel in an area less than 2 km across, but long-range movements of more than 250 km have been documented. Raccoons are well known for their intelligence and food-washing behavior, although they wash food only when near water. Front feet are well adapted for grasping and manipulating objects while the hind feet support the body. Raccoons may remain inactive for extended periods in winter, but they do not hibernate. Although this species frequently appears to be solitary, groups of adults, mothers and young, and unrelated individuals are common. Individuals or small groups often raid food supplies left by humans for domestic dogs and cats. Numerous vocalizations are made; most are between mothers and young. Olfactory communication is by scent-marking with odors from anal glands, urine, and feces.

LIFE HISTORY In Alabama, breeding may occur from early February to early August, but most raccoons breed during early March to late June, with a peak in mid-April. Gestation is about 63 days. Litters are born during early May to late August, with a peak in mid-June. Usually, 1 litter of 2–5 young is born each year. Weight at birth is 60–75 g. Although eyes and ears are closed at birth, the dark mask is visible against the pale hair

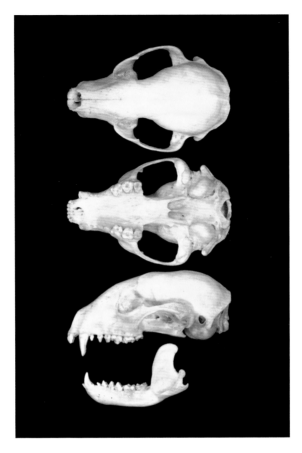

Dorsal, ventral, and lateral views of the cranium, and lateral view of the mandible of a raccoon of unknown sex. Greatest length of cranium is 111.5 mm.

on the face. Eyes and ears open at 2–3 weeks old and first exploration outside the den occurs when weight reaches about 1 kg. Weaning occurs at about 2–4 months old. Young may occupy a den with their mother until the spring following their birth. Both sexes are sexually mature at 1–2 years old. In the wild, most raccoons live fewer than 5 years, but some have lived up to 16 years. In captivity, life span has exceeded 17 years.

PARASITES AND DISEASES Ectoparasites include mites (*Androlaelaps, Echinonyssus, Eutrombicula, Hypoaspis, Laelaps, Ornithonyssus, Pygmephorus, Ursicoptes*), ticks (*Amblyomma, Dermacentor, Ixodes*), lice (*Neotrichodectes*), and fleas (*Ctenocephalides, Hoplopsyllus*). Endoparasites include protozoans (*Babesia, Eimeria, Hepatozoon, Isospora, Sarcocystis, Toxoplasma, Trypanosoma*), acanthocephalans (*Macracanthorhynchus*), cestodes (*Atriotaenia, Mesocestoides, Spirometra*), nematodes (*Ascaris, Baylisascaris, Capillaria, Crenosoma, Dirofilaria, Dracunculus, Filaroides, Gnathostoma, Gongylonema, Mansonella, Molineus, Physaloptera, Placoconus, Procyonostrongylus, Strongyloides, Synhimantus, Trichinella, Uncinaria*), and trematodes (*Ascocotyle, Brachylaemus, Carneophallus, Euryhelmis, Fibricola, Gynaecotyla, Heterobilharzia, Levinseniella, Lyperosomum, Maritrema, Mesostephanus, Microphallus, Parallelorchis, Pharyngostomoides, Ribeiroia, Stictodora*). Diseases include canine distemper (*Morbillivirus*), canine hepatitis (*Mastadenovirus*), Chagas disease (*Trypanosoma*), eastern equine encephalomyelitis (*Alphavirus*), encephalomyocarditis (*Cardiovirus*), Everglades virus (*Alphavirus*), leptospirosis (*Leptospira*), Lyme disease (*Borrelia*), parvoviruses (*Parvovirus*), rabies (*Lyssavirus*), ringworm (fungal dermatophytes), Saint Louis encephalitis (*Flavivirus*), salmonellosis (*Salmonella*), and tularemia (*Francisella*). Leptospirosis and tularemia can be transmitted to humans by direct contact or from water contaminated by urine and feces. More than one-third of the rabies reported in the United States each year is in raccoons. Canine distemper (*Morbillivirus*), which does not infect humans, is a common disease and can lead to the decimation of local populations of raccoons.

CONSERVATION STATUS Lowest conservation concern in Alabama.

COMMENTS *Procyon* is from the Greek prefix *pro-*, meaning "before," and *kynos*, meaning "dog" (ancestors of dogs were once believed to be raccoons); *lotor* is New Latin for "a washer," which refers to the habit of manipulating food in water.

REFERENCES Rausch and Tiner (1949), Lumsden and Zischke (1962), Lotze and Anderson (1979), Forrester (1992), Whitaker and Hamilton (1998), Oliver et al. (1999), Price et al. (2003), Best (2004*a*), Whitaker et al. (2007).

Even-toed Ungulates

Order Artiodactyla

Even-toed ungulates are represented by 10 families, 89 genera, and 240 species. Horns or antlers are present in 1 or both sexes of some species. Habitats are extremely varied, from tundra and grasslands to steep cliff faces, swamps, rivers, woodlands, and dense tropical forests. Artiodactyls are native worldwide except in Antarctica, the Australian region, and most oceanic islands. Three families (Bovidae, Cervidae, Suidae) are known from Alabama.

Swine

Family Suidae

There are 5 genera and 19 species of suids. They are medium to large and have coarse hair, relatively small eyes, and a mobile snout that is distally truncated and has terminal nostrils and disc-like cartilage in the tip. Swine tend to be gregarious and active throughout the day and night, and they make mud wallows. Most are omnivorous. They grub or root for foods such as snails, earthworms, fungi, leaves, fruits, roots, tubers, bulbs, rats, and snakes. They breed throughout the year and have 2–14 young/litter. Natural range is most of Europe, Asia, and Africa. One species (wild boar) has been introduced and is now widely distributed in Alabama.

Wild Boar
Sus scrofa

IDENTIFICATION A large, hoofed mammal with a muscular body tapering to a movable rostrum with nostrils on the flattened end. Head is wedge shaped and muscular. Hair is bristly and relatively sparse. Eyes, legs, and tail are small in comparison to overall size. Because wild boars are the same species as domestic swine, they readily interbreed with feral individuals. Thus, there is much variation in color, length and shape of rostrum, size and shape of ears, amount and distribution of hair, length and shape of tail, and size of body.

DENTAL FORMULA i 3/3, c 1/1, p 4/4, m 3/3, total = 44.

SIZE AND WEIGHT Range in size:
    head and body length, 900–1,800 mm / 36.0–72.0 inches
    shoulder height, 550–1,100 mm / 22.0–44.0 inches
    tail length, about 300 mm / 12.0 inches
    weight, 50–350 kg (110.0–770.0 pounds), but some domesticated
        animals may weigh 450 kg (990 pounds)
Males are larger than females.

DISTRIBUTION In Alabama, wild boars occur statewide, but the largest populations may be in southern areas of the state. Native range was northern Africa and most of Europe and Asia, but the species has been introduced around the world. In the 1500s, Spanish explorers introduced wild boars into what is now the southeastern United States.

Damage to agricultural crops in central Alabama from rooting by wild boars.

ECOLOGY Worldwide, habitats range from deserts to mountains to tropical regions but are usually in areas with vegetative cover. Wild boars are highly omnivorous and will consume almost any type of plant or animal materials, including agricultural crops, food discarded by humans, carrion, small mammals, birds, turtles, snakes, and amphibians. Competition for food probably occurs between wild boars and domestic livestock, white-tailed deer, wild turkeys (*Meleagris gallopavo*), American black bears, tree squirrels, chipmunks, striped and eastern spotted skunks, raccoons, Virginia opossums, gray and red foxes, bobcats, and waterfowl. Wild boars may cause extensive damage to fences, feed supplies, and watering sites. A group of wild boars can root up, trample, or consume several hectares of field or pasture in 1 night. White-tailed

deer avoid feeders, food plots, and natural foraging areas that are used by wild boars. Over a 2–3-month period, nightly movements were 2–15 km within 20–150 hectares; males had larger home ranges (1,000–2,000 hectares) than females (500–1,000 hectares). Wild boars are fast runners and good swimmers. In Alabama, humans are the only significant predators of adult wild boars, but young may also be taken by humans as well as by bobcats, foxes, and coyotes. Elsewhere in North America, humans, American black bears, and cougars may be significant predators of adult wild boars.

BEHAVIOR Senses of smell and taste are well developed. In some social situations with other adult males and females, a male may scent-mark with his tusk (canine teeth) glands by rubbing his lips against an object or by champing his jaws to produce foamy saliva, which contains the scent. Their short, powerful neck, muscular snout, and tusks are instrumental in their rooting type of foraging behavior. Wallowing aids thermoregulation and mud may discourage biting insects and prevent sunburn. Wild boars are most active at dusk and dawn and are usually active throughout

Distribution of the wild boar in Alabama and North America.

the night; hunted populations may be more active at night than in the day. Males may use their ever-growing tusks as weapons; tusks are sharpened as they move against each other. Males may fight to access females. Thick tissue in the thoracic region helps protect against piercing wounds during fights. Males are often solitary except during the breeding season. Groups of 10–20 individuals of mixed ages are common in Alabama; groups of wild boars are referred to as sounders. Vocalizations include snorts, grunts, and squeals.

LIFE HISTORY Gestation is 112–115 days. Usually, 1 litter of 4–8 young (range is 1–12) is produced each year in temperate regions of the world, but breeding may be year-round in tropical areas. Newborns weigh 500–1,500 g. Females are protective of young, which nurse for 3–4 months and become independent at about 1 year of age, before the next litter is born. Both sexes become sexually mature at 5–10 months old. Females usually breed at 18 months old but become sexually mature at 8–10 months of age. Males do not usually reach a large enough size to compete for females until about 2–5 years old. Although mortality is about 50% during the first months of life, life span is about 10 years; some have lived 27 years.

PARASITES AND DISEASES Ectoparasites include mites (*Demodex, Sarcoptes*), lice (*Haematopinus*), ticks (*Amblyomma, Dermacentor, Ixodes*), and larval flies (*Cochliomyia*). Endoparasites include protozoans (*Balantidium, Eimeria, Isospora, Sarcocystis, Toxoplasma*), acanthocephalans (*Macracanthorhynchus*), cestodes (*Spirometra, Taenia*), nematodes (*As-*

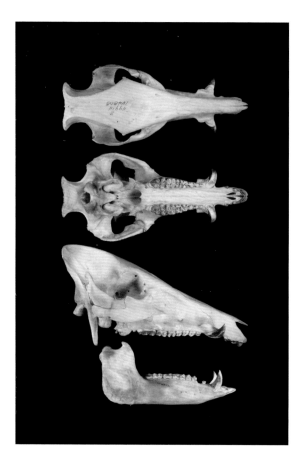

Dorsal, ventral, and lateral views of the cranium, and lateral view of the mandible of a male wild boar. Greatest length of cranium is 339.1 mm.

caris, Ascarops, Capillaria, Globocephalus, Gongylonema, Haemonchus, Hyostrongylus, Metastrongylus, Oesophagostomum, Physaloptera, Physocephalus, Stephanurus, Strongyloides, Trichinella, Trichostrongylus, Trichuris), and trematodes (*Brachylaima, Fasciola, Paragonimus*). Diseases include bovine tuberculosis (*Mycobacterium*), cysticercosis (*Taenia*), eastern equine encephalitis (*Alphavirus*), enterovirus (*Enterovirus*), foot-and-mouth disease (*Aphtae*), Japanese encephalitis (*Flavivirus*), leptospirosis (*Leptospira*), porcine parvovirus (*Parvovirus*), pseudorabies (*Varicellovirus*), swine brucellosis (*Brucella*), swine influenza, and toxoplasmosis (*Toxoplasma*).

CONSERVATION STATUS Exotic in Alabama; an invasive species.

COMMENTS Although commonly hunted for sport and for food in Alabama, this is one of the most destructive species of invasive mammals. They have significant adverse impacts on the land by destroying crops, property, native fauna, and natural vegetation and by rooting in large areas, which disrupts soils, seed banks, tree roots, and natural succession. *Sus* is Latin for "pig"; *scrofa* is Latin for "a sow."

REFERENCES Forrester (1992), Gingerich (1994), Tolleson et al. (1995), Whitaker and Hamilton (1998), Nowak (1999), Harveson et al. (2000), Sweeney et al. (2003), Best (2004*a*), Hampton et al. (2004), Gauss et al. (2005).

Deer, Elk, Caribou, and Moose

Family Cervidae

Cervids include 19 genera and 51 species. Color is usually a shade of brown, although some may be spotted (young are usually spotted). Antlers vary in size and shape but are usually present in males and are shed annually; both sexes of caribou (*Rangifer tarandus*) have antlers. Cervids are gregarious and some are seasonally migratory. These herbivores consume grasses and the tender bark, twigs, and shoots of trees. Habitats include Arctic tundra, dense woodlands, grassy plains, sparsely covered brush country, deserts, lowland swamps, and tropical rainforests. Geographic range of the family includes the Americas, Europe, Asia, and northwestern Africa. In Alabama, elk were present until about 2,500 years ago and were reintroduced and extirpated, fallow deer were introduced and possibly still occur in the wild, and white-tailed deer are widespread and common.

White-tailed Deer

Odocoileus virginianus

IDENTIFICATION The white-tailed deer is identified by its large size and by its tail, which is brown above and white below. Males have antlers with 1 main beam and tines that rise vertically. The back is reddish brown to tan in summer and gray brown in winter. The venter, underside of tail, insides of legs, and chin are white. Fawns have reddish-brown pelage with white dorsal spots that disappear at about 3–4 months of age.

DENTAL FORMULA i 0/3, c 0/1, p 3/3, m 3/3, total = 32.

SIZE AND WEIGHT Range in size of largest males of various subspecies:
    total length, 1,041–2,400 mm / 41.6–96.0 inches
    tail length, 100–365 mm / 4.0–14.6 inches
    hind foot length, 279–538 mm / 11.2–21.5 inches
    shoulder height, 533–1,067 mm / 21.3–42.7 inches
Weight of adult males was 90–135 kg (198.0–297.0 pounds) in northern subspecies to less than 23 kg (50.6 pounds) in the subspecies that occurs in the Florida Keys. Males are larger than females.

A female white-tailed deer (*Odocoileus virginianus*).

DISTRIBUTION Statewide in Alabama. White-tailed deer occur from central Canada across most of the United States and southward into much of northern South America.

ECOLOGY White-tailed deer inhabit a wide range of habitats, from semi-arid environments in the western United States to subtropics and rain-forests in Central and South America. This species is most numerous in the southeastern United States, but it is a prized game species throughout most of its geographic range. Carrying capacity varies greatly among regions, but abundance is directly related to the number and distribution of nonwooded areas within forests. In some intensively managed areas with abundant food and shelter, populations may reach 80 white-tailed deer/km^2. Populations have increased due to reintroductions into areas where overhunting once occurred and in response to the mosaic of habitats

produced by clearing forests, farming, and other agricultural activities. However, conversion of forests to even-aged monocultures of pine trees does not favor white-tailed deer. Diet includes agricultural crops (e.g., corn, wheat, soybeans, alfalfa, fruit trees), grasses, forbs, fruits, mushrooms and other fungi, and succulent leafy vegetation; acorns are readily consumed when available. White-tailed deer may cause significant damage to agricultural crops, orchards, nurseries, and ornamental plants in yards and gardens. Regulated hunting helps control overpopulation in Alabama. Historically, cougars and red wolves were the primary predators of white-tailed deer in Alabama; bobcats and coyotes often concentrate on fawns, which can reduce populations on a local scale. Currently, the most widespread nonhuman predators are domestic dogs.

For the first few days following birth, young white-tailed deer usually remain still and hide in vegetation while their mother forages.

BEHAVIOR Most activity is at dusk and dawn, but it varies with season, weather, and hunting pressure. The most common daily activity is foraging, which occurs even as individuals move to and from foraging sites. Home ranges are well defined and occupied year after year. White-tailed deer are not territorial, but they will defend bedding sites and limited resources, and tending males will defend females in estrus. Yearlings move

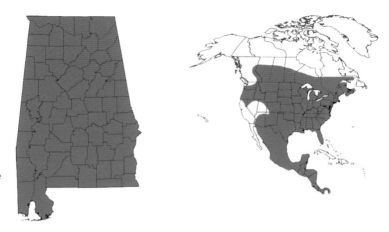

Distribution of the white-tailed deer in Alabama and North America.

farther and more often than other age classes. Males have larger home ranges and move farther than females, especially during the breeding season. Scrapes, rubs, and marking behaviors are important means of social communication by males; these provide visual and olfactory cues that establish dominance and facilitate intersexual interactions. Vocalizations include mews, whines, distress calls, alert-snorts, snort-wheezes, grunts, and bleats. Foot stamping is common when a predator or other intruder is detected. To avoid predators, white-tailed deer may remain motionless or flee with the tail erect and waving; this behavior exposes the white underside of the tail and the white rump patch.

LIFE HISTORY In Alabama, most breeding takes place in winter. Gestation is about 200 days. In summer, 1 young is born, but sometimes twins are born, rarely triplets. Young begin nursing immediately following birth, doubling their weight during the first 2 weeks, tripling their weight by 1 month old, and beginning to graze when a few weeks old. Males reach sexual maturity at about 18 months old and females can breed at 6–7 months old, but usually not until about 18 months old. White-tailed deer can potentially live for 20 years, but few live more than 10 years. Common causes of death are legal and illegal hunting, accidents with motorized vehicles, parasites, diseases, entanglement in fences, predation, and old age.

PARASITES AND DISEASES Ectoparasites include mites (*Demodex, Eutrombicula, Psoroptes*), ticks (*Amblyomma, Dermacentor, Ixodes, Rhipicephalus*), lice (*Solenopotes*), fleas (*Ctenocephalides*), and flies (*Cal-*

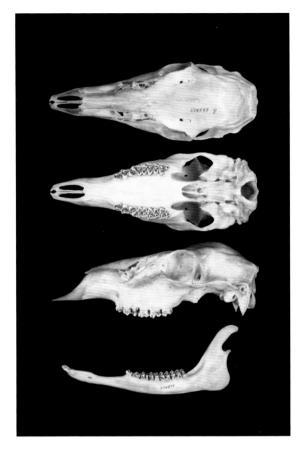

Dorsal, ventral, and lateral views of the cranium, and lateral view of the mandible of a female white-tailed deer. Greatest length of cranium is 272.5 mm.

liphora, Cephenamyia, Cochliomyia, Phaenicia, Sarcophaga). Endoparasites include protozoans (*Babesia, Eimeria, Sarcocystis, Toxoplasma, Trypanosoma*), cestodes (*Moniezia, Taenia*), nematodes (*Capillaria, Cooperia, Dictyocaulus, Elaeophora, Eucyathostomum, Gongylonema, Haemonchus, Mazamastrongylus, Monodontus, Nematodirus, Oesophagostomum, Ostertagia, Parelaphostrongylus, Setaria, Strongyloides, Trichostrongylus, Trichuris*), and trematodes (*Fascioloides, Paramphistomum*). Diseases and disorders include anaplasmosis (*Anaplasma*), anthrax (*Bacillus*), bluetongue virus (*Orbivirus*), bovine viral diarrhea (*Pestivirus*), brucellosis (*Brucella*), cutaneous fibroma, eastern equine encephalomyelitis (*Alphavirus*), Highlands J virus (*Alphavirus*), infectious bovine rhinotracheitis (*Varicellovirus*), leptospirosis (*Leptospira*), Lyme disease (*Borrelia*), parainfluenza (Paramyxoviridae), Saint Louis encephalitis (*Flavivirus*), tularemia (*Francisella*), and vesicular stomatitis (*Vesiculovirus*).

CONSERVATION STATUS In Alabama, this is the most economically important game species, with 300,000–500,000 white-tailed deer harvested each year. Lowest conservation concern in Alabama.

COMMENTS *Odocoileus* is from the Greek *odous*, meaning "tooth," and *koilos*, meaning "hollow," referring to prominent depressions in the molar teeth; *virginianus* is Latin for "of Virginia," referring to the place where the type specimen was obtained.

REFERENCES Smith (1991), Forrester (1992), Best (2004*a*).

Elk
Cervus elaphus

IDENTIFICATION This large ungulate is easily distinguished from the native white-tailed deer and from the introduced fallow deer by its much larger size, pale brown to chocolate color, and yellowish rump patch. Adult males have antlers and a dark mane on the neck.

DENTAL FORMULA i 0/3, c 1/1, p 3/3, m 3/3, total = 34.

SIZE AND WEIGHT Average range in size:
head and body length, 2,250–2,500 mm / 90.0–100.0 inches
tail length, 125–198 mm / 5.0–7.9 inches
shoulder height, 1,400–1,500 mm / 56.0–60.0 inches
weight, 200–350 kg / 440.0–770.0 pounds
Females do not have antlers and they average smaller than males.

DISTRIBUTION Also known as red deer, this species naturally occurred across much of the Northern Hemisphere in Europe, Asia, northern Africa, and North America. Red deer have been introduced into southern South America, Australia, and New Zealand. Although populations in

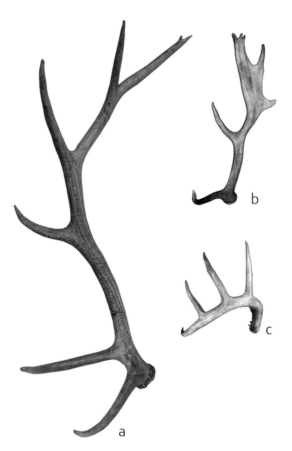

Mature elk usually have large antlers compared to those of other male Cervidae in Alabama: a) elk; b) fallow deer; and c) white-tailed deer. Greatest straight-line lengths of these antlers are 994, 495, and 304 mm, respectively.

North America are the same species as red deer, they are referred to as elk. Native populations of elk probably occurred in what is now Alabama until about 2,500 years ago. Historically, elk were present from northern Canada across most of the United States and into northern Mexico. Currently, elk have been widely reintroduced into Canada and the United States, and populations are well established in the Pacific Northwest and Rocky Mountains in western North America.

ECOLOGY Adapted to habitats that vary from dense coniferous forests to woodlands, nonforested valleys, and prairies, elk may seek refuge in wooded areas and graze in ecotones associated with forests, meadows, and grasslands. Many populations are migratory and move from their high-elevation range in summer to low-elevation sites in winter. Diet is

Distribution of the elk in Alabama and North America.

extremely variable, as elk occur in many types of vegetation across Canada and the United States; grasses, forbs, woody species, or other plants may be predominant in certain seasons or locations. Predators of adults include gray wolves (*Canis lupus*) and cougars; brown bears (*Ursus arctos*), American black bears, and coyotes may prey on newborns, young-of-the-year, or unhealthy elk. Although hunting is now regulated by wildlife agencies, humans hunted elk nearly to extinction throughout much of their native range before the 1900s.

BEHAVIOR Elk are gregarious, with groups varying in age and sex composition during the year. Usually, about 90% of the day is spent foraging and resting; the remainder is spent idling and in movement. Males maintain a harem of females and young during the mating season in autumn. At other times, herds are composed of mixed ages and sexes, and males may form bachelor herds. During the breeding season, males have a variety of dominance displays including posture and movements of antlers, wallowing, self-anointing with urine, and bugling, which is deeper and louder in larger males. Pitch varies noticeably during the bugle from low to high frequencies. As herds graze undisturbed in open habitats, individuals constantly emit a variety of vocalizations, which cease when a predator is detected. Thus, silence is a signal of danger and may be accompanied by a rigid stance, erect posture, a halting gait, and side-to-side movements of the head.

LIFE HISTORY Males 3 years of age or older are more sexually active than younger males. Although older males do most of the breeding, younger

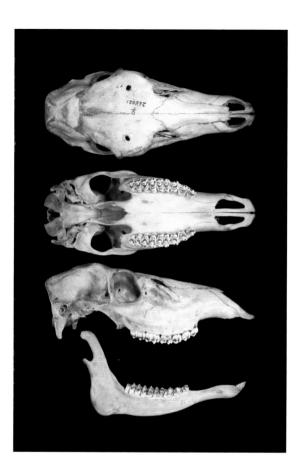

Dorsal, ventral, and lateral views of the cranium, and lateral view of the mandible of a female elk. Greatest length of cranium is 344.1 mm.

males with mature, hardened antlers have abundant sperm and they may contribute significantly to breeding when there are few older males. Females may become pregnant as yearlings, but usually not until 2 years of age. Gestation is about 250 days (range is 247–265 days). One young is born each year, rarely 2. Fertility is high in females 7–14 years old but declines after 14 years of age. Life span is usually 14–15 years in the wild, but males may have an average life span of only 3 years in heavily hunted populations.

PARASITES AND DISEASES Ectoparasites include mites (*Psoroptes*), ticks (*Dermacentor, Otobius, Ixodes*), lice (*Bovicola, Damalinia, Tricholipeurus*), and flies (*Cephenemyia, Chrysops, Haemotobia, Hybomitra, Lipoptena, Muscina, Symphoromyia, Tabanus*). Endoparasites include protozoans (*Eimeria, Sarcocystis, Trypanosoma*), cestodes (*Echinococ-*

cus, Moniezia, Taenia, Thysanosoma), nematodes (*Capillaria, Cooperia, Dictyocaulus, Elaeophora, Heamonchus, Marshallagia, Nematodirella, Nematodirus, Oesophagostomum, Orthostrongylus, Ostertagia, Parelaphostrongylus, Protostrongylus, Setaria, Trichostrongylus, Trichuris, Wehrdikmansia*), and trematodes (*Fasciola, Fascioloides*). Diseases and disorders may include anthrax (*Bacillus*), arthritis, botulism (*Clostridium*), brucellosis (*Brucella*), leptospirosis (*Leptospira*), listeriosis (*Listeria*), lump jaw (*Actinomyces*), necrotic stomatitis (*Fusobacterium*), neoplasms, rabies (*Lyssavirus*), and tetanus (*Clostridium*).

CONSERVATION STATUS No wild population is present in Alabama.

COMMENTS The elk is included here because it is an extant species, it occurred in what is now Alabama as recently as about 2,500 years ago, and it once inhabited much of the contiguous United States. In an attempt to establish a breeding population, 55 elk from Wyoming were released in Calhoun, Pickens, Sumter, and Tuscaloosa counties in 1916. Because of poaching, disease, and the unwillingness of landowners to tolerate damage to crops, the introduction was unsuccessful. The last elk was killed in 1921. *Cervus* is Latin for "deer"; *elaphus* is from the Greek *elaphos*, meaning "deer."

REFERENCES R. H. Allen (1965), Curren (1977), Bryant and Maser (1982), Geist (1982), Kistner et al. (1982), Nelson and Leege (1982), Skovlin (1982), Taber et al. (1982), Best (2004*a*).

Fallow Deer
Dama dama

IDENTIFICATION Adult males have palmate antlers, a prominent Adam's apple, and spotted pelage. Females have spotted pelage but usually no antlers. Color is the most variable of any species of cervid. A dark dorsal stripe extends from the nape of the neck to the tip of the tail.

DENTAL FORMULA i 0/3, c 0/1, p 3/3, m 3/3, total = 32.

SIZE AND WEIGHT Shoulder height of males is about 0.9–1 m (35.6–39.6 inches). In a captive population in Europe, average and range in weight of 20 males was 67 (46–80) kg / 147.4 (101.2–176.0) pounds and of 15 females was 44 (35–52) kg / 96.8 (77.0–114.4) pounds. Males are larger than females.

DISTRIBUTION During the early 1900s, attempts were made to establish this species in Alabama. Fallow deer were introduced into the Camden and Miller's Ferry area of Dallas County, where a small wild population may still exist. In 1932, 18–20 fallow deer escaped from their pens in Wilcox County and a feral population continued to exist in the area until at

Distribution of the fallow deer in Alabama and North America.

least the 1970s. Current status of this population is unknown. Natural range included the European region along the Mediterranean Sea, Asia Minor, and possibly northern Africa; breeding populations of fallow deer now exist in about 38 countries.

ECOLOGY Although most populations are in warm-humid climates, some occur in cool-humid and warm-dry areas. Fallow deer may occupy mixed forests, broad-leaved forests, subalpine vegetation, grasslands, woodlands, scrublands, and savannas. Home range varies depending on availability of food and shelter, degree of disturbance, density, and climate. Generally, males have larger home ranges and move more within home ranges than females. Diet is variable and includes grasses, forbs, mast, broad-leaved trees, and a variety of shrubs. In Alabama, potential predators include humans, bobcats, American black bears, and coyotes.

BEHAVIOR Except for some solitary males, fallow deer are gregarious and usually occur in groups all year. Size and composition of herds depend on the environment, amount of disturbance, abundance, season, and time of day. Peak foraging times are dusk and dawn, but fallow deer may feed throughout the day and night. During summer, adult males are often solitary, but they begin to join groups of females in early autumn. In the period before the mating season, males spend most of their time establishing a territory. Males mark territories by pawing the ground and urinating in the resulting scrape; they thrash understory vegetation with their antlers and deposit scent from suborbital glands. At this time, males produce low-pitched groans and grunts, belch, and spar with other males. During the breeding season, the male approaches the female several

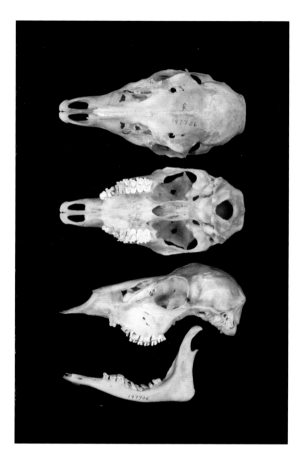

Dorsal, ventral, and lateral views of the cranium, and lateral view of the mandible of a female fallow deer. Greatest length of cranium is 247.5 mm

times, and usually she utters a high-pitched whine and moves away, but eventually she allows copulation, which may last 5 minutes but is usually only about 10 seconds. Several types of vocalizations occur; e.g., barking, bleating, mewing, and groaning. When fleeing, an adult female tends to lead, with the herd following her in single file in order of decreasing rank.

LIFE HISTORY Most mating takes place in October, but females are receptive from September to January. Gestation is 33–35 weeks. Prior to parturition, females become secretive and look for a hiding place to give birth. Immediately following birth, the mother licks the newborn, which aids in establishing the mother-offspring bond. Usually, 1 young is born each year during summer. Females usually return to the herd after 2–10 days. Newborns weigh about 4.5 kg. Fawns are born with 20 deciduous teeth. Eruption of molars begins at 3–4 months of age, permanent premolars erupt at 17–26 months, and the last molar is not completely functional until about 3 years of age. Mothers begin weaning young at about 20 days after birth, but young nurse until about 7 months old. Females can become pregnant at 6–7 months of age, but usually not until about 16 months old. Females are fully mature at 4–6 years of age. Although production of sperm may begin as early as 7 months of age, mature sperm appear at about 16 months, when testes and epididymides are much larger. Males become fully mature and reach their greatest mass at 5–9 years of age; they usually do not breed until about 4 years old. In captivity, life span may be more than 20 years.

PARASITES AND DISEASES Ectoparasites include mites (*Sarcoptes*), ticks (*Amblyomma, Dermacentor, Haemaphysalis, Ixodes*), lice (*Bovicola, Damalinia, Solenopotes*), and flies (*Cephenemyia, Hypoderma, Lipoptena, Neolipoptena*). Endoparasites include protozoans (*Eimeria, Sarcocystis*), cestodes (*Echinococcus, Moniezia, Taenia*), nematodes (*Apteragia, Artionema, Bicaulus, Bunostomum, Capillaria, Chabertia, Cooperia, Cutifilaria, Dictyocaulus, Gongylonema, Haemonchus, Muellerius, Nematodirus, Oesophagostomum, Onchocerca, Ostertagia, Parelaphostrongylus, Protostrongylus, Setaria, Skrjabinagia, Spiculopteragia, Strongyloides, Teladorsagia, Thelazia, Trichostrongylus, Trichuris, Wehrdikmansia*), and trematodes (*Dicrocoelium, Fasciola, Fascioloides, Paramphistomum*). Diseases and disorders include avian and bovine tuberculosis (*Mycobacterium*), brucellosis (*Brucella*), cerebrocortical necrosis, enzootic ataxia, erysipelas (*Erysipelothrix*), foot-and-mouth disease (*Aphtae*), leptospirosis (*Leptospira*), lump jaw (*Actinomyces*), lymphosarcoma, pseudotuberculosis (*Corynebacterium, Pasteurella*), rabies (*Lyssavirus*), and ringworm (fungal dermatophytes).

CONSERVATION STATUS Exotic in Alabama.

COMMENTS Fallow deer are commonly displayed in zoos and parks, and they are raised in many parts of the world for hunting, meat, and velvet from their antlers. *Dama* is Latin for "fallow deer."

REFERENCES R. H. Allen (1965), Brugh (1971), Espmark and Brunner (1974), Chapman and Chapman (1975), D. I. Chapman (1977), Feldhamer et al. (1988), Best (2004*a*).

Antelopes, Bison, Sheep, Goats, and Cattle

Family Bovidae

Bovidae contains 50 genera and 143 species. Color varies but is usually brown, black, or gray; size ranges from 25 cm at the shoulder in the royal antelope (*Neotragus pygmaeus*) to 2 m in the American bison. Bovids are primarily grazers, feeding by twisting leaves or grasses around the tongue and cutting them off against the lower incisors. Most inhabit grasslands, but they also occur in rocky habitats, mountainous areas, forests, swampy habitats, deserts, scrub, and tundra. Native range includes most of North America, Europe, Asia, and Africa, but domesticated species occur nearly worldwide. In recent times, only 1 bovid was native to what is now Alabama (American bison), but it is now extirpated in the wild in the state.

American Bison

Bison bison

IDENTIFICATION This is a massive species. The large head has rounded and pointed horns in both sexes, there is a prominent hump on the shoulders, hair is much longer on the head, neck, and shoulders than on the hindquarters, the neck is relatively short, and the tail is short and tufted.

DENTAL FORMULA i 0/3, c 0/1, p 3/3, m 3/3, total = 32.

SIZE AND WEIGHT Range in size for males and females, respectively:

    total length, 3,040–3,800, 2,130–3,180 mm / 10.1–12.7, 7.1–10.6 feet

    tail length, 330–910, 300–510 mm / 13.2–36.4, 12.0–20.4 inches

    hind foot length, 580–680, 500–530 mm / 23.2–27.2, 20.0–21.2 inches

    shoulder height, 1,670–1,860, 1,520–1,570 mm / 66.8–74.4, 60.8–62.8 inches

    Weight 544–907, 318–545 kg (1,196.8–1,995.4, 699.6–1,199.0 pounds), but 1 semidomesticated male weighed 1,724 kg (3,792.8 pounds). Males are significantly larger than females and differ in massiveness of horns, skull, and shoulders.

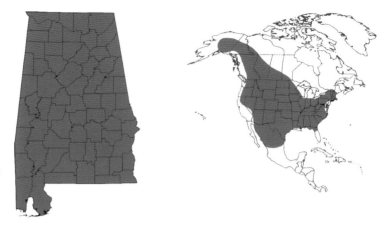

Distribution of the American bison in Alabama and North America.

DISTRIBUTION Extirpated in Alabama, but this species was probably present nearly statewide in open habitats. Although once widely distributed from central Alaska, western Canada, and most of the United States into central Mexico, and with a population numbering in the tens of millions, the American bison was hunted to near extinction in the late 1800s. This massive killing occurred with the passive blessing of the United States government, which viewed elimination of the bison as a way to subjugate western tribes of Native Americans. By the early 1900s, the only remaining wild populations were in Wood Buffalo National Park, Canada, and in Yellowstone National Park, Wyoming. Presently, isolated populations are maintained in parks and zoos and on government lands and private ranches.

ECOLOGY Historically, habitats occupied by American bison included grasslands and parklands of western North America and open habitats in eastern deciduous forests; American bison also occurred in habitats ranging from semideserts to boreal forests. Forested areas are used primarily for shade, escape from insects, and protection during inclement weather in winter. In all seasons, American bison are grazers that consume primarily grasses and sedges. Much of the geographic range of the American bison is in areas with heavy snowfall in winter. Musculature of the neck and shoulders may allow these animals to swing their head from side to side to clear away deep snow as they graze. In some parts of their geographic range, gray wolves (*Canis lupus*) may be significant predators of American bison, and humans hunt bison recreationally and harvest them commercially.

Dorsal, ventral, and lateral views of the cranium, and lateral view of the mandible of a female American bison. Greatest length of cranium is 484.0 mm.

BEHAVIOR Most activity is diurnal, with several grazing periods interspersed with loafing and ruminating. The sense of smell is acute and appears to be important in detecting danger. Depending on season and availability of forage, herds vary, but mixed herds of females of all ages, calves, males 2–3 years old, and 1 or a few older males occur together all year. Older males may be solitary or may aggregate into herds with other males, but during the breeding season, these males often join mixed herds. While herds change often, the bond between mother and calf appears to be the most stable relationship among American bison. Free-ranging populations are seasonally migratory, both directionally (up to 240 km) and elevationally, in response to changes in forage, weather conditions, temperature in spring, and snowfall in autumn. In summer, daily movements may average 3 km depending on many factors, including distribution of foraging sites, distance to water, and presence of large numbers of biting insects.

LIFE HISTORY Breeding is in summer, gestation is 285 days, and one 15–25-kg, reddish-tan calf is usually born during April or May. Twins are rare. Calves are precocious: they usually stand and begin nursing 10–30 minutes following birth; they may begin grazing within a week of birth; and they are weaned by 8–12 months of age. Both sexes usually attain sexual maturity at 2–4 years old. Greatest mass is reached at 10–12 years of age in males and by 3 years old in females. At 2–3 years of age, females usually conceive their first calf, and although sexually mature at about 3 years, males usually do not breed until 6 years old. Although life span is not well documented in the wild, old age is reached at 12–15 years; some survive more than 20 years, and a few live more than 40 years.

PARASITES AND DISEASES Ectoparasites include mites (*Demodex, Speleognathus*), ticks (*Dermacentor*), lice (*Damalinia*), and flies (*Aedes, Cochliomyia, Hypoderma, Musca, Simulium, Stomoxys, Symphoromyia, Tabanus*). In captive herds, more than 30 species of endoparasites have been reported, but the incidence of parasites in wild populations appears low. Endoparasites include protozoans (*Babesia, Eimeria, Sarcocystis, Toxoplasma*), cestodes (*Echinococcus, Moniezia*), nematodes (*Chabertia, Cooperia, Dictyocaulus, Haemonchus, Nematodirella, Nematodirus, Oesophagostomum, Ostertagia, Setaria, Strongylus, Trichostrongylus, Trichuris*), and trematodes (*Fasciola, Fascioloides, Paramphistomum*). Diseases include anaplasmosis (*Anaplasma*), anthrax (*Bacillus*), blackleg (*Clostridium*), brucellosis (*Brucella*), hemophilosis (*Histophilus*), and tuberculosis (*Mycobacterium*). Persons in the cattle industry are especially concerned that herds of American bison remain brucellosis-free because of the ease of transmission of brucellosis between cattle and bison.

CONSERVATION STATUS Extirpated in Alabama.

COMMENTS *Bison* is Greek for "wild ox."

REFERENCES J. A. Allen (1876), Hornaday (1889), Boyd (1936), Swanton (1938), Rostlund (1960), Hall (1981), Meagher (1986), Reynolds et al. (2003), Best (2004*a*).

Appendix 1

Cetaceans (Whales and Dolphins) that Occur, or May Occur, in the Gulf of Mexico and in Coastal Alabama

Below is an attempt to provide a complete listing of whales and dolphins that potentially occur in coastal Alabama. Although there is information about many kinds of cetaceans in other parts of the world, including the Gulf of Mexico, little is known about these animals in coastal Alabama. There are some records of strandings and observations, but no long-term comprehensive survey has been conducted. The Dauphin Island Sea Lab has an excellent program focusing on the endangered West Indian manatee and seeks to document occurrences of this species as well as cetaceans. A proper assessment of the conservation status of most cetaceans in the state must await verification of their occurrence and abundance. We hope this listing will inspire research on these fascinating species in coastal Alabama.

Order Cetacea
    Suborder Mysticeti, baleen whales
        Family Balaenidae, right and bowhead whales
            North Atlantic right whale, *Eubalaena glacialis*
        Family Balaenopteridae, rorquals
            Common minke whale, *Balaenoptera acutorostrata*
            Sei whale, *Balaenoptera borealis*
            Bryde's whale, *Balaenoptera edeni*
            Blue whale, *Balaenoptera musculus*
            Fin whale, *Balaenoptera physalus*
            Humpback whale, *Megaptera novaeangliae*
    Suborder Odontoceti, toothed whales and dolphins
        Family Delphinidae, ocean dolphins
            Long-beaked common dolphin, *Delphinus capensis*
            Short-beaked common dolphin, *Delphinus delphis*
            Pygmy killer whale, *Feresa attenuata*
            Short-finned pilot whale, *Globicephala macrorhynchus*
            Long-finned pilot whale, *Globicephala melas*
            Risso's dolphin, *Grampus griseus*
            Fraser's dolphin, *Lagenodelphis hosei*

Killer whale, *Orcinus orca*
Melon-headed whale, *Peponocephala electra*
False killer whale, *Pseudorca crassidens*
Pantropical spotted dolphin, *Stenella attenuata*
Clymene dolphin, *Stenella clymene*
Striped dolphin, *Stenella coeruleoalba*
Atlantic spotted dolphin, *Stenella frontalis*
Spinner dolphin, *Stenella longirostris*
Rough-toothed dolphin, *Steno bredanensis*
Bottlenose dolphin, *Tursiops truncatus*
Family Physeteridae, sperm whales
Pygmy sperm whale, *Kogia breviceps*
Dwarf sperm whale, *Kogia sima*
Sperm whale, *Physeter catodon*
Family Ziphiidae, beaked whales
Sowerby's beaked whale, *Mesoplodon bidens*
Blainville's beaked whale, *Mesoplodon densirostris*
Gervais' beaked whale, *Mesoplodon europaeus*
Cuvier's beaked whale, *Ziphius cavirostris*

REFERENCES Würsig et al. (2000), Wilson and Reeder (2005), R. H. Carmichael (in litt.).

Appendix 2

Metric to English Measurement Conversions

Length

Symbol	When you know	Multiply by	To obtain	Symbol
mm	millimeters	0.04	inches	in
cm	centimeters	0.4	inches	in
m	meters	3.3	feet	ft
m	meters	1.1	yards	yd
km	kilometers	0.6	miles	mi

Area

Symbol	When you know	Multiply by	To obtain	Symbol
cm^2	square centimeters	0.16	square inches	in^2
m^2	square meters	1.2	square yards	yd^2
km^2	square kilometers	0.4	square miles	mi^2
ha	hectares (10,000 m^2)	2.5	acres	ac

Weight

Symbol	When you know	Multiply by	To obtain	Symbol
g	grams	0.035	ounces	oz
kg	kilograms	2.2	pounds	lb

Volume

Symbol	When you know	Multiply by	To obtain	Symbol
ml	milliliters	0.03	fluid ounces	fl oz
l	liters	2.1	pints	pt
l	liters	1.06	quarts	qt
l	liters	0.26	gallons	gal
m^3	cubic meters	35	cubic feet	ft^3
m^3	cubic meters	1.3	cubic yards	yd^3

Temperature

$$°C = (°F - 32) \times 0.555$$
$$°F = (°C \times 1.8) + 32$$

Glossary

ACANTHOCEPHALAN A spiny-headed worm of the class Acanthocephala; an internal parasite occurring most often in fish, birds, and other vertebrates, but sometimes in mammals.

ADULT Usually, a sexually mature individual.

ADULT PELAGE Hair that is characteristic of adults of a species.

AERIAL Ability to fly or to leap high into the air. Bats are aerial mammals.

ALTRICIAL Newborns that are born with closed eyes and no hair; they are immobile and nearly helpless and require extended periods of parental care.

ANEMIA Reduced oxygen-carrying capacity of blood resulting from too few red blood cells, or other causes.

ANTERIOR Front end of an organism.

ANTHRAX A highly contagious bacterial disease caused by *Bacillus anthracis*; transmission may be by ingesting or inhaling bacterial endospores on vegetation or consuming tissues of infected animals.

ANTLER A frequently branched, paired, bony cranial projection present in deer.

AQUATIC Pertaining to life in water.

ARBOREAL Climbing in trees, living in trees, or both.

AUDITORY BULLA (*pl.* BULLAE) Rounded, thin-walled, bony capsule enclosing the middle and inner ear of most mammals.

BACHELOR COLONY In bats, a social organization where adult males and nonreproductive females occupy roosts that contain few or no pregnant or lactating females and their young; *see* maternity colony.

BLASTOCYST Early stage of development during which the embryo implants in the uterine wall.

BRUCELLOSIS Bacterial disease caused by species of *Brucella*; a common symptom is spontaneous abortion of a fetus; may be transmitted among bison, cattle, elk, and other mammals.

BULLAE *See* auditory bulla.

BURROW Tunnel excavated and inhabited by an animal.

CACHE Collection of stored items or food.

CALCAR Spur of cartilage that projects medially from the ankle in many species of bats; helps support the patagium; a keel may extend posteriorly along the calcar.

CALIFORNIA ENCEPHALITIS Viral disease caused by *Orthobunyavirus*; transmitted by mosquitoes.

CANINE DISTEMPER Common viral disease in members of the order Carnivora (except domestic cats); caused by *Morbillivirus*.

CANOPY Uppermost spreading, branchy layer of a forest.

CARNASSIALS In many Carnivora, the last upper premolar or the first lower molar; the largest of the molariform teeth that are adapted for shearing rather than for crushing or tearing.

CARNIVORE Any flesh-eating animal.

CATERWAUL Wailing sound made during breeding, especially in cats.

CESTODE A tapeworm; an internal parasite in the class Cestoda.

CHAGAS DISEASE Parasitic disease caused by the flagellate protozoan *Trypanosoma cruzi*; the protozoan is usually transmitted by triatomine bugs belonging to the family Reduviidae (also known as assassin bugs or kissing bugs).

CHEEK TEETH Premolars and molars; all teeth posterior to the canines; molariform teeth.

CLASS Taxonomic grouping that includes 1 or more orders; e.g., Mammalia.

COPROPHAGY Ingestion by an animal of its own undigested feces.

COPULATION Sexual coupling of 2 individuals.

CRANIUM Braincase of vertebrates.

DECIDUOUS DENTITION Juvenile or baby teeth of mammals.

DELAYED FERTILIZATION A condition where sperm remain viable in the uterus for an extended time before ovulation and then fertilization occur; e.g., in many bats, mating occurs in autumn and fertilization occurs the following spring.

DELAYED IMPLANTATION A condition where fertilization occurs soon after mating and the embryo develops to the blastula stage, but the blastula does not implant into the uterine wall; e.g., in some carnivores, mating occurs in summer and implantation occurs the following spring.

DEN Cave, hollow log, burrow, or other cavity used for shelter; includes shelters constructed by mammals from twigs, limbs, and other objects.

DENTAL FORMULA A convenient way of expressing the numbers of different kinds of teeth; e.g., i 3/3, c 1/1, p 4/4, m 3/3, total = 44. Letters designate incisors, canines, premolars, and molars, respectively. In shrews, "u" indicates unicuspid teeth. Numbers to the left of the diagonal line are the number of teeth of each kind on 1 side of the upper jaw; those to the right of the diagonal line indicate the number on 1 side of the lower jaw.

DIGIT Any finger or toe.

DISTAL Situated away from the base or area of attachment; e.g., tip of tail.

DIURNAL Active primarily during daylight hours; opposite of nocturnal.

DOMESTICATED A species that has been bred for use by humans. Examples are domestic dogs and cats.

DORSAL Referring to the back, dorsum, or upper surface.

DORSUM Back, dorsal, or upper surface.

EASTERN EQUINE ENCEPHALOMYELITIS A viral disease also known as sleeping sickness; it is caused by *Alphavirus* and is transmitted from birds to mammals by mosquitoes.

ECHOLOCATION Sonar; sensing objects and surfaces by emitting pulses of sound and receiving and evaluating echoes reflected by the objects or surfaces. Used by most bats.

ECTOPARASITE Parasite on or in the integument of an animal; e.g., mites, ticks, lice, fleas, and some species of crustaceans, bugs, flies, and beetles.

EMERGENT VEGETATION Those parts of aquatic plants that grow toward, or extend above, the surface of water.

ENCEPHALITIS Inflammation of the brain.

ENCEPHALOMYELITIS Inflammation of the brain and spinal cord.

ENDANGERED Population, subspecies, or species that is in danger of extinction throughout its entire geographic range or within a specific part of its range.

ENDOPARASITE Parasite occurring within the body of an animal; e.g., protozoans, acanthocephalans, cestodes, nematodes, and trematodes.

ESTRUS Usually an annual period, from hours to weeks in duration depending on species, when a female is sexually receptive to a male and reproductively fertile.

EXTANT Currently living; not extinct.

EXTINCTION Complete and irrevocable disappearance of all individuals constituting a taxon.

EXTIRPATION Extinction of a species within a particular geographic region, such as American bison in Alabama, but not from its total worldwide distribution.

FAMILY Taxonomic category just below order in scientific classification; contains 1 or more genera.

FECES Animal excrement; guano; scat.

FENESTRA (*pl.* FENESTRAE) A labyrinth of small openings; e.g., in the rostrum of rabbits.

FERAL Domestic animals that have reverted to a wild state and are no longer dependent on humans.

FLEAS Small, wingless insects in the order Siphonaptera that have mouthparts adapted for piercing skin and sucking blood; common ectoparasites of mammals.

FLOODPLAIN Usually dry land adjacent to a stream or river that is covered with water during floods.

FOOT-AND-MOUTH DISEASE Infectious viral disease caused by *Aphtae epizooticae*; also known as hoof-and-mouth disease.

FORAGE To obtain food, such as by grazing or searching for insects.

FOSSORIAL Living under the surface of the ground; adapted for digging and burrowing.

FUR Soft pelage, especially dense underhair with definitive growth; serves primarily as insulation.

GENUS (*pl.* GENERA) Classification category within families; contains 1 or more species.

GESTATION Period from fertilization to birth.

GLEAN To capture nonflying prey by collecting them from a surface, such as from the ground, a leaf, or a branch.

GREGARIOUS Living in groups; e.g., Brazilian free-tailed bats.

GUANO Feces, especially of bats or birds; may be harvested for its nitrates and phosphates and used as an agricultural fertilizer; has been used in the manufacture of gunpowder.

GUARD HAIRS Long, usually coarser outer hairs that lie over the shorter underfur of most mammals.

HABITAT Any place that is occupied by, or has the potential to support, a population of a species.

HAIR Cylindrical, filamentous outgrowth of the epidermis that consists of cornified epidermal cells. Present only in mammals.

HALLUX Medial digit on the hind limb of many vertebrates; the big toe.

HAMMOCK Elevated island of unflooded habitat within marshy or swampy areas.

HANTAVIRUS Group of viral diseases caused by *Hantavirus*; usually transmitted by exposure to urine, feces, or saliva of infected rodents.

HECTARE Area of ground (abbreviated ha) equal to 2.47 acres.

HERBIVORE An animal that consumes plant material as the primary component of its diet.

HIBERNACULUM (*pl.* HIBERNACULA) A place occupied by an animal during winter.

HIBERNATION Period of inactivity usually induced by cold temperatures; usually characterized by lowered body temperature and a depressed metabolic rate.

HISTOPLASMOSIS Fungal disease of the respiratory system that is caused by *Histoplasma capsulatum*, which occurs in accumulations of bird and bat guano.

HOOF-AND-MOUTH DISEASE *See* foot-and-mouth disease.

IMPLANTATION Process by which the blastula embeds in the wall of the uterus.

INSECTIVORE An animal that consumes insects and other small invertebrates as the primary component of its diet.

INTRODUCTION Intentional, negligent, or accidental placement of individuals of a species into a location where that species did not exist previously; e.g., releasing wild boars from Eurasia into North America.

INVERTEBRATE An animal without a spinal column, such as insects, spiders, millipedes, snails, and worms.

JUVENILE An individual that is immature or undeveloped.

LACTATION The formation and secretion of milk by mammary glands for nourishing the developing young after birth. Present only in mammals.

LARVA (*pl.* LARVAE) Developmental form of many higher insects (as well as mites and ticks), e.g., grubs, maggots, or caterpillars, which is fundamentally unlike the adult in appearance and undergoes thorough reorganization at metamorphosis.

LATERAL Away from the midline, or toward the side.

LEPROSY Chronic bacterial disease caused by *Mycobacterium*; affects nervous tissue, upper respiratory tract, and skin.

LEPTOSPIROSIS Bacterial disease caused by *Leptospira*; transmitted in water and urine of an infected animal; also known as Weil's disease.

LITTER One or more young resulting from a single pregnancy.

LOUSE (*pl.* LICE) Small, wingless insect in the order Phthiraptera; common ectoparasite of mammals.

LYME DISEASE Tick-borne bacterial disease caused by various species of *Borrelia*.

MAMMA (*pl.* MAMMAE) *See* mammary glands; may also refer to nipples.

MAMMALIA Class of vertebrates usually distinguished by the presence of a jaw articulation between the squamosal and dentary bones, hair, mammary glands, muscular diaphragm, and nonnucleated red blood cells.

MAMMARY GLANDS Milk-producing glands unique to mammals and believed to be specialized sudoriferous (sweat) glands.

MANGE Group of persistent, contagious diseases of the skin caused by parasitic mites.

MARSUPIUM External pouch formed by folds of skin on the abdominal wall; e.g., in Virginia opossums.

MAST Accumulation on the forest floor of fruits of various trees, such as acorns, beechnuts, and hickory nuts; an important source of food for many mammals.

MATERNITY COLONY In bats, a social organization where pregnant or lactating females and their young occupy roosts that contain few or no adult males or nonreproductive females; *see* bachelor colony.

MIDDEN A refuse heap; a pile of guano, food, or other items such as plant materials, bones, etc., deposited by woodrats, or stored cones or other foods of squirrels; may also refer to woodrat houses.

MIST NET Fine-meshed net used to capture bats.

MITES Small arthropods in the subclass Acari and class Arachnida; common parasites of mammals.

MOLARIFORM TEETH Premolars and molars; all teeth posterior to the canines; cheek teeth.

MOLT Periodic shedding and replacing of all or much of the hair, typically once or twice a year; also, the shedding of hair by juveniles or subadults and its replacement with hair typical of adults.

MUSCULAR DIAPHRAGM Muscular septum between thoracic and abdominal cavities of mammals.

NEMATODE Roundworm of the class Nematoda; many are internal parasites of mammals.

NEONATE A newborn.

NEST Structure made of grass, leaves, or some other material built by a mammal for shelter.

NOCTURNAL Active during nighttime; opposite of diurnal.

OMNIVORE Animal that consumes plants, animals, and other organisms as food.

ORDER Category of scientific classification that is below class; contains 1 or more families; e.g., Rodentia.

OSTEOMALACIA Disorder in which bones are inadequately mineralized.

PARASITE Any organism that spends all or part of its life cycle on or in the living body of another species (the host) and that obtains its food from tissues or the digestive tract of its host.

PARTURITION Process of giving birth.

PATAGIUM (*pl.* PATAGIA) Web of skin for flight or gliding, such as the gliding membranes of a flying squirrel or the membranes on the wing of a bat.

PELAGE All hairs on a mammal.

PELAGIC Refers to the open ocean, especially as distinguished from coastal waters.

PILOERECTION Elevating the ends of hairs away from the skin; e.g., to assist in regulating body temperature or to enhance a threatening posture.

PINNA (*pl.* PINNAE) External fleshy flap of skin around the opening of the ear canal in most mammals; absent in many fossorial and aquatic species.

PISCIVORE Animal that consumes fish.

PLAGUE Bacterial disease that is caused by *Yersinia pestis* and is usually transmitted to mammals by fleas.

POPULATION All of the individuals that form a single interbreeding group within a species.

POSTORBITAL PROCESS A projection of the frontal bone directly behind the eye socket.

PRECOCIAL A newborn that is fully furred, open eyed, and able to move about immediately; e.g., newborn white-tailed deer.

PREDATOR An animal that lives, at least in part, by killing and consuming other animals.

PREHENSILE Adapted for grasping by curling or wrapping; e.g., tail of the Virginia opossum.

PREY Any animal seized or hunted by another for food.

PROMISCUOUS Mating indiscriminately and perhaps often.

PROTOZOAN One of a group of single-celled microscopic organisms including amoebas, ciliates, flagellates, and sporozoans.

PROXIMAL Situated nearer to, or nearest to, the main part of the body; e.g., the proximal end of the tail is attached to the body.

Q FEVER Bacterial disease caused by *Coxiella burnetii*; infection usually comes from contact with urine, feces, milk, vaginal mucus, or semen.

QUILL A hardened, hollow, pointed, and barbed hair of a North American porcupine.

RABIES Deadly viral disease (*Lyssavirus*) that infects the central nervous system; usually transmitted by a bite from an infected animal.

RADIORECEIVER *See* radiotelemetry.

RADIOTELEMETRY Use of radiotransmitters, which are attached to animals, and radioreceivers to track movements and to determine locations of dens and roosts of individual animals.

RADIOTRANSMITTER *See* radiotelemetry.

RANGE Geographic area inhabited by a particular group of organisms, e.g., a population or a taxonomic group such as a species or family.

RINGWORM Also known as dermatophytosis, a fungal infection caused by several species of dermatophytes; these fungi survive on keratin in skin and hair.

RIPARIAN Habitat along the banks and floodplain of a waterway.

ROCKY MOUNTAIN SPOTTED FEVER Tick-borne disease caused by the bacterium *Rickettsia rickettsii*.

ROSTRUM Facial region that extends forward from the eyes to the tip of the nose.

RUNWAY A worn, cut, or otherwise detectable pathway produced by and repeatedly used by small mammals.

SAGITTAL CREST A medial dorsal ridge on the braincase.

SAINT LOUIS ENCEPHALITIS Viral infection caused by *Flavivirus* and transmitted by mosquitoes from birds to mammals.

SALTATORIAL Adapted for hopping or jumping; e.g., meadow jumping mouse.

SCENT GLANDS Various types of glands that are modified for production of odoriferous secretions and may be used to mark territories, defend a resource, identify individuals, or attract the opposite sex.

SCHISTOSOMIASIS Parasitic disease caused by trematodes in the genus *Schistosoma*.

SEBACEOUS GLAND Epidermal gland that secretes a fatty substance and usually opens onto a hair follicle.

SEXUAL DIMORPHISM Apparent differences between males and females of a species; e.g., size or color.

SHELTER FORM A place where rabbits rest and hide from predators that is adjacent to, within, or under a clump of grass or other vegetation or debris; floor may be a shallow excavation.

SNOUT *See* rostrum.

SPECIES Actually or potentially interbreeding populations that are reproductively isolated from other kinds of organisms; also, a taxonomic classification between genus and subspecies.

SUBADULT Young individual, generally not fully grown, that may be a young-of-the-year and may or may not be in reproductive condition.

SUBSPECIES A relatively uniform and genetically distinct portion of a species.

SUBTERRANEAN Living beneath the surface of the ground; fossorial.

SUCCESSION Gradual and natural replacement of a biotic community (chiefly vegetation) by another, such as the replacement of grassland by shrubland, and eventually by forest.

SUCCULENT VEGETATION Fleshy plants that store large amounts of water in specially adapted, thickened, or swollen stems, leaves, or roots; often in aquatic habitats.

SWARMING (BY BATS) Large numbers of bats (1 or more species) gathering in late summer and autumn and flying outside cave openings; this behavior may be related to mating before bats enter hibernation.

SWEAT GLAND Long, tubular gland that extends from the dermis to the surface of the skin and secretes perspiration or scent.

TACTILE Pertaining to the sense of touch.

TERRESTRIAL Living primarily on or in the ground.

THREATENED At risk of becoming endangered.

TICKS Small arthropods in the subclass Acari and class Arachnida; common ectoparasites of mammals.

TORPID *See* torpor.

TORPOR Type of adaptive dormancy in which heart rate, body temperature, and respiration are reduced.

TOXOPLASMOSIS Parasitic disease caused by the protozoan *Toxoplasma gondii*; infects many mammals but occurs primarily in cats.

TRAGUS In most bats, the projection from the lower medial margin of the pinna; also present in some other mammals.

TREMATODE Fluke, or internal parasite, of the invertebrate class Trematoda.

TULAREMIA Bacterial disease that is caused by *Francisella tularensis* and can be spread through water and the bites of ticks, lice, fleas, and flies that infest rabbits and rodents.

TYPE LOCALITY Site where a type specimen was obtained.

TYPE SPECIMEN The specimen used in the original description of a new species or subspecies.

ULTRASONIC Sound waves that are above the range that can be heard by humans.

UNDERFUR Thick, soft fur beneath the longer and coarser guard hairs.

UNDERSTORY Layer of herbs, shrubs, and smaller trees beneath a canopy.

UNICUSPID In shrews, 1 of the 3–5 small teeth between the large anterior 2-cusped upper incisors and the large cheek teeth.

UROPATAGIUM (*pl.* UROPATAGIA) Membrane of skin between the legs of bats.

VASOCONSTRICTION Decrease in diameter of blood vessels.

VENTER Belly or underside.

VENTRUM Belly or underside.

VERTEBRATE An animal with a backbone, or spinal column; i.e., fish, amphibians, reptiles, birds, and mammals.

VIBRISSAE Long, stiff hairs that serve primarily as touch receptors.

WEIL'S DISEASE *See* leptospirosis.

WHITE-NOSE SYNDROME A disease caused by a fungal pathogen (*Geomyces destructans*) that is devastating populations of cave-roosting bats in the United States and Canada.

YOUNG-OF-THE-YEAR An animal that was born in the most recent breeding season and is less than a year old.

REFERENCES Kistner et al. (1982), Forrester (1992), Martin et al. (2001), Whitaker and Hamilton (1998), Whitaker and Mumford (2009), Vaughan et al. (2011).

References

Agosta, S. J., and D. Morton. 2003. Diet of the big brown bat, *Eptesicus fuscus*, from Pennsylvania and western Maryland. Northeastern Naturalist 10:89–104.

Allen, J. A. 1876. The American bisons, living and extinct. Memoirs of the Museum of Comparative Zoology 4:1–246.

Allen, R. H., Jr. 1965. History and results of deer restocking in Alabama. Alabama Department of Conservation, Division of Game and Fish Bulletin 6:1–50.

Anderson, A. J., E. C. Greiner, C. T. Atkinson, and M. E. Roelke. 1992. Sarcocysts in the Florida bobcat (*Felis rufus floridanus*). Journal of Wildlife Diseases 28:116–120.

Anderson, S., and J. K. Jones, Jr. 1967. Recent mammals of the world: a synopsis of families. Ronald Press Company, New York.

[Anonymous]. 1994. Managing Iowa wildlife: moles. Iowa State University, University Extension, Ames PM1302B:1–6.

Arlton, A. V. 1936. An ecological study of the mole. Journal of Mammalogy 17:349–371.

Audubon, J. J., and J. Bachman. 1846. The viviparous quadrupeds of North America. J. J. Audubon, New York.

Babero, B. B., and J. W. Lee. 1961. Studies on the helminths of nutria, *Myocastor coypus* (Molina), in Louisiana with check-list of other worm parasites from this host. Journal of Parasitology 47:378–390.

Baird, S. F. 1857. Reports of explorations and surveys, to ascertain the most practicable and economical route for a railroad from the Mississippi River to the Pacific Ocean. Part I. General report upon the zoology of the several Pacific railroad routes. A. O. P. Nicholson Printer, Washington, D.C. 8:1–737.

Baker, R. H., and R. W. Dickerman. 1956. Daytime roost of the yellow bat in Veracruz. Journal of Mammalogy 37:443.

Barbour, R. W., and W. H. Davis. 1969. Bats of America. University Press of Kentucky, Lexington.

Barclay, R. M. R., P. A. Faure, and D. R. Farr. 1988. Roosting behavior and roost selection by migrating silver-haired bats (*Lasionycteris noctivagans*). Journal of Mammalogy 69:821–825.

Barigye, R., E. Schamber, T. K. Newell, and N. W. Dyer. 2007. Hepatic lipidosis and other test findings in two captive adult porcupines (*Erethizon dorsatum*) dying from a "sudden death syndrome." Journal of Veterinary Diagnostic Investigation 19:712–716.

Barkalow, F. S., Jr. 1961. The porcupine and fisher in Alabama archaeological sites. Journal of Mammalogy 42:544–545.

Barkalow, F. S., Jr. 1972. Vertebrate remains from archeological sites in the Tennessee Valley of Alabama. Southern Indian Studies 24:3–41.

Barnett, S. A. 1963. The rat: a study in behaviour. Aldine Publishing Company, Chicago, Illinois.

Barr, T. R. B. 1963. Infectious diseases in the opossum: a review. Journal of Wildlife Management 27:53–71.

Bartram, W. 1791. Travels through North and South Carolina, Georgia, East and West Florida, the Cherokee country, the extensive territories of the Muscogulges or Creek Confederacy, and the country of the Chactaws. Containing an account of the soil and natural productions of those regions; together with observations on the manners of the Indians. Embellished with copper-plates. Printed by James and Johnson, Philadelphia, Pennsylvania.

Beckett, J. V., and V. Gallicchio. 1967. A survey of helminths of the muskrat, *Ondatra z. zibethica* Miller, 1912, in Portage County, Ohio. Journal of Parasitology 53:1169–1172.

Bekoff, M. 1977. *Canis latrans*. Mammalian Species 79:1–9.

Beolens, B., M. Watkins, and M. Grayson. 2009. The eponym dictionary of mammals. Johns Hopkins University Press, Baltimore, Maryland.

Best, T. L., compiler. 2004a. Mammals. Pp. 185–204 in Alabama wildlife: a checklist of vertebrates and selected invertebrates: aquatic mollusks, fishes, amphibians, reptiles, birds, and mammals (R. E. Mirarchi, ed.). University of Alabama Press, Tuscaloosa 1:1–209.

Best, T. L. 2004b. Red wolf *Canis rufus* Audubon and Bachman. Pp. 171–172 in Alabama wildlife: imperiled amphibians, reptiles, birds, and mammals (R. E. Mirarchi, M. A. Bailey, T. M. Haggerty, and T. L. Best, eds.). University of Alabama Press, Tuscaloosa 3:1–225.

Best, T. L., S. D. Carey, K. G. Caesar, and T. H. Henry. 1993. Distribution and abundance of bats (Mammalia: Chiroptera) in the Coastal Plain caves of southern Alabama. Journal of Cave and Karst Studies 54:61–65.

Best, T. L., and K. N. Geluso. 2003. Summer foraging range of Mexican free-tailed bats (*Tadarida brasiliensis mexicana*) from Carlsbad Cavern, New Mexico. Southwestern Naturalist 48:590–596.

Best, T. L., and E. B. Hart. 1976. Swimming ability of pocket gophers (Geomyidae). Texas Journal of Science 27:361–366.

Best, T. L., B. Hoditschek, and H. H. Thomas. 1981. Foods of coyotes (*Canis latrans*) in Oklahoma. Southwestern Naturalist 26:67–69.

Best, T. L., and M. K. Hudson. 1996. Movements of gray bats (*Myotis grisescens*) between roost sites and foraging areas. Journal of the Alabama Academy of Science 67:6–14.

Best, T. L., and J. B. Jennings. 1997. *Myotis leibii*. Mammalian Species 547:1–6.

Best, T. L., and M. L. Kennedy. 1972. The porcupine (*Erethizon dorsatum* Linnaeus) in the Texas Panhandle and adjacent New Mexico and Oklahoma. Texas Journal of Science 24:351.

Best, T. L., B. A. Milam, T. D. Haas, W. S. Cvilikas, and L. R. Saidak. 1997. Variation in diet of the gray bat (*Myotis grisescens*). Journal of Mammalogy 78:569–583.

Blair, W. F. 1936. The Florida marsh rabbit. Journal of Mammalogy 17:197–207.

Blair, W. F. 1948. Population density, life span, and mortality rates of small mammals in the blue-grass meadow and blue-grass field associations of southern Michigan. American Midland Naturalist 40:395–419.

Blankespoor, H. D., and M. J. Ulmer. 1970. Helminths from six species of Iowa bats. Proceedings of the Iowa Academy of Science 77:200–206.

Bounds, D. L., M. H. Sherfy, and T. A. Mollett. 2003. Nutria *Myocastor coypus*. Pp. 1119–1147 in Wild mammals of North America: biology, management, and conservation (G. A. Feldhamer, B. C. Thompson, and J. A. Chapman, eds.). 2nd ed. Johns Hopkins University Press, Baltimore, Maryland.

Bowers, M. A., and H. D. Smith. 1979. Differential habitat utilization by sexes of the deermouse, *Peromyscus maniculatus*. Ecology 60:869–875.

Boyce, K. A., and R. E. Barry. 2007. Seasonal home range and diurnal movements of *Sylvilagus obscurus* (Appalachian cottontail) at Dolly Sods, West Virginia. Northeastern Naturalist 14:99–110.

Boyd, I. L., C. Lockyer, and H. D. Marsh. 1999. Reproduction in marine mammals. Pp. 218–286 in Biology of marine mammals (J. E. Reynolds, III and S. A. Rommel, eds.). Smithsonian Institution Press, Washington, D.C.

Boyd, M. F. 1936. The occurrence of the American bison in Alabama and Florida. Science 84(2174):203.

Brack, V., Jr., and J. O. Whitaker, Jr. 2001. Foods of the northern myotis, *Myotis septentrionalis*, from Missouri and Indiana, with notes on foraging. Acta Chiropterologica 3:203–210.

Bradley, R. D., D. D. Henson, and N. D. Durish. 2008. Re-evaluation of the geographic distribution and phylogeography of the *Sigmodon hispidus* complex based on mitochondrial DNA sequences. Southwestern Naturalist 53:301–310.

Brannon, P. A. 1923. Cacomixtl in Alabama. Journal of Mammalogy 4:54.

Brown, R. N., M. W. Gabriel, G. Wengert, S. Matthews, J. M. Higley, and J. E. Foley. 2006. Fecally transmitted viruses associated with Pacific fishers (*Martes pennanti*) in northwestern California. Transactions of the Western Section of the Wildlife Society 42:40–46.

Brugh, T. H., Jr. 1971. A survey of the internal parasites of a feral herd of fallow deer (*Dama dama*) in Alabama. M.S. thesis, Auburn University, Alabama.

Bryant, L. D., and C. Maser. 1982. Classification and distribution. Pp. 1–59 in Elk of North America: ecology and management (J. W. Thomas and D. E. Toweill, eds.). Stackpole Books, Harrisburg, Pennsylvania.

Buechner, H. K. 1944. Helminth parasites of the gray fox. Journal of Mammalogy 25:185–188.

Burgdorfer, W., and K. L. Gage. 1987. Susceptibility of the hispid cotton rat (*Sigmodon hispidus*) to the Lyme disease spirochete (*Borrelia burgdorferi*). American Journal of Tropical Medicine and Hygiene 37:624–628.

Caceres, M. C., and R. M. R. Barclay. 2000. *Myotis septentrionalis*. Mammalian Species 634:1–4.

Callahan, E. V., R. D. Drobney, and R. L. Clawson. 1997. Selection of summer roosting sites by Indiana bats (*Myotis sodalis*) in Missouri. Journal of Mammalogy 78:818–825.

Cameron, G. N., and S. R. Spencer. 1981. *Sigmodon hispidus*. Mammalian Species 158:1–9.

Carroll, D. S., L. L. Peppers, and R. D. Bradley. 2004. Molecular systematics and phylogeography of the *Sigmodon hispidus* species group. Pp. 85–98 in Contribuciones mastozoologicas en homenaje a Bernardo Villa (V. Sanchez-Cordero and R. A. Medellín, eds.). Instituto de Biología e Instituto de Ecología, Universidad Nacional Autónoma de México, México, Distrito Federal, México.

Castleberry, S. B., M. T. Mengak, and W. M. Ford. 2006. *Neotoma magister*. Mammalian Species 789:1–5.

Chandler, A. C., and D. M. Melvin. 1951. A new cestode, *Oochoristica pennsylvanica*, and some new or rare helminth host records from Pennsylvania mammals. Journal of Parasitology 37:106–109.

Chapman, D., and N. Chapman. 1975. Fallow deer: their history, distribution and biology. Terence Dalton Limited, Lavenham, United Kingdom.

Chapman, D. I. 1977. Fallow deer. Pp. 429–437 in The handbook of British mammals (G. B. Corbet and H. N. Southern, eds.). 2nd ed. Blackwell Scientific Publishing, London, United Kingdom.

Chapman, J. A. 1975. *Sylvilagus transitionalis*. Mammalian Species 55:1–4.

Chapman, J. A., K. L. Cramer, N. J. Dippenaar, and T. J. Robinson. 1992. Systematics and biogeography of the New England cottontail, *Sylvilagus transitionalis* (Bangs, 1895), with the description of a new species from the Appalachian Mountains. Proceedings of the Biological Society of Washington 105:841–866.

Chapman, J. A., and G. A. Feldhamer. 1981. *Sylvilagus aquaticus*. Mammalian Species 151:1–4.

Chapman, J. A., J. G. Hockman, and M. M. Ojeda. 1980. *Sylvilagus floridanus*. Mammalian Species 136:1–8.

Chapman, J. A., and J. A. Litvaitis. 2003. Eastern cottontail *Sylvilagus floridanus* and allies. Pp. 101–125 in Wild mammals of North America: biology, management, and conservation (G. A. Feldhamer, B. C. Thompson, and J. A. Chapman, eds.). 2nd ed. Johns Hopkins University Press, Baltimore, Maryland.

Chapman, J. A., and J. R. Stauffer, Jr. 1981. The status and distribution of the New England cottontail. Pp. 973–983 in Proceedings of the World Lagomorph Conference (K. Meyers and C. D. McInnes, eds.). University of Guelph, Guelph, Ontario, Canada.

Chapman, J. A., and G. R. Willner. 1981. *Sylvilagus palustris*. Mammalian Species 153:1–3.

Cook, F. A. 1942. *Sorex longirostris longirostris* in Mississippi. Journal of Mammalogy 23:218.

Coultrip, R. L., R. W. Emmons, L. J. Legters, J. D. Marshall, Jr., and K. F. Murray. 1973. Survey for the arthropod vectors and mammalian hosts of Rocky Mountain spotted fever and plague at Fort Ord, California. Journal of Medical Entomology 10:303–309.

Crawford, R. L., and W. W. Baker. 1981. Bats killed at a North Florida television tower: a 25-year record. Journal of Mammalogy 62:651–652.

Cryan, P. M. 2008. Mating behavior as a possible cause of bat fatalities at wind turbines. Journal of Wildlife Management 72:845–849.

Cryan, P. M., C. U. Meteyer, J. G. Boyles, and D. S. Blehert. 2010. Wing pathology of white-nose syndrome in bats suggests life-threatening disruption of physiology. BMC Biology 8(135):1–8.

Cudmore, W. W. 1986. Nest associates and ectoparasites of the eastern wood rat, *Neotoma floridana*, in Indiana. Canadian Journal of Zoology 64:353–357.

Curren, C. B., Jr. 1977. Prehistoric range extension of the elk: *Cervus canadensis*. American Midland Naturalist 97:230–232.

Currier, M. J. P. 1983. *Felis concolor*. Mammalian Species 200:1–7.

Daggett, P. M., and D. R. Henning. 1974. The jaguar in North America. American Antiquity 39:465–469.

Davis, J. R. 1955. Food habits of the bobcat in Alabama. M.S. thesis, Alabama Polytechnic Institute, Auburn.

Decher, J., and J. R. Choate. 1995. *Myotis grisescens.* Mammalian Species 510:1–7.

Dellinger, J. A. 2011. Foraging and spatial ecology of red wolves (*Canis rufus*) in northeastern North Carolina. M.S. thesis, Auburn University, Alabama.

de Oliveira, T. G. 1998. *Herpailurus yagouaroundi.* Mammalian Species 578:1–6.

de S. Pinto, I., J. R. Botelho, L. P. Costa, Y. L. R. Leite, and P. M. Linardi. 2009. Siphonaptera associated with wild mammals from the Central Atlantic Forest Biodiversity Corridor in southeastern Brazil. Journal of Medical Entomology 46:1146–1151.

Dick, T. A., and R. D. Leonard. 1979. Helminth parasites of fisher *Martes pennanti* (Erxleben) from Manitoba, Canada. Journal of Wildlife Diseases 15:409–412.

Dolan, P. G., and D. C. Carter. 1977. *Glaucomys volans.* Mammalian Species 78:1–6.

Dragoo, J. W., and R. L. Honeycutt. 1997. Systematics of mustelid-like carnivores. Journal of Mammalogy 78:426–443.

Duchamp, J. E., D. W. Sparks, and J. O. Whitaker, Jr. 2004. Foraging-habitat selection by bats in an urban-rural interface: comparison between a successful and a less successful species. Canadian Journal of Zoology 82:1157–1164.

Durden, L. A. 1995. Bot fly (*Cuterebra fontinella fontinella*) parasitism of cotton mice (*Peromyscus gossypinus*) on St. Catherines Island, Georgia. Journal of Parasitology 81:787–790.

Durden, L. A., M. W. Cunningham, R. McBride, and B. Ferree. 2006. Ectoparasites of free-ranging pumas and jaguars in the Paraguayan chaco. Veterinary Parasitology 137:189–193.

Durden, L. A., and D. J. Richardson. 2003. Ectoparasites of the striped skunk, *Mephitis mephitis*, in Connecticut, U.S.A. Comparative Parasitology 70:42–45.

Dusi, J. L. 1959. *Sorex longirostris* in eastern Alabama. Journal of Mammalogy 40:438–439.

Dyer, W. G. 1969. Helminths of the striped skunk, *Mephitis mephitis*, in North America. American Midland Naturalist 82:601–605.

Eads, R. B., and G. C. Menzies. 1950. Fox ectoparasites collected incident to a rabies control program. Journal of Mammalogy 31:78–80.

Ellis, L. L., Jr. 1955. A survey of the ectoparasites of certain mammals in Oklahoma. Ecology 36:12–18.

Erickson, A. B. 1944*a.* Helminths of Minnesota Canidae in relation to food habits, and a host list and key to the species reported from North America. American Midland Naturalist 32:358–372.

Erickson, A. B. 1944b. Parasites of beavers, with a note on *Paramphistomum castori* Kofoid and Park, 1937 a synonym of *Stichorchis subtriquetrus*. American Midland Naturalist 31:625–630.

Erickson, A. B. 1947. Helminth parasites of rabbits of the genus *Sylvilagus*. Journal of Wildlife Management 11:255–263.

Esher, R. J., J. L. Wolfe, and J. N. Layne. 1978. Swimming behavior of rice rats (*Oryzomys palustris*) and cotton rats (*Sigmodon hispidus*). Journal of Mammalogy 59:551–558.

Espmark, Y., and W. Brunner. 1974. Observations on rutting behaviour in fallow deer, *Dama dama*. Säugetierkundliche Mitteilungen 22:135–142.

Fauquier, D., F. Gulland, M. Haulena, M. Dailey, R. L. Rietcheck, and T. P. Lipscomb. 2004. Meningoencephalitis in two stranded California sea lions (*Zalophus californianus*) caused by aberrant trematode migration. Journal of Wildlife Diseases 40:816–819.

Favorov, M. O., M. Y. Kosoy, S. A. Tsarev, J. E. Childs, and H. S. Margolis. 2000. Prevalence of antibody to hepatitis E virus among rodents in the United States. Journal of Infectious Diseases 181:449–455.

Feldhamer, G. A., L. C. Drickamer, S. H. Vessey, J. F. Merritt, and C. Krajewski. 2007. Mammalogy: adaptation, diversity, ecology. Johns Hopkins University Press, Baltimore, Maryland.

Feldhamer, G. A., K. C. Farris-Renner, and C. M. Barker. 1988. *Dama dama*. Mammalian Species 317:1–8.

Feldhamer, G. A., R. S. Klann, A. S. Gerard, and A. C. Driskell. 1993. Habitat partitioning, body size, and timing of parturition in pygmy shrews and associated soricids. Journal of Mammalogy 74:403–411.

Felix, Z., L. J. Gatens, Y. Wang, and C. J. Schweitzer. 2009. First records of the smoky shrew (*Sorex fumeus*) in Alabama. Southeastern Naturalist 8:750–753.

Fenton, M. B., and R. M. R. Barclay. 1980. *Myotis lucifugus*. Mammalian Species 142:1–8.

Fitch, H. S., P. Goodrum, and C. Newman. 1952. The armadillo in the southeastern United States. Journal of Mammalogy 33:21–37.

Fitzgerald, S. D., M. R. White, and K. R. Kazakos. 1991. Encephalitis in two porcupines due to *Baylisascaris* larval migration. Journal of Veterinary Diagnostic Investigation 3:359–362.

Flynn, J. J., J. A. Finarelli, S. Zehr, J. Hsu, and M. A. Nedbal. 2005. Molecular phylogeny of the Carnivora (Mammalia): assessing the impact of increased sampling on resolving enigmatic relationships. Systematic Biology 54:317–337.

Forrester, D. J. 1992. Parasites and diseases of wild mammals in Florida. University Press of Florida, Gainesville.

Forys, E. A., and R. D. Dueser. 1993. Inter-island movements of rice rats (*Oryzomys palustris*). American Midland Naturalist 130:408–412.

French, T. W. 1980*a*. *Sorex longirostris*. Mammalian Species 143:1–3.

French, T. W. 1980*b*. Natural history of the southeastern shrew, *Sorex longirostris* Bachman. American Midland Naturalist 104:13–31.

Frey, J. K., and J. N. Stuart. 2009. Nine-banded armadillo (*Dasypus novemcinctus*) records in New Mexico, USA. Edentata 8–10:54–55.

Fritzell, E. K., and K. J. Haroldson. 1982. *Urocyon cinereoargenteus*. Mammalian Species 189:1–8.

Fujita, M. S., and T. H. Kunz. 1984. *Pipistrellus subflavus*. Mammalian Species 228:1–6.

Gash, S. L., and W. L. Hanna. 1973. Occurrence of some helminth parasites in the muskrat, *Ondatra zibethicus*, from Crawford County, Kansas. Transactions of the Kansas Academy of Science 75:251–254.

Gauss, C. B. L., J. P. Dubey, D. Vidal, F. Ruiz, J. Vicente, I. Marco, S. Lavin, C. Gortazar, and S. Almería. 2005. Seroprevalence of *Toxoplasma gondii* in wild pigs (*Sus scrofa*) from Spain. Veterinary Parasitology 131:151–156.

Geist, V. 1982. Adaptive behavioral strategies. Pp. 219–277 in Elk of North America: ecology and management (J. W. Thomas and D. E. Toweill, eds.). Stackpole Books, Harrisburg, Pennsylvania.

Genoways, H. H., and J. R. Choate. 1998. Natural history of the southern short-tailed shrew, *Blarina carolinensis*. Occasional Papers, Museum of Southwestern Biology 8:1–43.

George, J. E., and R. W. Strandtmann. 1960. New records of ectoparasites on bats in West Texas. Southwestern Naturalist 5:228–229.

George, S. B., J. R. Choate, and H. H. Genoways. 1986. *Blarina brevicauda*. Mammalian Species 261:1–9.

Gingerich, J. L. 1994. Florida's fabulous mammals. World Publications, Tampa Bay, Florida.

Glass, G. E., et al. 1998. Black Creek Canal virus infection in *Sigmodon hispidus* in southern Florida. American Journal of Tropical Medicine and Hygiene 59:699–703.

Goodpaster, W. W., and D. F. Hoffmeister. 1952. Notes on the mammals of western Tennessee. Journal of Mammalogy 33:362–371.

Gosse, P. H. 1859. Letters from Alabama, (U.S.) chiefly relating to natural history. Morgan and Chase, London, United Kingdom.

Greer, K. R. 1955. Yearly food habits of the river otter in the Thompson Lakes region, northwestern Montana, as indicated by scat analysis. American Midland Naturalist 54:299–313.

Grizzell, R. A., Jr. 1955. A study of the southern woodchuck, *Marmota monax monax*. American Midland Naturalist 53:257–293.

Guillot, J., T. Petit, F. Degorce-Rubiales, E. Guého, and R. Chermette. 1998. Dermatitis caused by *Malassezia pachydermatis* in a California sea lion (*Zalophus californianus*). Veterinary Record 142:311–312.

Gunter, G. 1968. The status of seals in the Gulf of Mexico with a record of feral otariid seals off the United States Gulf Coast. Gulf Research Reports 2:301–308.

Hall, E. R. 1981. The mammals of North America. 2nd ed. John Wiley and Sons, New York.

Hamilton, W. J., Jr. 1934. The life history of the rufescent woodchuck, *Marmota monax rufescens* Howell. Annals of the Carnegie Museum 23:85–178.

Hamilton, W. J., Jr. 1940. The biology of the smoky shrew (*Sorex fumeus fumeus* Miller). Zoologica 25:473–491.

Hamilton, W. J., Jr. 1958. Life history and economic relations of the opossum (*Didelphis marsupialis virginiana*) in New York State. Memoirs of Cornell University Agricultural Experiment Station 354:1–48.

Hamilton, W. J., Jr., and A. H. Cook. 1955. The biology and management of the fisher in New York. New York Fish and Game Journal 2:13–35.

Hampton, J. O., P. B. S. Spencer, D. L. Alpers, L. E. Twigg, A. P. Woolnough, J. Doust, T. Higgs, and J. Pluske. 2004. Molecular techniques, wildlife management and the importance of genetic population structure and dispersal: a case study with feral pigs. Journal of Applied Ecology 41:735–743.

Hanson, J. D., J. L. Indorf, V. J. Swier, and R. D. Bradley. 2010. Molecular divergence within the *Oryzomys palustris* complex: evidence for multiple species. Journal of Mammalogy 91:336–347.

Harkema, R. 1936. The parasites of some North Carolina rodents. Ecological Monographs 6:151–232.

Hartman, A. C., and R. E. Barry. 2010. Survival and winter diet of *Sylvilagus obscurus* (Appalachian cottontail) at Dolly Sods, West Virginia. Northeastern Naturalist 17:505–516.

Hartman, D. S. 1979. Ecology and behavior of the manatee (*Trichechus manatus*) in Florida. Special Publication, American Society of Mammalogists 5:1–153.

Harveson, L. A., M. E. Tewes, N. J. Silvy, and J. Rutledge. 2000. Prey use by mountain lions in southern Texas. Southwestern Naturalist 45:472–476.

Harvey, M. J., J. S. Altenbach, and T. L. Best. 2011. Bats of the United States and Canada. Johns Hopkins University Press, Baltimore, Maryland.

Hayes, J. P., and M. E. Richmond. 1993. Clinal variation and morphology of woodrats (*Neotoma*) of the eastern United States. Journal of Mammalogy 74:204–216.

Heath, C. B. 2002. California, Galapagos, and Japanese sea lions: *Zalophus californianus, Z. wollebaeki,* and *Z. japonicus.* Pp. 180–186 in Encyclopedia of marine mammals (W. F. Perrin, B. Würsig, and J. G. M. Thewissen, eds.). Academic Press, San Diego, California.

Hein, C. D., S. B. Castleberry, and K. V. Miller. 2005. Winter roost-site selection by Seminole bats in the lower coastal plain of South Carolina. Southeastern Naturalist 4:473–478.

Henry, T. H., T. L. Best, and C. D. Hilton. 2000. Body size, reproductive biology, and sex ratio of a year-round colony of *Eptesicus fuscus fuscus* and *Tadarida brasiliensis cynocephala* in eastern Alabama. Occasional Papers of the North Carolina Museum of Natural Sciences and the North Carolina Biological Survey 12:50–56.

Herman, C. M., and L. J. Goss. 1940. Trichinosis in an American badger, *Taxidea taxus taxus.* Journal of Parasitology 26:157.

Hester, L. C., T. L. Best, and M. K. Hudson. 2007. Rabies in bats from Alabama. Journal of Wildlife Diseases 43:291–299.

Hill, E. P., III. 1967. Notes on the life history of the swamp rabbit in Alabama. Proceedings of the Southeastern Association of Game and Fish Commissioners 21:117–123.

Hilton, C. D., and T. L. Best. 2000. Gastrointestinal helminth parasites of bats in Alabama. Occasional Papers of the North Carolina Museum of Natural Sciences and the North Carolina Biological Survey 12:57–66.

Hirt, S. J. 2008. Analysis of stable isotopes of hydrogen to determine migrational source of silver-haired bats (*Lasionycteris noctivagans*) in Alabama. M.S. thesis, Auburn University, Alabama.

Hoberg, E. P., C. J. Henny, O. R. Hedstrom, and R. A. Grove. 1997. Intestinal helminths of river otters (*Lutra canadensis*) from the Pacific Northwest. Journal of Parasitology 83:105–110.

Hodgson, A. 1824. Letters from North America, written during a tour in the United States and Canada. Hurst, Robinson, & Co., Edinburgh, United Kingdom 1:1–405.

Hoditschek, B., J. F. Cully, Jr., T. L. Best, and C. Painter. 1985. Least shrew (*Cryptotis parva*) in New Mexico. Southwestern Naturalist 30:600–601.

Holliman, D. C. 1963. The mammals of Alabama. Ph.D. dissertation, University of Alabama, Tuscaloosa.

Hornaday, W. T. 1889. The extermination of the American bison. Pp. 367–548 in Annual Report 1887, part 2. Smithsonian Institution, Washington, D.C.

Howell, A. H. 1909. Notes on the distribution of certain mammals in the southeastern United States. Proceedings of the Biological Society of Washington 22:55–68.

Howell, A. H. 1921. A biological survey of Alabama: 1. Physiography and life zones. 2. The mammals. North American Fauna 45:1–88.

Hubbard, C. A. 1947. Fleas of western North America: their relation to the public health. Iowa State College Press, Ames.

Hunt, R. H., and J. J. Ogden. 1991. Selected aspects of the nesting ecology of American alligators in the Okefenokee Swamp. Journal of Herpetology 25:448–453.

Hurst, T. E., and M. J. Lacki. 1997. Food habits of Rafinesque's big-eared bat in southeastern Kentucky. Journal of Mammalogy 78:525–528.

Hurst, T. E., and M. J. Lacki. 1999. Roost selection, population size and habitat use by a colony of Rafinesque's big-eared bats (*Corynorhinus rafinesquii*). American Midland Naturalist 142:363–371.

Husar, S. L. 1978. *Trichechus manatus*. Mammalian Species 93:1–5.

Jaeger, E. C. 1955. A source-book of biological names and terms. 3rd ed. Charles Thomas Publisher, Springfield, Illinois.

Jaffe, G., D. A. Zegers, M. A. Steele, and J. F. Merritt. 2005. Long-term patterns of botfly parasitism in *Peromyscus maniculatus*, *P. leucopus*, and *Tamias striatus*. Journal of Mammalogy 86:39–45.

Jenkins, S. H., and P. E. Busher. 1979. *Castor canadensis*. Mammalian Species 120:1–8.

Jones, C. 1977. *Plecotus rafinesquii*. Mammalian Species 69:1–4.

Jones, C., and R. W. Manning. 1989. *Myotis austroriparius*. Mammalian Species 332:1–3.

Jones, J. K., Jr., R. P. Lampe, C. A. Spenrath, and T. H. Kunz. 1973. Notes on the distribution and natural history of bats in southeastern Montana. Occasional Papers of the Museum, Texas Tech University 15:1–12.

Kalkan, A., and M. F. Hansen. 1966. *Ancylostoma taxideae* sp. n. from the American badger, *Taxidea taxus taxus*. Journal of Parasitology 52:291–294.

Keppner, E. J. 1969a. *Filaria taxideae* n. sp. (Filarioidea: Filariidae) from the badger, *Taxidea taxus taxus* from Wyoming. Transactions of the American Microscopical Society 88:581–588.

Keppner, E. J. 1969b. Occurrence of *Atriotaenia procyonis* and *Molineus mustelae* in the badger, *Taxidea taxus* (Schreber, 1778), in Wyoming. Journal of Parasitology 55:1161.

Kilgore, C. H. 2008. Ecological associations of bats (Mammalia: Chiroptera) in the upper Mobile-Tensaw River Delta, Alabama. M.S. thesis, Auburn University, Alabama.

Kinlaw, A. 1995. *Spilogale putorius*. Mammalian Species 511:1–7.

Kinsella, J. M. 1974. Comparison of helminth parasites of the cotton rat, *Sig-*

modon hispidus, from several habitats in Florida. American Museum Novitates 2540:1–12.

Kinsey, K. P. 1976. Social behaviour in confined populations of the Allegheny woodrat, *Neotoma floridana magister*. Animal Behaviour 24:181–187.

Kistner, T. P., K. R. Greer, D. E. Worley, and O. A. Brunetti. 1982. Diseases and parasites. Pp. 181–217 in Elk of North America: ecology and management (J. W. Thomas and D. E. Toweill, eds.). Stackpole Books, Harrisburg, Pennsylvania.

Komarek, E. V. 1939. A progress report on southeastern mammal studies. Journal of Mammalogy 20:292–299.

Koprowski, J. L. 1994*a*. *Sciurus niger*. Mammalian Species 479:1–9.

Koprowski, J. L. 1994*b*. *Sciurus carolinensis*. Mammalian Species 480:1–9.

Kosoy, M. Y., L. H. Elliott, T. G. Ksiazek, C. F. Fulhorst, P. E. Rollin, J. E. Childs, J. N. Mills, G. O. Maupin, and C. J. Peters. 1996. Prevalence of antibodies to arenaviruses in rodents from the southern and western United States: evidence for an arenavirus associated with the genus *Neotoma*. American Journal of Tropical Medicine and Hygiene 54:570–576.

Kosoy, M. Y., R. L. Regnery, T. Tzianabos, E. L. Marston, D. C. Jones, D. Green, G. O. Maupin, J. G. Olson, and J. E. Childs. 1997. Distribution, diversity, and host specificity of *Bartonella* in rodents from the southeastern United States. American Journal of Tropical Medicine and Hygiene 57:578–588.

Kunz, T. H. 1982. *Lasionycteris noctivagans*. Mammalian Species 172:1–5.

Kurta, A., and R. H. Baker. 1990. *Eptesicus fuscus*. Mammalian Species 356:1–10.

Kwiecinski, G. G. 1998. *Marmota monax*. Mammalian Species 591:1–8.

Lackey, J. A. 1978. Reproduction, growth, and development in high-latitude and low-latitude populations of *Peromyscus leucopus* (Rodentia). Journal of Mammalogy 59:69–83.

Lackey, J. A., D. G. Huckaby, and B. G. Ormiston. 1985. *Peromyscus leucopus*. Mammalian Species 247:1–10.

Lacki, M. J. 2000. Effect of trail users at a maternity roost of Rafinesque's big-eared bats. Journal of Cave and Karst Studies 62:163–168.

Lacki, M. J., and K. M. Ladeur. 2001. Seasonal use of lepidopteran prey by Rafinesque's big-eared bats (*Corynorhinus rafinesquii*). American Midland Naturalist 145:213–217.

Laerm, J., L. Lepardo, T. Gaudin, N. Monteith, and A. Szymczak. 1996. First records of the pygmy shrew, *Sorex hoyi winnemana* Preble (Insectivora: Soricidae), in Alabama. Journal of the Alabama Academy of Science 67:43–48.

Laerm, J., M. A. Menzel, D. M. Krishon, and J. L. Boone. 1999. Morphological discrimination between the eastern red bat, *Lasiurus borealis*, and Seminole bat, *Lasiurus seminolus* (Chiroptera: Vespertilionidae), in the southeastern United States. Journal of the Elisha Mitchell Scientific Society 115:131–139.

Larivière, S. 1999. *Mustela vison*. Mammalian Species 608:1–9.

Larivière, S. 2001. *Ursus americanus*. Mammalian Species 647:1–11.

Larivière, S., and M. Pasitschniak-Arts. 1996. *Vulpes vulpes*. Mammalian Species 537:1–11.

Larivière, S., and L. R. Walton. 1997. *Lynx rufus*. Mammalian Species 563:1–8.

Larivière, S., and L. R. Walton. 1998. *Lontra canadensis*. Mammalian Species 587:1–8.

La Val, R. K. 1967. Records of bats from the southeastern United States. Journal of Mammalogy 48:645–648.

Leiby, P. D., P. J. Sitzmann, and D. C. Kritsky. 1971. Studies on helminths of North Dakota. 2. Parasites of the badger, *Taxidea taxus* (Schreber). Proceedings of the Helminthological Society of Washington 38:225–228.

Lewis, J. A. 2004. West Indian manatee *Trichechus manatus* Linneaus. P. 197 in Alabama wildlife: conservation and management recommendations for imperiled wildlife (R. E. Mirarchi, M. A. Bailey, J. T. Garner, T. M. Haggerty, T. L. Best, M. F. Mettee, and P. O'Neil, eds.). University of Alabama Press, Tuscaloosa 4:1–221.

Lindzey, F. G. 2003. Badger: *Taxidea taxus*. Pp. 683–691 in Wild mammals of North America: biology, management, and conservation (G. A. Feldhamer, B. C. Thompson, and J. A. Chapman, eds.). 2nd ed. Johns Hopkins University Press, Baltimore, Maryland.

Linzey, D. W., and A. V. Linzey. 1969. First record of the yellow bat in Alabama. Journal of Mammalogy 50:845.

Linzey, D. W., and R. L. Packard. 1977. *Ochrotomys nuttalli*. Mammalian Species 75:1–6.

Long, C. A. 1973. *Taxidea taxus*. Mammalian Species 26:1–4.

Long, C. A. 1974. *Microsorex hoyi* and *Microsorex thompsoni*. Mammalian Species 33:1–4.

López-González, C., and T. L. Best. 2006. Current status of wintering sites of Mexican free-tailed bats *Tadarida brasiliensis mexicana* (Chiroptera: Molossidae) from Carlsbad Cavern, New Mexico. Vertebrata Mexicana 18:13–22.

Lotz, J. M., and W. F. Font. 1991. The role of positive and negative interspecific associations in the organization of communities of intestinal helminths of bats. Parasitology 103:127–138.

Lotze, J. H., and S. Anderson. 1979. *Procyon lotor*. Mammalian Species 119:1–8.

Lubelczyk, C. B., T. Hanson, E. H. Lacombe, M. S. Holman, and J. E. Keirans. 2007. First U.S. record of the hard tick *Ixodes* (*Pholeoixodes*) *gregsoni* Lindquist, Wu, and Redner. Journal of Parasitology 93:718–719.

Lumsden, R. D., and J. A. Zischke. 1962. Seven trematodes from small mammals in Louisiana. Tulane Studies in Zoology 9:87–98.

Macy, R. W. 1933. A review of the trematode family Urotrematidae with the description of a new genus and two new species. Transactions of the American Microscopical Society 52:247–254.

Mager, K. J., and T. A. Nelson. 2001. Roost-site selection by eastern red bats (*Lasiurus borealis*). American Midland Naturalist 145:120–126.

Margolis, L., and M. D. Dailey. 1972. Revised annotated list of parasites from sea mammals caught off the West Coast of North America. United States Department of Commerce, National Oceanic and Atmospheric Administration, National Marine Fisheries Service, NOAA Technical Report NMFS SSRF-647:1–23.

Martin, R. E., R. H. Pine, and A. F. DeBlase. 2001. A manual of mammalogy with keys to families of the world. 3rd ed. McGraw Hill Companies, New York.

McAllister, C. T., C. R. Bursey, and A. D. Burns. 2005. Gastrointestinal helminths of Rafinesque's big-eared bat, *Corynorhinus rafinesquii* (Chiroptera: Vespertilionidae), from southwestern Arkansas, U.S.A. Comparative Parasitology 72:121–123.

McAllister, C. T., and S. J. Upton. 1989. *Eimeria cryptotis* n. sp. (Apicomplexa: Eimeriidae) from the least shrew, *Cryptotis parva* (Insectivora: Soricidae), in north-central Texas. Journal of Parasitology 75:212–214.

McBee, K., and R. J. Baker. 1982. *Dasypus novemcinctus*. Mammalian Species 162:1–9.

McCay, T. S. 2001. *Blarina carolinensis*. Mammalian Species 673:1–7.

McIntyre, N. E., Y. -K. Chu, R. D. Owen, A. Abuzeineh, N. de la Sancha, C. W. Dick, T. Holsomback, R. A. Nisbett, and C. Jonsson. 2005. A longitudinal study of Bayou virus, hosts, and habitat. American Journal of Tropical Medicine and Hygiene 73:1043–1049.

McKenzie, C. E., and H. E. Welch. 1979. Parasite fauna of the muskrat, *Ondatra zibethica* (Linnaeus, 1766), in Manitoba, Canada. Canadian Journal of Zoology 57:640–646.

McManus, J. J. 1974. *Didelphis virginiana*. Mammalian Species 40:1–6.

McWilliams, L. A. 2005. Variation in diet of the Mexican free-tailed bat (*Tadarida brasiliensis mexicana*). Journal of Mammalogy 86:599–605.

Meagher, M. 1986. *Bison bison*. Mammalian Species 266:1–8.

Menzel, J. M., M. A. Menzel, W. M. Ford, J. W. Edwards, S. R. Sheffield, J. C. Kilgo, and M. S. Bunch. 2003. The distribution of the bats of South Carolina. Southeastern Naturalist 2:121–152.

Menzel, M. A., T. C. Carter, B. R. Chapman, and J. Laerm. 1998. Quantitative comparison of tree roosts used by red bats (*Lasiurus borealis*) and Seminole bats (*L. seminolus*). Canadian Journal of Zoology 76:630–634.

Menzel, M. A., T. C. Carter, L. R. Jablonowski, B. L. Mitchell, J. M. Menzel, and B. R. Chapman. 2001. Home range size and habitat use of big brown bats (*Eptesicus fuscus*) in a maternity colony located on a rural-urban interface in the Southeast. Journal of the Elisha Mitchell Scientific Society 117:36–45.

Menzel, M. A., D. M. Krishon, T. C. Carter, and J. Laerm. 1999. Notes on tree roost characteristics of the northern yellow bat (*Lasiurus intermedius*), the Seminole bat (*L. seminolus*), the evening bat (*Nycticeius humeralis*), and the eastern pipistrelle (*Pipistrellus subflavus*). Florida Scientist 62:185–193.

Menzel, M. A., J. M. Menzel, J. C. Kilgo, M. W. Ford, T. C. Carter, and J. W. Edwards. 2003. Bats of the Savannah River Site and vicinity. United States Department of Agriculture, Forest Service, Southern Research Station, General Technical Report SRS-68:1–69.

Metzgar, L. H. 1967. An experimental comparison of screech owl predation on resident and transient white-footed mice (*Peromyscus leucopus*). Journal of Mammalogy 48:387–391.

Meyer, M. C., and B. G. Chitwood. 1951. Helminths from fisher (*Martes p. pennanti*) in Maine. Journal of Parasitology 37:320–321.

Meyer, M. C., and J. R. Reilly. 1950. Parasites of muskrats in Maine. American Midland Naturalist 44:467–477.

Milazzo, C., J. G. de Bellocq, M. Cagnin, J. -C. Casanova, C. Di Bella, C. Feliu, R. Fons, S. Morand, and F. Santalla. 2003. Helminths and ectoparasites of *Rattus rattus* and *Mus musculus* from Sicily, Italy. Comparative Parasitology 70:199–204.

Miller, G. C., and R. Harkema. 1968. Helminths of some wild mammals in the southeastern United States. Proceedings of the Helminthological Society of Washington 35:118–125.

Miller, S. D., and D. W. Speake. 1980. Prey utilization by bobcats on quail plantations in southern Alabama. Proceedings of the Annual Conference of the Southeastern Association of Fish and Wildlife Agencies 32:100–111.

Mirarchi, R. E. (ed.). 2004. Alabama wildlife: a checklist of vertebrates and selected invertebrates: aquatic mollusks, fishes, amphibians, reptiles, birds, and mammals. University of Alabama Press, Tuscaloosa 1:1–209.

Mirarchi, R. E., M. A. Bailey, J. T. Garner, T. M. Haggerty, T. L. Best, M. F.

Mettee, and P. O'Neil (eds.). 2004a. Alabama wildlife: conservation and management recommendations for imperiled wildlife. University of Alabama Press, Tuscaloosa 4:1–221.

Mirarchi, R. E., M. A. Bailey, T. M. Haggerty, and T. L. Best (eds.). 2004b. Alabama wildlife: imperiled amphibians, reptiles, birds, and mammals. University of Alabama Press, Tuscaloosa 3:1–225.

Mormann, B. M., and L. W. Robbins. 2007. Winter roosting ecology of eastern red bats in Southwest Missouri. Journal of Wildlife Management 71:213–217.

Morzunov, S. P., J. E. Rowe, T. G. Ksiazek, C. J. Peters, S. C. St. Jeor, and S. T. Nichol. 1998. Genetic analysis of the diversity and origin of hantaviruses in *Peromyscus leucopus* mice in North America. Journal of Virology 72:57–64.

Mount, R. H. (ed.). 1984. Vertebrate wildlife of Alabama. Alabama Agricultural Experiment Station, Auburn University.

Mount, R. H. 1986. Vertebrate animals of Alabama in need of special attention. Alabama Agricultural Experiment Station, Auburn University.

Negus, N. C., and H. A. Dundee. 1965. The nest of *Sorex longirostris*. Journal of Mammalogy 46:495.

Neill, W. T. 1961. On the trail of the jaguarundi. Florida Wildlife 15:10–13.

Nelson, E. W. 1909. The rabbits of North America. North American Fauna 29:1–314.

Nelson, J. R., and T. A. Leege. 1982. Nutritional requirements and food habits. Pp. 323–367 in Elk of North America: ecology and management (J. W. Thomas and D. E. Toweill, eds.). Stackpole Books, Harrisburg, Pennsylvania.

NeSmith, C. C., and J. Cox. 1985. Red-winged blackbird nest usurpation by rice rats in Florida and Mexico. Florida Field Naturalist 13:35–36.

Neubaum, D. J., K. R. Wilson, and T. J. O'Shea. 2007. Urban maternity-roost selection by big brown bats in Colorado. Journal of Wildlife Management 71:728–736.

Nielsen, L. T., D. K. Eaton, D. W. Wright, and B. Schmidt-French. 2006. Characteristic odors of *Tadarida brasiliensis mexicana* Chiroptera: Molossidae. Journal of Cave and Karst Studies 68:27–31.

Niewiesk, S., and G. Prince. 2002. Diversifying animal models: the use of hispid cotton rats (*Sigmodon hispidus*) in infectious diseases. Laboratory Animals 36:357–372.

Nims, T. N., L. A. Durden, C. R. Chandler, and O. J. Pung. 2008. Parasitic and phoretic arthropods of the oldfield mouse (*Peromyscus polionotus*) from burned habitats with additional ectoparasite records from the eastern harvest mouse (*Reithrodontomys humulis*) and southern short-tailed shrew (*Blarina carolinensis*). Comparative Parasitology 75:102–106.

Nowak, R. M. 1972. The mysterious wolf of the South. Natural History 81:50–53, 74–77.

Nowak, R. M. 1999. Walker's mammals of the world. 6th ed. Johns Hopkins University Press, Baltimore, Maryland 2:837–1936.

Nugent, R. F., and J. R. Choate. 1970. Eastward dispersal of the badger, *Taxidea taxus*, into the northeastern United States. Journal of Mammalogy 51:626–627.

Odell, D. K. 2003. West Indian manatee *Trichechus manatus*. Pp. 855–864 in Wild mammals of North America: biology, management, and conservation (G. A. Feldhamer, B. C. Thompson, and J. A. Chapman, eds.). 2nd ed. Johns Hopkins University Press, Baltimore, Maryland.

Oliver, J. H., Jr., F. W. Chandler, Jr., M. P. Luttrell, A. M. James, D. E. Stallknecht, B. S. McGuire, H. J. Hutcheson, G. A. Cummins, and R. S. Lane. 1993. Isolation and transmission of the Lyme disease spirochete from the southeastern United States. Proceedings of the National Academy of Sciences of the United States of America 90:7371–7375.

Oliver, J. H., Jr., L. A. Magnarelli, H. J. Hutcheson, and J. F. Anderson. 1999. Ticks and antibodies to *Borrelia burgdorferi* from mammals at Cape Hatteras, NC and Assateague Island, MD and VA. Journal of Medical Entomology 36:578–587.

Oswald, V. H. 1958. Helminth parasites of the short-tailed shrew in central Ohio. Ohio Journal of Science 58:325–334.

Owen, J. G. 1984. *Sorex fumeus*. Mammalian Species 215:1–8.

Pabody, C. M., R. H. Carmichael, L. Rice, and M. Ross. 2009. A new sighting network adds to 20 years of historical data on fringe West Indian manatee (*Trichechus manatus*) populations in Alabama waters. Gulf of Mexico Science 1:52–61.

Paradiso, J. L., and R. M. Nowak. 1972. *Canis rufus*. Mammalian Species 22:1–4.

Paradiso, J. L., and R. M. Nowak. 1973. New data on the red wolf in Alabama. Journal of Mammalogy 54:506–509.

Parsons, H. J., D. A. Smith, and R. F. Whittam. 1986. Maternity colonies of silver-haired bats, *Lasionycteris noctivagans*, in Ontario and Saskatchewan. Journal of Mammalogy 67:598–600.

Pembleton, E. F., and S. L. Williams. 1978. *Geomys pinetis*. Mammalian Species 86:1–3.

Peterson, R. L. 1966. Recent mammal records from the Galapagos Islands. Mammalia 30:441–445.

Peurach, S. C. 2003. High-altitude collision between an airplane and a hoary bat, *Lasiurus cinereus*. Bat Research News 44:2–3.

Poglayen-Neuwall, I., and D. E. Toweill. 1988. *Bassariscus astutus*. Mammalian Species 327:1–8.

Powell, R. A. 1981. *Martes pennanti*. Mammalian Species 156:1–6.

Price, R. D., R. A. Hellenthal, R. L. Palma, K. P. Johnson, and D. H. Clayton. 2003. The chewing lice: world checklist and biological overview. Illinois Natural History Survey, Special Publication 24:1–501.

Pruitt, L., and L. TeWinkel (eds.). 2007. Indiana bat (*Myotis sodalis*) draft recovery plan: first revision. United States Fish and Wildlife Service, Fort Snelling, Minnesota.

Rageot, R. H. 1955. A new northernmost record of the yellow bat, *Dasypterus floridanus*. Journal of Mammalogy 36:456.

Rausch, R. 1947. A redescription of *Taenia taxidiensis* Skinker, 1935. Proceedings of the Helminthological Society of Washington 14:73–75.

Rausch, R., and J. D. Tiner. 1948. Studies on the parasitic helminths of the north central states. 1. Helminths of Sciuridae. American Midland Naturalist 39:728–747.

Rausch, R., and J. D. Tiner. 1949. Studies on the parasitic helminths of the north central states. 2. Helminths of voles (*Microtus* spp.) preliminary report. American Midland Naturalist 41:665–694.

Rausch, R. L. 1975. Cestodes of the genus *Hymenolepis* Weinland, 1858 (sensu lato) from bats in North America and Hawaii. Canadian Journal of Zoology 53:1537–1551.

Ray, C. E. 1964. The jaguarundi in the Quaternary of Florida. Journal of Mammalogy 45:330–332.

Reimer, J. P., E. R. Baerwald, and R. M. R. Barclay. 2010. Diet of hoary (*Lasiurus cinereus*) and silver-haired (*Lasionycteris noctivagans*) bats while migrating through southwestern Alberta in late summer and autumn. American Midland Naturalist 164:230–237.

Reynolds, H. W., C. C. Gates, and R. D. Glaholt. 2003. Bison *Bison bison*. Pp. 1009–1060 in Wild mammals of North America: biology, management, and conservation (G. A. Feldhamer, B. C. Thompson, and J. A. Chapman, eds.). 2nd ed. Johns Hopkins University Press, Baltimore, Maryland.

Rice, D. W. 1957. Life history and ecology of *Myotis austroriparius* in Florida. Journal of Mammalogy 38:15–32.

Riedman, M. 1990. The pinnipeds: seals, sea lions, and walruses. University of California Press, Berkeley.

Riley, G. A., and R. T. McBride. 1975. A survey of the red wolf (*Canis rufus*). Pp. 263–277 in The wild canids: their systematics, behavioral ecology and evolution (M. W. Fox, ed.). Van Nostrand Reinhold Company, New York.

Rollin, P. E., et al. 1995. Isolation of Black Creek Canal virus, a new hanta-virus from *Sigmodon hispidus* in Florida. Journal of Medical Virology 46:35–39.

Rosatte, R., and S. Larivière. 2003. Skunks: genera *Mephitis*, *Spilogale*, and *Conepatus*. Pp. 692–707 in Wild mammals of North America: biology, management, and conservation (G. A. Feldhamer, B. C. Thompson, and J. A. Chapman, eds.). 2nd ed. Johns Hopkins University Press, Baltimore, Maryland.

Rose, R. K. 1980. The southeastern shrew, *Sorex longirostris*, in southern Indiana. Journal of Mammalogy 61:162–164.

Rostlund, E. 1960. The geographic range of the historic bison in the Southeast. Annals of the Association of American Geographers 50:395–407.

Ryan, J. M. 1986. Dietary overlap in sympatric populations of pygmy shrews, *Sorex hoyi*, and masked shrews, *Sorex cinereus*, in Michigan. Canadian Field-Naturalist 100:225–228.

Saugey, D. A., D. R. Heath, and G. A. Heidt. 1989. The bats of the Ouachita Mountains. Proceedings of the Arkansas Academy of Science 43:71–77.

Schmidt, C. A., and M. D. Engstrom. 1994. Genic variation and systematics of rice rats (*Oryzomys palustris* species group) in southern Texas and northeastern Tamaulipas, Mexico. Journal of Mammalogy 75:914–928.

Schubert, B. W., R. C. Hulbert, Jr., B. J. Macfadden, M. Searle, and S. Searle. 2010. Giant short-faced bears (*Arctodus simus*) in Pleistocene Florida USA, a substantial range extension. Journal of Paleontology 84:79–87.

Schwartz, C. W., and E. R. Schwartz. 2001. The wild mammals of Missouri. 2nd rev. ed. University of Missouri Press, Columbia.

Scott, A. F., and T. French. 1974. Rediscovery of *Zapus hudsonius* (Zimmerman) in Alabama. Journal of the Alabama Academy of Science 46:77–78.

Serfass, T. L., L. M. Rymon, and R. P. Brooks. 1992. Ectoparasites from river otters in Pennsylvania. Journal of Wildlife Diseases 28:138–140.

Seymour, K. L. 1989. *Panthera onca*. Mammalian Species 340:1–9.

Sharp, W. M. 1959. A commentary on the behavior of free-running gray squirrels. Proceedings of the Southeastern Association of Game and Fish Commissioners 13:382–387.

Sheffield, S. R., and H. H. Thomas. 1997. *Mustela frenata*. Mammalian Species 570:1–9.

Sherman, H. B. 1930. Birth of the young of *Myotis austroriparius*. Journal of Mammalogy 11:495–503.

Shump, K. A., Jr., and A. U. Shump. 1982a. *Lasiurus borealis*. Mammalian Species 183:1–6.

Shump, K. A., Jr., and A. U. Shump. 1982*b*. *Lasiurus cinereus*. Mammalian Species 185:1–5.

Skovlin, J. M. 1982. Habitat requirements and evaluations. Pp. 369–413 in Elk of North America: ecology and management (J. W. Thomas and D. E. Toweill, eds.). Stackpole Books, Harrisburg, Pennsylvania.

Smith, K. E., J. R. Fischer, and J. P. Dubey. 1995. Toxoplasmosis in a bobcat (*Felis rufus*). Journal of Wildlife Diseases 31:555–557.

Smith, W. P. 1991. *Odocoileus virginianus*. Mammalian Species 388:1–13.

Smolen, M. J. 1981. *Microtus pinetorum*. Mammalian Species 147:1–7.

Snyder, D. P. 1982. *Tamias striatus*. Mammalian Species 168:1–8.

Sperry, C. C. 1933. Opossum and skunk eat bats. Journal of Mammalogy 14:152–153.

Stafford, K. C., III, R. F. Massung, L. A. Magnarelli, J. W. Ijdo, and J. F. Anderson. 1999. Infection with agents of human granulocytic ehrlichiosis, Lyme disease, and babesciosis in wild white-footed mice (*Peromyscus leucopus*) in Connecticut. Journal of Clinical Microbiology 37:2887–2892.

Stalling, D. T. 1990. *Microtus ochrogaster*. Mammalian Species 355:1–9.

Stalling, D. T. 1997. *Reithrodontomys humulis*. Mammalian Species 565:1–6.

Stone, J. E., and D. B. Pence. 1977. Ectoparasites of the bobcat from West Texas. Journal of Parasitology 63:463.

Sullivan, E. G. 1956. Gray fox reproduction, denning, range, and weights in Alabama. Journal of Mammalogy 37:346–351.

Sumner, P. W., E. P. Hill, and J. B. Wooding. 1984. Activity and movements of coyotes in Mississippi and Alabama. Proceedings of the Annual Conference of the Southeastern Association of Fish and Wildlife Agencies 38:174–181.

Swanson, G., and A. B. Erickson. 1946. *Alaria taxideae* n. sp., from the badger and other mustelids. Journal of Parasitology 33:17–19.

Swanton, J. R. 1938. Notes on the occurrence of bison near the Gulf of Mexico. Journal of Mammalogy 19:379–380.

Sweeney, J. R., J. M. Sweeney, and S. W. Sweeney. 2003. Feral hog: *Sus scrofa*. Pp. 1164–1179 in Wild mammals of North America: biology, management, and conservation (G. A. Feldhamer, B. C. Thompson, and J. A. Chapman, eds.). 2nd ed. Johns Hopkins University Press, Baltimore, Maryland.

Taber, F. W. 1939. Extension of the range of the armadillo. Journal of Mammalogy 20:489–493.

Taber, R. D., K. Raedeke, and D. A. McCaughran. 1982. Population characteristics. Pp. 279–298 in Elk of North America: ecology and management (J. W. Thomas and D. E. Toweill, eds.). Stackpole Books, Harrisburg, Pennsylvania.

Talmage, R. V., and G. D. Buchanan. 1954. The armadillo (*Dasypus novem-cinctus*): a review of its natural history, ecology, anatomy and reproductive physiology. Rice Institute Pamphlet, Monograph in Biology 41:1–135.

Thomas, D. P., and T. L. Best. 2000. Radiotelemetric assessment of movement patterns of the gray bat (*Myotis grisescens*) at Guntersville Reservoir, Alabama. Occasional Papers of the North Carolina Museum of Natural Sciences and the North Carolina Biological Survey 12:27–39.

Thomson, C. E. 1982. *Myotis sodalis*. Mammalian Species 163:1–5.

Tiekotter, K. L. 1985. Helminth species diversity and biology in the bobcat, *Lynx rufus* (Schreber), from Nebraska. Journal of Parasitology 71:227–234.

Timm, R. M. 1985. Parasites. Pp. 455–534 in Biology of New World *Microtus* (R. H. Tamarin, ed.). Special Publication, American Society of Mammalogists 8:1–893.

Tiner, J. D. 1953. Fatalities in rodents caused by larval *Ascaris* in the central nervous system. Journal of Mammalogy 34:153–167.

Tolleson, D. R., W. E. Pinchak, D. Rollins, and L. J. Hunt. 1995. Feral hogs in the rolling plains of Texas: perspectives, problems, and potential. Great Plains Wildlife Damage Control Workshop Proceedings 12:124–128.

Trousdale, A. W., and D. C. Beckett. 2004. Seasonal use of bridges by Rafinesque's big-eared bat, *Corynorhinus rafinesquii*, in southern Mississippi. Southeastern Naturalist 3:103–112.

Turner, R. W. 1974. Mammals of the Black Hills of South Dakota and Wyoming. Miscellaneous Publications, Museum of Natural History, University of Kansas 60:1–178.

Vaughan, T. A., J. M. Ryan, and N. J. Czaplewski. 2011. Mammalogy. 5th ed. Jones and Bartlett Publishers, Sudbury, Massachusetts.

Veilleux, J. P., J. O. Whitaker, Jr., and S. L. Veilleux. 2003. Tree-roosting ecology of reproductive female eastern pipistrelles, *Pipistrellus subflavus*, in Indiana. Journal of Mammalogy 84:1068–1075.

Verts, B. J. 1967. The biology of the striped skunk. University of Illinois Press, Urbana.

Wade-Smith, J., and B. J. Verts. 1982. *Mephitis mephitis*. Mammalian Species 173:1–7.

Ward, J. W. 1934. A study of some parasites of rabbits of central Oklahoma. Proceedings of the Oklahoma Academy of Science 14:31–32.

Watkins, L. C. 1972. *Nycticeius humeralis*. Mammalian Species 23:1–4.

Watson, T. G., V. F. Nettles, and W. R. Davidson. 1981. Endoparasites and selected infectious agents in bobcats (*Felis rufus*) from West Virginia and Georgia. Journal of Wildlife Diseases 17:547–554.

Webster, W. A., and G. A. Casey. 1973. Studies on the parasites of Chiroptera.

3. Helminths from various bat species collected in British Columbia. Canadian Journal of Zoology 51:633–636.

Webster, W. D. 1987. Kyphosis in the marsh rice rat (*Oryzomys palustris*). Journal of Wildlife Diseases 23:171–172.

Webster, W. D., J. K. Jones, Jr., and R. J. Baker. 1980. *Lasiurus intermedius.* Mammalian Species 132:1–3.

Wehinger, K. A., M. E. Roelke, and E. C. Greiner. 1995. Ixodid ticks from panthers and bobcats in Florida. Journal of Wildlife Diseases 31:480–485.

Whitaker, J. O., Jr. 1968. Parasites. Pp. 254–311 in Biology of *Peromyscus* (Rodentia) (J. A. King, ed.). Special Publication, American Society of Mammalogists 2:1–593.

Whitaker, J. O., Jr. 1972. *Zapus hudsonius.* Mammalian Species 11:1–7.

Whitaker, J. O., Jr. 1974. *Cryptotis parva.* Mammalian Species 43:1–8.

Whitaker, J. O., Jr. 1982. Ectoparasites of mammals of Indiana. Indiana Academy of Science Monograph 4:1–240.

Whitaker, J. O., Jr., and P. Clem. 1992. Food of the evening bat, *Nycticeius humeralis*, from Indiana. American Midland Naturalist 127:211–214.

Whitaker, J. O., Jr., and T. W. French. 1984. Foods of six species of sympatric shrews from New Brunswick. Canadian Journal of Zoology 62:622–626.

Whitaker, J. O., Jr., and R. Goff. 1979. Ectoparasites of wild Carnivora of Indiana. Journal of Medical Entomology 15:425–430.

Whitaker, J. O., Jr., and W. J. Hamilton, Jr. 1998. Mammals of the eastern United States. 3rd ed. Comstock Publishing Associates, Ithaca, New York.

Whitaker, J. O., Jr., and R. E. Mumford. 2009. Mammals of Indiana. Revised and enlarged ed. Indiana University Press, Bloomington.

Whitaker, J. O., Jr., B. L. Walters, L. K. Castor, C. M. Ritzi, and N. Wilson. 2007. Host and distribution lists of mites (Acari), parasitic and phoretic, in the hair or on the skin of North American wild mammals north of Mexico: records since 1974. Faculty Publications from the Harold W. Manter Laboratory of Parasitology, University of Nebraska, Lincoln.

White, J. A., P. R. Moosman, Jr., C. H. Kilgore, and T. L. Best. 2006. First record of the eastern pipistrelle (*Pipistrellus subflavus*) from southern New Mexico. Southwestern Naturalist 51:420–422.

Wiley, R. W. 1980. *Neotoma floridana.* Mammalian Species 139:1–7.

Wilkins, K. T. 1987. *Lasiurus seminolus.* Mammalian Species 280:1–5.

Wilkins, K. T. 1989. *Tadarida brasiliensis.* Mammalian Species 331:1–10.

Williams, D. F., and J. S. Findley. 1979. Sexual size dimorphism in vespertilionid bats. American Midland Naturalist 102:113–126.

Williams, R. R. 1962. Trematodes from the cave bat, *Myotis sodalis* Miller and Allen. Ohio Journal of Science 62:273.

Willner, G. R., G. A. Feldhamer, E. E. Zucker, and J. A. Chapman. 1980. *Ondatra zibethicus*. Mammalian Species 141:1–8.

Wilson, D. E., and D. M. Reeder (eds.). 2005. Mammal species of the world: a taxonomic and geographic reference. 3rd ed. Johns Hopkins University Press, Baltimore, Maryland.

Wilson, D. E., and S. Ruff (eds.). 1999. The Smithsonian book of North American mammals. Smithsonian Institution Press, Washington, D.C.

Winhold, L., A. Kurta, and R. Foster. 2008. Long-term change in an assemblage of North American bats: are eastern red bats declining? Acta Chiropterologica 10:359–366.

Wittrock, D. D., and G. L. Hendrickson. 1979. Helminths of shrews, *Blarina brevicauda* and *Sorex cinereus*, in Iowa. Journal of Parasitology 65:985–986.

Wolfe, J. L. 1968. Armadillo distribution in Alabama and Northwest Florida. Quarterly Journal of the Florida Academy of Science 31:209–212.

Wolfe, J. L. 1982. *Oryzomys palustris*. Mammalian Species 176:1–5.

Wolfe, J. L., and A. V. Linzey. 1977. *Peromyscus gossypinus*. Mammalian Species 70:1–5.

Woods, C. A. 1973. *Erethizon dorsatum*. Mammalian Species 29:1–6.

Woods, C. A., L. Contreras, G. Willner-Chapman, and H. P. Whidden. 1992. *Myocastor coypus*. Mammalian Species 398:1–8.

Würsig, B., T. A. Jefferson, and D. J. Schmidly. 2000. The marine mammals of the Gulf of Mexico. Texas A&M University Press, College Station.

Yabsley, M. J., T. N. Nims, M. Y. Savage, and L. A. Durden. 2009. Ticks and tick-borne pathogens and putative symbionts of black bears (*Ursus americanus floridanus*) from Georgia and Florida. Journal of Parasitology 95:1125–1128.

Yates, T. L., and D. J. Schmidly. 1978. *Scalopus aquaticus*. Mammalian Species 105:1–4.

Young, R. A., and E. A. H. Sims. 1979. The woodchuck, *Marmota monax*, as a laboratory animal. Laboratory Animal Science 29:770–780.

Young, S. P. 1958. The bobcat of North America: its history, life habits, economic status and control, with list of currently recognized subspecies. Stackpole Company and Wildlife Management Institute, Washington, D.C.

Young, S. P., and E. A. Goldman. 1944. The wolves of North America. American Wildlife Institute, Washington, D.C.

Young, S. P., and E. A. Goldman. 1946. The puma: mysterious American cat. American Wildlife Institute, Washington, D.C.

Young, S. P., and H. H. T. Jackson. 1951. The clever coyote. Stackpole Company and Wildlife Management Institute, Washington, D.C.

About the Authors

TROY L. BEST was born on 30 August 1945 in Fort Sumner, New Mexico. His fascination with wildlife and natural history began during his childhood, which was spent among the farming and ranching communities of eastern New Mexico. He attended public schools in Clayton, Clovis, Lovington, Portales, and Tatum, New Mexico. After graduating from high school in Clayton in 1963, he entered Eastern New Mexico University in Portales, where he completed B.S. degrees in biology, anthropology, and secondary education in 1967. He began graduate studies at New Mexico State University in 1967; then he transferred to The University of Oklahoma in Norman, where he received an M.S. in zoology in 1971 and a Ph.D. in zoology with a minor in medical entomology in 1976. During 1974–1976, he served as assistant professor of biology at Northeastern University in Boston, Massachusetts. In 1976–1986, he held a variety of professorial appointments at Eastern New Mexico University and The University of New Mexico in Albuquerque; also in 1983, he was employed as a wildlife biologist by the New Mexico Department of Game and Fish, where he conducted research on rare and endangered species. He was an assistant professor at The University of New Mexico during 1983–1988 and he accepted a position as assistant professor of zoology and wildlife science at Auburn University in 1988. He continues to serve as a professor of biological sciences and as curator of the Auburn University Julian L. Dusi Collection of Mammals. He has conducted research in Argentina, Canada, Kenya, Mexico, and many countries in Europe, and he has published numerous research articles on mammals, birds, reptiles, and other organisms in Alabama, the southwestern United States, and Mexico. In addition, he has served in a variety of editorial positions for the *Journal of Mammalogy*, *Mammalian Species*, and *The Southwestern Naturalist*.

JULIAN L. DUSI was born on 10 November 1920 in Columbus, Ohio. He attended public schools in Columbus, and while growing up, he learned carpentry and other builder's trades from his father, who was a building contractor. In his spare time, he worked with his father on carpentry jobs. For his freshman year in college, he attended Capital University in Bexley, Ohio; then he transferred to The Ohio State University, receiving his B.S. in wildlife conservation in 1943. During this period, he also enrolled in a civilian pilot training program, and in 1943 he entered the United States Army Air Corps to be trained as a pilot. He graduated as a P-38 high-altitude photo reconnaissance pilot at the rank of second lieutenant. At the end of World War II, he was released from active duty and returned to pursue graduate studies at The Ohio State University. He received his M.S. degree in 1946 and his Ph.D. in 1949. During this period, he met Rosemary T. Dearth and married her on 22 November 1947. In December 1949, he accepted a temporary position at the Alabama Polytechnic Institute (Auburn University) as assistant professor in the Department of Zoology-Entomology and taught courses in mammalogy and ornithology for the next 45 years. During this time, he conducted research on mammals and birds, assisted by Rosemary, and published numerous research papers, some jointly with Rosemary. They also established an environmental consulting company and conducted studies of construction sites for power plants, pipelines, and electrical transmission lines. Both Julian and Rosemary were active in bird banding, various bird counts, and scientific organizations. They helped establish the Alabama Ornithological Society, edited the resulting journal, and participated in other activities of the organization. Julian retired from Auburn University in June 1993 and died 28 August 2012.

Illustration Credits

Photographs Courtesy of:

J. Scott Altenbach: 209, 215, 222, 227, 232, 236, 240, 244, 248, 253, 258, 263, 269, 273, 277, 281

Rodger W. Barbour: 71, 76, 80, 107, 121, 134

Troy L. Best: viii, 5, 6, 7, 8 (top), 9, 10, 11, 12, 13, 14, 15, 24, 25, 37, 47, 61, 62, 67, 103, 147, 162, 203, 216, 217, 218, 228, 249, 259, 264, 265, 266, 282, 292, 316, 327, 339, 349, 357, 363, 397, 402, 407

Steven B. Castleberry: 66, 94

C. Scott Clem: 93

Tanya Dewey: 297

Candice M. Dunning: 325

Julian L. Dusi: 184, 202, 223, 348

Thomas W. French: 197

Keith Geluso: 88, 89, 112, 129, 204

L. Michelle Gilley: 46

Joseph W. Hinton: iv, 51, 152, 153, 288, 307, 311, 312, 313, 326, 337, 338, 378, 379

Chris Jaworowski: 384, 385, 386

Michael L. Kennedy: 125

Scott A. Kincaid: 358, 398

Damon B. Lesmeister: 368

Jill C. Medford: 332

Lisa H. Moates: 30, 36, 41, 287, 289, 306, 391, 392

Tom Murray: 55, 60, 84, 108, 138, 171, 320, 343, 393

James F. Parnell: 116, 142, 176, 180, 189, 353

Jeffrey S. Pippen: 19, 167

Edward B. Pivorun: 193

Aaron K. Poole: 8 (bottom)

Denise I. Stetson: 98, 99, 100:

U.S. Fish and Wildlife Service, National Digital Library: 117

Chris Wemmer: 374

Jeff Whitlock: 158, 163, 301

Used with permission:

Harvey et al. (2011): 208

Skull Plates

Prepared and photographed by Scott A. Kincaid

Maps

Prepared by Troy L. Best and Scott A. Kincaid

Index

and porcupine, 148

and procyonid, 375, 380

and rabbit, 159, 163, 168, 172

and rat, 90, 95, 130

and shrew, 177, 190

and skunk, 365, 370

and squirrel, 38, 42, 48, 52, 56

Boehmiella, 40, 44, 92, 155

Borrelia, 22, 106, 111, 132, 137, 319, 323, 382, 395

botfly. *See* fly

Bothriocephalus, 44

botulism, 356, 401

Bovicola, 400, 405

Bovidae, 383, 406–10

box elder. *See* elder, box

Brachylaema, 111, 115, 366

Brachylaeme, 22, 128

Brachylaemus, 33, 356, 382

Brachylaima, 22, 137, 179, 183, 388

Brachylecithum, 183

bramble, 81

Branta canadensis, 308

Brevisterna, 110

brier, 81, 95, 99, 100, 159, 172, 173, 181

bromeliad, 298

Brucella, 388, 395, 401, 405, 410

brucellosis. See *Brucella*

Bubo, 177

virginianus, 90, 95, 148, 172, 185, 190, 211, 224, 241, 255, 349, 354, 364, 370, 375

bug, 118, 191, 210, 216, 224

Cicadellidae, 255

Cimex, 219, 226, 243, 256, 276, 280, 284

Coreidae, 241

Corixidae, 255

Hemiptera, 234, 249, 255, 265

Isopoda, 190

Lygaeidae, 185, 241

Pentatomidae, 241

Bunostomum, 405

burdock, 274

Buteo, 177, 182, 194

jamaicensis, 72, 130, 172, 203, 211

lagopus, 185

lineatus, 153, 203

cactus, 211, 375

caddisfly, 249, 255, 265, 278, 282

calicivirus, 295

Callinectes, 126

Calliphora, 394

Callithrix, 298

Calodium, 137, 145

cancer, 300, 310

Canidae, 305–23

Canis

latrans, 306–10

lupus, 310, 311, 399, 408

rufus, 311–15

rufus floridanus, 312

rufus gregoryi, 312

virginianus, 1

cantaloupe, 308

Cantella repanda, 172

Cantheridae, 190

Capillaria

bat, 219, 243, 247, 251, 256, 267, 276

bear, 328

beaver, 63

boar, 388

canid, 310, 319, 323

coypu, 155

deer, 395, 401, 405

felid, 290, 295, 304

gopher, 69

diabetes, 304

diarrhea, 395

Dicrocoelium, 155, 212, 219, 243, 405

Dictyocaulus, 395, 401, 405, 410

Didelphidae, 18–22

Didelphilichus, 22

Didelphimorphia, 18–22, 286

Didelphis
    virginiana, 19–22
    virginiana pigra, 20
    virginiana virginiana, 20

Dioctophyma, 310, 346, 356

Diospyros virginiana, 317

Dipetalonema, 22, 40, 150, 155, 334, 366

diphtheria, 132

Diphyllobothrium, 304, 323, 328, 334

diplomonad, 74

Dipodidae, 70–74

Diptera, 234, 245, 249, 255, 265, 278

dipteran, 194

Dipylidium, 323

Dirofilaria
    and bear, 328
    and canid, 310, 314, 319, 323
    and felid, 290, 295
    and muskrat, 87
    and otter, 341
    and porcupine, 150
    and rabbit, 165, 170, 174
    and raccoon, 382
    and sealion, 334

disease, xvi, 7, 27, 35, 135, 333, 401
    Aleutian, 356
    Chagas, 22, 33, 132, 145, 213, 319, 367, 382
    distemper, 22, 310, 314, 319, 323, 342, 346, 356, 360, 367, 371, 382
    foot-and-mouth, 388, 405
    heart, 329

    hemorrhagic, 87
    leprosy-like, 33
    Lyme, 22, 106, 111, 132, 137, 319, 323, 382, 395
    neurological, 97
    periodontal, 128, 329
    renal, 33
    respiratory tract, 342
    Tyzzer's, 87, 165, 319
    viral, 243
    Weil's, 141, 145
    yellow-fat, 87

Distoma, 226, 231

dock, 81

dog, 7, 14, 20, 33, 38, 42, 43, 52, 62, 118, 135, 139, 143, 148, 153, 159, 172, 177, 182, 185, 198, 288, 293, 305–23, 339, 349, 365, 370, 379, 380, 382, 393

dogwood, 42, 90, 94, 99, 190, 237

Dolichopodidae, 237

dolphin, 2, 333, 411–12

Doratopsylla, 83, 101, 110, 115, 179, 183, 187, 192

Dracunculus, 87, 295, 346, 366, 382

dragonfly, 229, 234

Drosophilidae, 241

duck, 198, 340

dugongs, manatees, and sea cows, 23–28

Dujardinia, 334

Dytiscidae, 234

eagle, 85, 309, 380

Echidnophaga
    and armadillo, 33
    and bobcat, 290
    and fox, 319
    and gopher, 69
    and mouse, 110, 115, 120, 137
    and mustelid, 352, 360

Rocky Mountain spotted, 54, 115, 161, 319

tick, 150

Fibricola, 22, 87, 106, 128, 356, 366, 382

fibroma, 40, 54, 395

Filaria, 155, 205, 360, 366

Filaroides, 295, 310, 352, 356, 366, 382

fir, 344

fish, 25, 85, 126, 135, 139, 177, 214, 286, 288, 294, 298, 302, 308, 321, 338, 340, 379

anchovy, 333

cages, 354

cray, 85, 379

gar, 153

mackerel, 333

rock, 333

salmon, 333

sardine, 7, 14

shark, 333

sucker, 27

sun, 182

fisher, 2, 148, 163, 286, 336, 343–47

Flavivirus

and armadillo, 33

and bat, 213, 219

and bear, 329

and boar, 388

and deer, 395

and mouse, 106

and opossum, 22

and raccoon, 382

and rat, 128, 132

and squirrel, 40

and weasel, 352

flea, 256

Atyphloceras, 83, 115

Catallagia, 110, 115

Cediopsylla, 165, 174, 310, 319, 323

Ceratophyllus, 44, 54, 58, 83, 87, 115, 150, 154, 165

Chaetopsylla, 310, 328, 351, 366

Conorhinopsylla, 50, 92, 115

Corrodopsylla, 74, 110, 115, 187, 192

Ctenocephalides, 22, 44, 106, 132, 141, 145, 165, 170, 174, 187, 295, 300, 319, 366, 371, 381, 394

Ctenophthalmus, 58, 74, 79, 83, 101, 106, 110, 115, 120, 137, 165, 179, 183, 187, 192, 205, 352, 356, 366

Doratopsylla, 83, 101, 110, 115, 179, 183, 187, 192

Echidnophaga, 22, 33, 40, 44, 69, 110, 115, 120, 132, 137, 141, 145, 290, 319, 352, 360, 366, 371

Epitedia, 50, 58, 79, 83, 92, 97, 101, 110, 115, 132, 187, 192, 352, 366

Euhoplopsyllus, 290

Hoplopsyllus, 40, 44, 115, 132, 161, 165, 174, 310, 319, 352, 366, 371, 377, 381

Hystrichopsylla, 79, 83, 110, 115, 352

Leptopsylla, 44, 50, 106, 132, 137, 141, 145

Megabothris, 58, 74, 79, 83, 110, 115, 352, 356, 366

Monopsyllus, 58, 79, 110, 352

Myodopsylla, 219, 267, 276, 284

Nearctopsylla, 79, 115, 192

Nosopsyllus, 58, 79, 110, 115, 141, 145, 352, 356

and bear, 328
and boar, 388
and bobcat, 290
and deer, 395, 405
and mole, 205
and mouse, 101, 106, 115, 120, 137
and raccoon, 382
and rat, 132
and skunk, 366
goose, Canada, 308
gopher
   pocket, 293, 349, 369
   southeastern pocket, 65–69
   tortoise, 104, 317, 369
Gopherus polyphemus, 104, 317, 369
Gordonia lasianthus, 237
Gossypium, 104
grackle, 216
Grahamella, 101
grape, 100, 317
Graphidioides, 155
Graphidium, 161, 165
Graptemys, 126
grass, 67, 135, 159, 163, 168, 182, 190, 194, 198, 203, 390, 393, 399, 403, 406, 408
   blue, 163
   bluestem, 118
   bur, 211
   field, 81, 99, 108, 122, 130, 185
   fruit, 72
   nest, 73, 78, 82, 90, 95, 100, 108, 117, 122, 126, 131, 163, 169, 174, 182, 186, 187, 199, 203, 349, 375
   orchard, 163
   pasture, 237, 313
   roost, 224, 241
   rye, 153

   seed, 72, 123
   shelter form, 159, 163, 173, 174
   slope, 339
   stem, 81
   swamp, 159
   timothy, 163
grasshopper, 52, 122, 185, 216, 229, 317, 364, 375, 379
grassland, 70, 77, 89, 113, 148, 163, 169, 175, 177, 181, 194, 288, 298, 349, 354, 358, 362, 383, 398, 403, 406, 408
Gryllus assimilis, 237
gum, 42, 128, 132, 172, 224, 237, 382

Haemaphysalis, 44, 58, 92, 101, 145, 161, 165, 170, 174, 290, 377, 405
Haematopinoides, 205
Haematopinus, 387
Haemogamasus
   and badger, 359
   and mole, 205
   and mouse, 74, 79, 83, 106, 110, 120, 137
   and opossum, 22
   and rat, 97, 132, 141
   and shrew, 179, 183, 187, 192, 196, 200
   and skunk, 371
   and squirrel, 39, 43, 49, 58
   and weasel, 351
Haemolaelaps, 192
Haemonchus, 319, 388, 395, 405, 410
Haemotobia, 400
Hamanniella, 22, 33
Hammondia, 304
hamster, 75–132
Hantavirus, 106, 111, 115, 128, 132, 141, 145
hare, snowshoe, 344

necrosis, cerebrocortical, 405

nectar, 375

nematode, 262, 284

 Aelurostrongylus, 304

 Allintoshius, 219, 243, 267

 Anafilaroides, 290, 323

 Anatrichosoma, 22

 Ancylostoma, 290, 295, 300, 304, 310, 314, 319, 323, 328, 360

 Angiocaulus, 360

 Angiostrongylus, 128, 179, 290, 323

 Anisakis, 334, 341

 Anoplostrongylus, 212

 Apteragia, 405

 Arthrocephalus, 346, 366

 Artionema, 405

 Ascaris, 40, 44, 54, 87, 110, 165, 346, 360, 366, 382, 387–88

 Ascarops, 33, 388

 Aspicularis, 106, 110, 115, 137

 Aspidodera, 33

 Baylisascaris, 97, 155, 165, 328, 346, 356, 382

 Bicaulus, 405

 Boehmiella, 40, 44, 92, 155

 Bunostomum, 405

 Calodium, 137, 145

 Capillaria, 22, 40, 44, 50, 58, 63, 69, 79, 83, 87, 92, 106, 110, 115, 128, 132, 137, 141, 145, 155, 165, 179, 183, 192, 200, 219, 243, 247, 251, 256, 267, 276, 290, 295, 304, 310, 319, 323, 328, 341, 346, 352, 356, 366, 371, 382, 388, 395, 401, 405

 Carolinensis, 101, 106, 120

 Castorstrongylus, 63

 Chabertia, 405, 410

 Citellina, 54

 Citellinema, 40, 44, 50, 54, 290

 Citellinoides, 74

 Cnathostoma, 341

 Contracaecum, 334, 341

 Cooperia, 395, 401, 405, 410

 Crenosoma, 323, 328, 341, 346, 366, 382

 Cruzia, 22

 Cutifilaria, 405

 Cyathospirura, 290

 Cylicospirura, 290

 Cyrnea, 219, 290

 Delicata, 33

 Dermatoxys, 165, 174

 Dictyocaulus, 395, 401, 405, 410

 Dioctophyma, 310, 346, 356

 Dipetalonema, 22, 40, 150, 155, 334, 366

 Dirofilaria, 87, 150, 165, 170, 174, 290, 295, 310, 314, 319, 323, 328, 334, 341, 382

 Dracunculus, 87, 295, 346, 366, 382

 Dujardinia, 334

 Elaeophora, 395, 401

 Enterobius, 44, 50

 Eucoleus, 319

 Eucyathostomum, 395

 Eustrongylides, 341

 Filaria, 155, 205, 360, 366

 Filaroides, 295, 310, 352, 356, 366, 382

 Globocephalus, 388

 Gnathostoma, 22, 295, 328, 341, 356, 366, 382

 Gongylonema, 101, 106, 115, 120, 132, 137, 205, 290, 328, 366, 382, 388, 395, 405

 Graphidioides, 155

 Graphidium, 161, 165

salmonellosis. See *Salmonella*

Salvelinus, 177

Sarcocystis

    and bison, 410

    and boar, 387

    and coypu, 155

    and deer, 395, 400, 405

    and felid, 290, 295

    and fox, 323

    and mink, 356

    and raccoon, 381

    and vole, 83

Sarcophaga, 395

Sarcoptes, 44, 150, 319, 323, 387, 405

    scabei, 314

sarcosporidiosis, 155

Sassafras albidum, 194

Scalopacarus, 205, 323, 351

Scalopus aquaticus, 202–6

Scaphiostomum, 58, 106

Scarabaeidae, 185, 190, 216, 241, 255

Schistocephalus, 341

Schistosoma, 33, 155

Schistosomatium, 74, 87

schistosomiasis, 33

Schizocarpus, 63, 87

Schizotaenia, 155

Sciuridae, 35–58

Sciurocoptes, 58

Sciurus

    capistratus, 1

    carolinensis, 36–40

    carolinensis carolinensis, 37

    carolinensis fuliginosis, 37

    niger, 1, 41–45

    niger bachmani, 42

    niger niger, 42

Scomber, 333

Scutacarus, 183

sea cow, 23–28

Steller's, 23

sealion, California, 2, 286, 331–35

seals, eared, fur seals, and sealions, 331–35

Sebastes, 333

Sebekia, 341

sedge, 67, 123, 127, 159, 174, 198, 408

seed, 47, 56, 77, 81, 100, 108, 113, 114, 122, 124, 126, 130, 139, 143, 148, 168, 308, 321, 379

    bank, 389

    blackberry, 99

    bluestem, 118

    cherry, 99

    dogwood, 99

    grass, 72, 123

    ivy, 99

    locust, 90

    oats, 118

    sedge, 123

    sumac, 99

    tree, 42, 172

    wind-deposited, 118

seedling, 56, 153, 159

Sellacotyle, 319, 367

Setaria, 395, 401, 405, 410

Seuratum, 212, 219

shark, 25, 333

sheep, 293, 406–10

shrew, xv, 90, 114, 175–206, 344, 349

    American pygmy, 3, 175, 193–96

    North American least, 175, 184–88

    northern short-tailed, 175–80, 190

    smoky, 3, 175, 189–92

    southeastern, 175, 197–200

    southern short-tailed, 175, 176, 180–83

shrews, moles, desmans, and relatives, 175–206